CAMBRIDGE LIBRARY COLLECTION

Books of enduring scholarly value

Earth Sciences

In the nineteenth century, geology emerged as a distinct academic discipline. It pointed the way towards the theory of evolution, as scientists including Gideon Mantell, Adam Sedgwick, Charles Lyell and Roderick Murchison began to use the evidence of minerals, rock formations and fossils to demonstrate that the earth was older by millions of years than the conventional, Bible-based wisdom had supposed. They argued convincingly that the climate, flora and fauna of the distant past could be deduced from geological evidence. Volcanic activity, the formation of mountains, and the action of glaciers and rivers, tides and ocean currents also became better understood. This series includes landmark publications by pioneers of the modern earth sciences, who advanced the scientific understanding of our planet and the processes by which it is constantly re-shaped.

New Zealand

Inhabited by Polynesians since the thirteenth century and discovered by Europeans in the seventeenth, New Zealand is a geologically diverse island group where active volcanoes and frequent earthquakes have resulted in a rich variety of rock formations and geothermal activity. In 1859–60, the geologist Ferdinand von Hochstetter (1829–84) was employed by Auckland's government to undertake the first systematic geological survey of the islands, the results of which were first published in German in 1863 and translated into this English version in 1867. Hochstetter describes his travels across New Zealand, his encounters with native people and his scientific observations. He analyses plants, wildlife and fossils, describes mountains, rocks and boiling springs, and evaluates evidence of glaciers and tectonic activity. As a result of Hochstetter's work, several species in New Zealand were named after him. This book remains a valuable resource in the history of Australasian natural science.

Cambridge University Press has long been a pioneer in the reissuing of out-of-print titles from its own backlist, producing digital reprints of books that are still sought after by scholars and students but could not be reprinted economically using traditional technology. The Cambridge Library Collection extends this activity to a wider range of books which are still of importance to researchers and professionals, either for the source material they contain, or as landmarks in the history of their academic discipline.

Drawing from the world-renowned collections in the Cambridge University Library and other partner libraries, and guided by the advice of experts in each subject area, Cambridge University Press is using state-of-the-art scanning machines in its own Printing House to capture the content of each book selected for inclusion. The files are processed to give a consistently clear, crisp image, and the books finished to the high quality standard for which the Press is recognised around the world. The latest print-on-demand technology ensures that the books will remain available indefinitely, and that orders for single or multiple copies can quickly be supplied.

The Cambridge Library Collection brings back to life books of enduring scholarly value (including out-of-copyright works originally issued by other publishers) across a wide range of disciplines in the humanities and social sciences and in science and technology.

New Zealand

Its Physical Geography, Geology and Natural History,
with Special Reference to the Results of Government Expeditions
in the Provinces of Auckland and Nelson

FERDINAND VON HOCHSTETTER
TRANSLATED BY EDWARD SAUTER

CAMBRIDGE
UNIVERSITY PRESS

CAMBRIDGE UNIVERSITY PRESS

Cambridge, New York, Melbourne, Madrid, Cape Town,
Singapore, São Paolo, Delhi, Mexico City

Published in the United States of America by Cambridge University Press, New York

www.cambridge.org
Information on this title: www.cambridge.org/9781108059657

© in this compilation Cambridge University Press 2013

This edition first published 1867
This digitally printed version 2013

ISBN 978-1-108-05965-7 Paperback

Ko Paora Matutaera.

(Paul Marshall.)

Maori chief at Kapanga.

Coromandel Harbour Prov. Auckland.

printed by Meermann, Munich.

NEW ZEALAND

ITS PHYSICAL GEOGRAPHY, GEOLOGY AND NATURAL HISTORY

WITH SPECIAL REFERENCE

TO THE RESULTS OF GOVERNMENT EXPEDITIONS IN THE PROVINCES OF

AUCKLAND AND NELSON

BY

DR. FERDINAND VON HOCHSTETTER,

Professor of Mineralogy and Geology at the Polytechnic Institution of Vienna, late member of the Austrian
Novara-Expedition, President of the J. R. Geographical Society of Vienna, Honorary member of the
New Zealand Society at Wellington, and of the Philosophical Institute of Canterbury N. Z. etc. etc.

———————

TRANSLATED FROM THE GERMAN ORIGINAL PUBLISHED IN 1863
BY EDWARD SAUTER, A.-M., PRINCIPAL LITTLE ROCK ACADEMY, ARKANSAS.
WITH ADDITIONS UP TO 1866 BY THE AUTHOR.
ILLUSTRATED WITH TWO MAPS, SEVEN PLATES IN TINTS, TEN LARGE WOODCUTS, PAGE-SIZE,
AND NINTY-THREE WOOD ENGRAVINGS IN THE TEXT.

———————

STUTTGART, J. G. COTTA.

1867.

Printed by J. G. Cotta, Stuttgart.

Preface to the German Edition,
published in 1863.

An Austrian man-of-war, *His Majesty's* frigate *Novara*, conveyed me towards the end of the year 1858 to the shores of New Zealand. In the capacity of geologist, I was a member of the Expedition fitted out under the orders of His Imperial Highness the Archduke Ferdinand Maximilian for a voyage round the world.

By the kind arrangements of the Commander of this Expedition, Vice-Admiral, then Commodore, Baron von Wüllerstorf-Urbair and by the timely preparations of the Colonial Government and the Colonists of New Zealand I was placed in such a position that I could fully devote myself for the space of nine months to the exploration of one of the most remarkable countries of the world, a beautiful country, which Albions enterprising sons, it's occupiers, looking forward to a rich and blooming future, are wont to call "the Great-Britain of the South Sea."

Was it the amiable disposition of the inhabitants? Was it the ties of friendship, that I formed there? Or was it the grandeur and peculiarity of the natural features of a country, appearing in its isolation like a world of its own, that attracted me so strongly? I cannot say; but I still look back with enthusiasm to my stay in the Antipodes.

On returning to my home in the beginning of the year 1860, it became my duty to work up the rich and copious materials I had brought with me in the shape of observations and collections, and to publish them as one of the results of the Novara-Expedition. A scientific work, accompanied by an atlas with numerous illustrations of newly discovered fossils, with views of the country and geological maps were to comprise the results of my geological researches. In a second work of a more general character I intended to present the results of my observations in a form better suited to the general reader and, in the proposed English translation, more especially accessible to the Colonists of New Zealand.

With heartfelt gratitude I may be allowed to state here, that the Imperial Austrian Government accorded me every assistance necessary to enable me to carry out both designs. While for the publication of the scientific work the Government itself has provided in a most liberal manner, I am indebted to the kindness of Baron von Cotta, for the publication of the Book of Travels in a form worthy to stand side by side with the Narrative of the Novara-Expedition as published in the Imperial Printing Office at Vienna.

By letters and other communications from numerous friends in New Zealand I have been enabled to follow from this hemisphere also, the course of events on those distant islands. Amongst these my friend and former fellow traveller, Dr. Julius Haast, the present Government Geologist of the Province of Canterbury, has contributed most amply to the completion of the present work by the important and interesting information obtained by him during his travels in the Alpine regions of the Southern Island.

Should I have been so fortunate as to have afforded any new information regarding the youngest and most distant colony of the British crown, I venture to hope, that I may at least have partly cancelled a debt of gratitude I owe to a nation, from the members of which I have experienced in all parts of the world the most friendly hospitality and the most energetic assistance in my labours.

My own countrymen, on the other hand, I hope will be gratified by my having presented to them for the first time on a larger scale the wonders and peculiarities of a country of which Carl Ritter, our great geographer, said already in 1842 in enthusiastic language, that it seems destined before all other countries to become a mother of civilized nations.

Vienna, December, 1862.

Dr. F. von Hochstetter.

Preface to the English Edition.

Already when the German original was published, it was my most earnest desire, that this work should also appear in an English edition for the purpose of facilitating its circulation in the Colony and in England. The accomplishment of this wish has now been rendered possible by the intervention of the General Assembly of New Zealand, which, during their session in 1866, voted in a most liberal manner a contribution towards the cost of the English work by the purchase of 500 copies of the same. I trust that I may be allowed to express here my most sincere thanks to the General Assembly and Government of New Zealand, for this farther proof of their kind interest in the undertaking.

The English edition, as now presented, is not a mere translation of the German original. A great portion of the matter in the German work, such as the chapter on the History of the Colonisation, on the Maori·war, on the Maori poetry, and on the statistics of New Zealand, was intended exclusively for German readers, to whom the numerous English works on New Zealand, treating at length upon these subjects, are often inaccessible. In these chapters I could have offered nothing new to the English public. I have therefore entirely omitted them in the English edition, and have instead rewritten and enlarged the chapters on the Physical Geography and Geology. In the same way, also, the chapter on the Southern Alps had, in consequence of the discoveries and explorations of the latter years, to be entirely rewritten, and likewise in the other chapters additions up to the year 1866 have been made. Also the sequence of the chapters has been altered in the English edition. Thus, the first part of the book contains now the general matter, whilst the second part, beginning chapter XI, consists of accounts of travels and descriptions of single districts and landscapes on both Islands. A number of the former illustrations have likewise been replaced here by new ones.

In delivering this book into the hands of the English public, I beg to request them not to subject the same to the severest criticism as regards its manner of expression and style, but to be lenient in their judgment, taking into consideration that neither author nor translator have written in their native language. May even the severest critic acknowledge, that the German author and translator, as well as the German publisher, have done their best to render this work as correct as possible. I for my part certainly feel myself greatly indebted to Mr. **Edward Sauter**, *A. M.*, *Principal Little Rock Academy*, *Arkansas*, who, during his stay in Europe in 1865, took the trouble to translate the German original, and to the Publisher for their aid in overcoming all those difficulties, which the publication of an English work in Germany necessarily entails.

Vienna, August, 1867.

Dr. F. von Hochstetter.

Contents.

Chapter I. Nine Months in New Zealand.

Chapter II. New Zealand. Sketch of its Physical Structure.

Chapter III. Geology and Palæontology.

Chapter IV. The Mineral Riches of New Zealand. Mineral Coal.

Chapter V. Gold.

Chapter VI. The Flora.

Chapter VII. Kauri and Harakeke. The New Zealand Pine and the New Zealand flax plant.

Chapter VIII. The Fauna.

Chapter IX. Kiwi and Moa, the wingless birds of New Zealand.

Chapter X. The Maoris.

Chapter XI. The Isthmus of Auckland.

Chapter XII. The North Shore.

Chapter XIII. Round the Manukau Harbour and to the Mouth of Waikato River.

Chapter XIX. The East Coast from Maketu to Tauranga, and Return to Auckland.

Chapter XX. Nelson.

Chapter XXI. The Southern Alps.

Illustrations.

Maps.

Coloured Engravings.

Woodcuts.

Wood engravings in the text.

CHAPTER I.

Nine Months in New Zealand.

THE ARRIVAL. On the 7th of December 1858, after a stay of four weeks on the coast of Australia, the *Novara* set sail from Port Jackson, the harbour of Sydney. The frigate, which on the long voyage from China to Australia had suffered great damage from a hurricane in the Chinese sea, and subsequently from stormy weather in the neighbourhood of New Caledonia, had undergone a thorough repair in the famous docks of Sydney, and was from masthead to keel as good as new; and the wind and waves, as though they had conspired to test at once the workmanship of carpenters, caulkers, and sail-makers, began to use us very roughly as soon as we had passed the "Sydney-Heads" at 9 o'clock a. m., and were steering our course towards New Zealand.

The ship bore up most gallantly. The keener the South wind blew, and the more boisterous the sea, the faster she moved ahead. About noon the coast of Australia had already vanished from the horizon. Less agreeable, however, was the sudden change of scenery for us "Naturalists," who found ourselves, — together with the hundred different things we had collected and brought on board with us, a doubly helpless sport at the mercy of a boisterous sea. I was, indeed, fortunate enough, never to suffer from actual sea-sickness, yet, at every such sudden transfer from land to the upheaving sea, an invincible drowsiness would come over me, and not until I had paid the God of Sleep double or treble the usual tribute, did I feel myself entirely acclimatized on board.

After two days the wind and waves subsided, and from this time, favoured by fair weather and alternate breezes, we had a pleasant, if not a very speedy passage. In the night from the 18th to the 19th of December we passed along the North Cape of New Zealand, without, however, coming in sight of the "Three Kings," three small rocky isles, which for the navigation from Sydney to the North Island of New Zealand form, so to say, the corner-stones, behind which ships bearing South-East, are wont to pass to the numerous harbours on the East-coast of the Northern peninsula.

On the 19th of December, shaping our course South East, we sailed along the East-coast, but at such a distance from the land, that we could only just see the very prominent Cape Brett near the Bay of Islands. However, a whale, which had followed the frigate close to the larboard-side for upwards of an hour, afforded us ample amusement. Every two minutes it would appear on the surface of the water to inhale fresh air, and thus exhibit its colossal dimensions. It was from 50 to 60 feet long. At eventide, a calm having set in, boats were lowered and a party was arranged for shooting albatrosses and other sea-birds (*puffinus, procellaria*), which swarmed round the frigate, and from sheer curiosity flew towards the boats in greater numbers as the firing increased.

On the 20th December we stood in front of the entrance to the Hauraki Gulf, the South-West bay of which forms the harbour of Auckland. Great and Little Barrier Island, or Otea and Hou-touru, as the natives call them, with their peaks about 2000 feet high, lay before us. In these parts certainly but very few vessels had occasion to complain of too fair weather and perfect calm; yet, here again we made an exception to the general rule. It was a beautiful day; but not a breath of air was stirring to waft us to-wards our destination. The weather was much the same on the 21th December; currents and slightly adverse winds had driven us out of the common route into the Hauraki Gulf past the "Hen and Chickens," thence between Little Barrier Island and Rodney Point as far as the East-coast of Great Barrier Island. Hence Commodore v. Wüllerstorf decided on entering by the Southern Channel between Cape Barrier and Cape Colville, a navigable passage ten miles wide, and we sailed slowly along the East-coast of Great Barrier Island.

This island, about 25 miles long, consists of a chain of steep-rising, serrated rocky mountains with many summits and sharp peaks. The highest point, in the middle of the island, named Mount Hobson after the first Governor of New Zealand, is indica-ted on the maps as 2330 feet above the level of the sea. Its Northern extremities are very remarkably indented rocks, called the "Needles," and the South angle is formed by the round summit of the rocky Cape Barrier. While the West-side of the island has a great many deeply excavated bays affording excellent moorings, on the shores of which both natives and Europeans have settled, the East-coast appears as a naked, uninhabited rocky shore with but one bay of any size, which is partially screened by "Aride Island," a lone, barren, and utterly inaccessible rock, so named by Cook. On the Northwest-side of Great Barrier there are some rich copper-mines; and in the woods of the island herds of wild horned cattle are said to roam.

[1] In the central part of the island, to the eastward of Wangaparapara, hot springs have been recently discovered. They spring up in the bed of a creek, which

During the night we passed the straits between Cape Barrier and Cape Colville, and in the morning of the 22^d, favoured by a North-Eastern breeze, we sailed down the spacious Hauraki Gulf. The weather had entirely changed by this time. The mountains were shrouded in fogs, and the horizon was so misty and murky, that we could hardly see the little islands surrounding us, or the land we were making for. Suddenly, however, the wind veered round to the opposite direction, the haze cleared away, and we now saw before us the entrance to the Waitemata Bay; — we had arrived in front of Auckland Harbour. The South-West wind had so suddenly withdrawn the veil of mist and clouds, in which we had hitherto been wrapped, that we were perfectly amazed at the first sight presented to us.

We were surrounded on all sides by islands, peninsulas and main-land, Tiritirimatangi, Wangaparoa, and the outlines of the North-shore; a low, undulating country destitute of woods, with steep shores, exhibiting regular layers of sandstone and shale, with small, sandy bays, the beach of which was dotted with small, isolated wood-huts. — Before us, in the direction of the sporadic groups of houses composing Auckland City, there lay numerous small truncated cones of hills, the form of which at the very first glance betrayed their volcanic nature. Pre-eminent among all the rest, as it were the leader of the whole host, who alone had ventured out ,into the sea, and here proudly reared his lofty head, arose the Rangitoto, an island mountain, 900 feet high, — the true prognostic of Auckland.

Attractive as the view of this volcanic island was to me, with

---

takes its rise under a dome-shaped hill at the back of Wangaparapara and flows towards the east coast into a large swamp near the sea. The water of the springs is very clear, and the temperature ranges from 50 deg. to 212 deg., boiling point. The water is strongly saline, and gives forth a strong sulphuric smell like burnt gunpowder. A large quantity of white sediment accumulates rapidly on the stones about the springs, and quickly becomes hard and cemented to them. In the neighbourhood of these springs, about half-a-mile to the east of them, and flowing from the side of a heavily wooded hill, are some clear waters very cold, and exceedingly salt.

its black streams of lava, with its strangely formed summit, where one small cone seemed to be set in the crater of another larger one: the first view of Auckland, I must confess, equalled by no means my brillant anticipations of New Zealand.

Entrance to Auckland Harbour.

Is that Auckland? — I said to myself, the farfamed capital of the *"Great Britain of the South sea?"* Where is the New Zealand Thames? Where the steaming, seething geysers and boiling springs? Where are all the volcanic cones of which I had read, the ever-steaming Tongariro; the Ruapahu covered with perpetual snow and ice; the Taranaki rearing its lofty head to the very clouds; where the New Zealand Alps? The picture my imagination had created of New Zealand was quite different to that now presented to my view. The stupendous conical mountains in reality seemed to me shrunk up into little insignificant conical eruptions from 500 to 600 feet high. Although I knew full well, that those gigantic volcanoes, and the snow-clad mountains of the South-Island were no fables, but that they lay at such a distance from this coast as to render them invisible; yet my eye searched inquiringly after them, and I felt quite disappointed that not even the last trace of them was to be descried.

However, I always felt so, whenever I first set foot on a land, about which I had read a great deal; and every traveller, I think, will experience the same. The reality of the spot which he first steps upon in a new country, never corresponds with the picture created by the imagination. After a long voyage he approaches the new coast with a feeling of impatience and utmost curiosity in the full belief of finding all that is attractive and remarkable collected on the very spot he happens first to set foot on, ready and waiting for him, who has come so far over the waters to see, with his own eyes, all he had read and heard of. But as it is with the traveller, who would like to see and experience at once all on one and the same spot, so it is, on the other hand, with others in regard to the traveller. He, in his turn, is expected to have seen all and every thing, to have experienced and passed through every thing, especially if he happens to be a so-called "circumnavigator." And if he moreover should happen to have just visited the gold-fields of Australia, — why, nothing seems more natural than that he should have brought home with him all his pockets and coffers stuffed with gold dust. It is the imagination which ever speculates, brings the most distant objects near, and would fain comprise all in one grasp.

Should my friends in Auckland require any further apology after this my candid avowal, that the impression made on me on the 22^d of December 1858, on my first viewing the scenery of that country, did not realise the grand picture my imagination had drawn of New Zealand, I can only assure them, that as Auckland and New Zealand live in my memory at present, all my former expectations and anticipations have been surpassed by far, and should I live to be permitted a second view of that panorama, and to greet once more the Rangitoto, my heart would leap for joy.

On a nearer approach we could perceive, that the signal had

already been hoisted on the flagstaff upon Mount Victoria, announcing to the inhabitants the arrival of an Austrian man-of-war, and at 2 o'clock p. m. Captain Burgess came on board as pilot. I had no idea at that time, that I should make many a trip yet on those waters in the Captain's neat, fast-sailing cutter, and that to this same Captain Burgess, who tendered us so hearty a welcome, saying that we had long since been announced and expected in Auckland, — I should be indebted afterwards for many a kind favour received at his hands.

Although but a few miles distant from the harbour, the vessel laboured long and hard against the contrary wind to pass up the narrow channel between the Rangitoto and the Northshore into the Waitemata. The nearer we approached, the more enlivening was the scenery Boats came rowing up to us; natives paddled along in their canoes, and from the deck of a ship just leaving the harbor we were hailed with loud shouts of welcome, as she passed close by us; but it was not until 6 o'clock p. m. that we arrived at the anchoring place in front of the city.

We met five other ships in the harbour; alongside of them now lay the Novara, the largest man-of-war that had ever anchored here. The whole population of Auckland appeared to have gathered together on the shore, when our frigate cast anchor, firing twenty one salutes in honour of the British flag upon New Zealand. The salutes were responded to from the fort. The Governor bade the expedition through his secretary and adjutant a most hearty welcome; messengers arrived to hail our coming also in the name of the colonists and inhabitants of Auckland; and the very first reception betokened that genuine cordiality, that amiability and complaisance shown to the expedition in so eminent a degree by the generous inhabitants of Auckland, and which I met with every where among the colonists with whom I chanced to become acquainted on my subsequent wanderings.

NEGOTIATIONS AND RESOLUTION. I little thought, on my first arrival in Auckland, that after a fortnight's stay the Novara would weigh anchor, and that I should wave from the shore a last farewell to my companions, and henceforth continue my travels alone. First of all, the duty seems to devolve upon me to relate, how unexspectedly this came to pass.

My first plans and hopes of being able to travel through the interior of the North Island of New Zealand, so remarkable on account of its volcanic features, date from the stay of the Novara Expedition at the Cape of Good Hope, in November 1857. *Sir George Grey*, then Governor of the Cape Colony, — formerly, in the commencement of his brillant career, from 1847 to 1853, Governor of New Zealand, and 1861 recalled to it — gave the first encouragement. He pictured to the commander of the expedition, Commodore *von Wüllerstorf-Urbair*, the grand natural curiosities of that unexplored country, its volcanoes and boiling springs, which he himself had seen on various journeys through the interior, in such lively colours, pointing at the same time to the great advantages, that must necessarily result from exploring the interior of the North Island, especially to geology and geography: — that the Commodore, who never for a moment lost sight of the scientific task of the expedition, resolved on remaining, if possible, longer in New Zealand, than had been laid down in the original plan of travel. Therefore it was at that time already, that I consulted Sir George Grey on the expediency of my undertaking an overland journey from Auckland to Wellington during the stay of the Novara in those two ports which she was to visit. Sir George Grey also very kindly placed his very select library at my disposal for my further guidance, and furnished the naturalists of the Novara with the kindest letters of recommendation to influential men of that conntry. It is with a sense of heartfelt gratitude I remember the cordiality and the friendly encouragements of that noble-minded man, who, wherever he was, in Australia, New Zealand or the Cape of Good Hope, invariably made use of his influential position to the furtherance of science.

However, the fine plans we had made at the Cape of Good Hope, at the very commencement of our travels round the world, would hardly have been put into execution, had not the question concerning the exploration of New Zealand, by a singular concurrence of circumstances, been solved for me otherwise than it had been originally proposed. In November 1858, about the time of the stay of our frigate in the harbour of Sydney, *Sir William Denison*, Governor-General of Australia, to whom the Novara Expedition is greatly indebted for a vigorous furtherance of its purpose, and for liberal contributions to its collections, had, shortly before our arrival on the coast of Australia, received a document from the New Zealand Government, requesting the services of a geologist to examine a newly discovered coal-field near Auckland. Sir William Denison, on becoming acquainted with Commodore Wüllerstorf's intention to visit the harbour of Auckland, requested him, that he would permit the geologist on board the Novara to examine those coal-fields more closely during the stay of the vessel in the harbour of Auckland, and to deliver reports thereon to the Governor of New Zealand. Commodore Wüllerstorf gladly embraced this opportunity, of rendering the Government of an English colony at least some service, however slight; in order to prove by it our unfeigned gratitude for the cordial welcome and the vigorous assistance we had received everywhere on English ground.

Consequently, on my arrival in Auckland, I was commissioned by the commander of the expedition to undertake a close examination of the coal-field in question, to give my opinion as to the quality of the coal and extent of the coal-fields, and also to point out the place best suited for working a mine.

With pleasure I accepted this commission, and owing to the various excellent preparations, which the Government at Auckland had made for this purpose previous to our arrival, I was enabled within the short space of time from December 24th to January 2d, to carry my researches so far as to arrive at definite results, which I presented in a special report "On the coal-field in the Drury and Hunua Districts in the province of

Auckland." This report, urging most emphatically the esta-
blishing of mines, was delivered by Commodore Wüllerstorf to
His Excellency, the Governor of New Zealand, Colonel *Th. Gore
Brown*, while our frigate was yet at anchor in the harbour of
Auckland; — it called forth further proceedings and negotiations.

In a country, as yet perfectly unknown in a geological point
of view, in the various parts of which various mineral treasures,
such as gold, copper, iron, coal, promised most essentially to raise
the natural resources of the young and fast-rising colony, the necessity of a more extensive geological exploration seemed so press-
ing; it appeared to be so much the universal wish of the colonists,
that the opportunity for carrying such an exploration into effect,
now offered by the presence of a geologist, should be eagerly embraced: — that the Government of New Zealand applied to the
Commodore with the request, that he would consent to a longer stay
in New Zealand of the geologist of the expedition, for the purpose
of making geological surveys in that country, but especially in the
province of Auckland.

Commodore *von Wüllerstorf*, on reflecting that during the rest
of the voyage no more such unexplored countries would be touched,
was quite inclined to comply with the request, provided I could
make up my mind to remain behind alone. Indeed, he was the
more willing from the conviction, that the geological exploration
of a country so little known as New Zealand, would lead to results, which through all generations to come would secure, even
at the Antipodes, a lasting memento to the Novara Expedition,
which had first been projected by His Imperial Highness, the Archduke *Ferdinand Max*, and which, in consequence of being sanc-
tioned by His Majesty, the Emperor, has proved so important
to Austria.

---

[1] Report of a Geological Exploration of the Coalfield in the Drury and Hunua
Districts, in the Province of Auckland (N. Z.) by Dr. F. Hochstetter, Geologist on
board the Austrian frigate Novara. It appeared first on January 14. 1858 in the
"New Zealand Government Gazette," and in a supplement to the "New Zealander;"
afterwards in several other New Zealand papers.

I must confess, it was no easy matter for me to make up my mind. It had, indeed, never been my intention to put, while separated from the expedition and from my fellow-travellers, those plans into execution, which had engaged my thoughts ever since our stay at the Cape of Good Hope, and but too well do I remember the pain the decision cost me. It was on the 5th of January. Accompanied by my friend, the Rev. Mr. *A. G. Purchas*, with whom I afterwards became so intimate, I entered the Council-Chamber of the Colonial Government-Office, resolved to state to the assembled ministry my special reasons for declining the honourable proposition made me by the New Zealand Government, and preferring to continue my travels on board the Novara. My ignorance of the language of the natives, the extraordinary territorial difficulties which the country, at no great distance from the capital, with its gloomy, pathless forests seemed to present, the want of any sort of topographical map of the interior, without which I deemed a geological exploration useless if not impracticable — these and various other circumstances made me doubt a successful solution of the grand problem before me.

However, the eloquence and amiability of my friends, the Rev. Mr. *Purchas* and Dr. *C. Fischer*, as well as the obliging disposition of the ministers present overcame my scruples, and in cheerful anticipation of an interesting, time of travel; and of happy results, I finally assented. The ministry, consisting of Attorney-General *F. Whitacker*, Colonial-Treasurer *C. W. Richmond*, and Postmaster-General *H. T Tancred*, as also Superintendent *John Williamson*, the head of the Provincial-Government, promised me the most vigorous assistance and the supply of every means within their power.

Consequently, in an agreement between the Commander of the Novara Expedition, B. von Wüllerstorf-Urbair and His Excellency, the Governor of New Zealand, Colonel Gore Brown, the conditions were laid down under which I was to perform my new and honourable task. [1]

---

[1] See the official documents in the Appendix to Ch. I.

On the 7th of January, during very stormy weather, I dis-
embarked. I had all my things brought ashore, including the
instruments and the apparatus necessary for my future operations.
I had to leave my cabin, which had so long been my shelter,
and to which I had become so much attached; and one of my
colleagues occupied it after me.

The departure of the Novara for Tahiti was fixed for the 8th of
January. At break of day I was on board. After the several stormy
days, which had delayed her departure, this was the first calm,
bright morning. The frigate was ready to set sail, and was only
waiting for a breeze to spring up, and for a change of the current
at the time of low-water. About 8 a. m. the command was given
to weigh anchor, and for me the sad moment of parting had arrived. To separate myself suddenly and for good from a ship,
that had been my home for nearly two years, and the fate of which
had been so closely linked to mine, cost me many a bitter pang.
I tried to return thanks to the noble Commodore, and the gallant
Commandant, but my voice failed me; I pressed the hands of my
late fellow-travellers, my companions in joy and sorrow, who I
saw were as sad at heart as myself. However, the band commenced
playing; up went the anchor, the sails were unfurled, I hastened
down the rope-ladder into my boat, — and made for the shore.
Before I had reached the land, the Novara with studding sails was
gliding along the placid, watery mirror, buoyed up by a gentle
breeze. I stood gazing after her for a long long while, wishing her a
safe voyage and a happy return to my native home. After the
body of the vessel had disappeared behind the Northshore, I still
saw her mast-heads; for a moment the whole ship was once more visible above the low country; many a last farewell was waved to her
by her friends on land, unobserved, however, by those on board; the
breeze grew fresher, and the Novara had vanished from the horizon.

It was not till then, that I awoke to a full sense of the novelty and change of my situation. The Novara was a piece of my
native home; on her I had felt at home even while abroad in far
distant countries. In my previous travels I had been among friends

and acquaintances, and lived after the good old home-fashion; the language we spoke was our mother-tongue; — only the scenery changed during our travels; and even among people of different complexion, and on the most distant shores did I feel at home as long as the Novara lay in the harbour. Hence my travels in foreign countries, and among strange people might truly be said to date from now. I was alone and left to shift for myself.

Such were my thoughts, when I wrote in my diary:

"Alone among the Antipodes!"

————

SOJOURN. Now, after having returned to my native hemisphere, when I recall the pleasant times spent in the Antipodes, I may well say, that I had every reason to be satisfied with my fate. New Zealand was altogether an extremely remunerative field of exploration for me. In a geological point of view, every step was attended with new results, which, whether in accordance with or adverse to the sanguine expectations of the colonists, were generally connected with questions of material and practical value for the young colony. I had thus the satisfaction of finding the whole population taking a lively interest in my labours. This cheering and truly honouring sympathy was to me the most welcome reward for the many privations and hardships, which had to be necessarily incurred. The kind and valuable services rendered me by word and deed, wherever my scientific peregrinations led me; the numerous and attentive audience in Auckland and Nelson at my evening lectures on the geology of New Zealand; the honours and distinctions, with which I was overwhelmed at my departure, — all this imparted to me the soothing consciousness and the cheering certainty, that I had not laboured only for myself or for a few initiated in the science, but happily for a whole nation, who with their lively interest and national energy participated in the results of geological and geographical researches, and most vigorously endeavoured to turn them to account. Perhaps I have also aroused a dormant taste for natural sciences in many a distant friend, and

may therefore rejoice in new explorations and enterprises, the interesting results of which my friends communicate to me by letter;
thus maintaining a connection between myself and a country, the
grand nature and the amiable inhabitants of which I shall always
recall to my memory with feelings of warmest interest.

The consequence of these happy and pleasant relations was,
that instead of remaining in New Zealand only four or five months,
as was my original intention, I remained nine, and only decided
on returning to my native country on receiving tidings of the safe
arrival home of the Novara.

After the departure of the Novara I had taken up spacious
and comfortable quarters in the so-called Clermont-House, Princes-
Street, with Mr. Winchy, an ever obliging and complaisant host.
A large, saloon-like apartment with a commanding view over a
large portion of the lower town with the harbour as far as the wood-
clad Titirangi-Range extending along the West-coast, served me
both for a study, and for a cabinet for my collections. Zoology and
botany having been the special fields assigned to my colleagues of
the Novara Expedition, I thought it my duty now — left behind as
I was, all by myself, — to avail myself of every opportunity on my
travels, of making at the same time also zoological and botanical
collections, and I arranged at once everything necessary for this
purpose. At the same time I put advertisements in the papers requesting the public to forward to me all objects in any way per-
taining to natural history. In doing this I had a double object in
view; first, the chance of one day receiving with the objects forwarded also information concerning the nature of districts which in
consequence of my very limited time I should not be able to explore myself; and secondly to be enabled to contribute to a museum of natural history in the town of Auckland. I, therefore,
requested the forwarding of duplicates of the same objects; and collected also myself multiplied specimens of every object of interest,
in order to leave a portion of my collections behind for the Auckland Museum. The colonists manifesting a lively sympathy for all
my undertakings, my summons was crowned with success, and

together with my own contributions, the collections gradually grew
so numerous, that at last I was scarcely able to find room for them
in my lodgings. Most readily, therefore, the Government fitted
up a neat little house close by for my museum. It was open to
the public at all times on my return from excursions, and I was
always favoured there with numerous kind visitors, who were desirous
of seeing what there was new and remarkable in New Zealand.

It is with the deepest sense of gratitude that I state, that
both the Colonial, and the Provincial Governments of Auckland did
all in their power to further my plans. Numerous friends, amongst
them the most influential and experienced men of the colony —
men, whose names I shall have frequent occasions to refer to —
were ever eady to aid me by word and deed. Their kind recommendations accompanying me on my various excursions, were always certain to procure me the most hospitable reception wherever
I went. But I was particularly fortunate here in meeting with a
true and trusty German, who became henceforth my inseparable
travelling companion, the faithful participant of all the toil and
troubles as well as my pleasures during my peregrinations through
New Zealand; I am speaking of my friend, *Julius Haast.* A singular chance had brought him in an emigrant-ship to the coast of
New Zealand the very day before the arrival of the Novara. He
had come for the special purpose of becoming acquainted with the
country and its inhabitants, and sounding to what extent New
Zealand was adapted to German immigration. With youthful
enthusiasm he entered at once into all my plans, duly apprecia-
ting the importance of the task before me; with unfeigned friendly
devotion and an unwavering cheerfulness of mind he stood ever
at my side, aiding and furthering my projects and labours until
we separated on my departure from Nelson. He remained behind,
and has since acquired a well-earned fame and distinction by his
bold and persevering explorations of the wild, mountainous districts
of the South Island. [1]

---

[1] Dr. J. Haast is at present Government geologist of the Canterbury Province.

My first field of action was the Province of *Auckland*. By the ample means with which the most worthy Superintendent of the Province, the *Hon. J. Williamson*, furnished me, I was enabled in the short space of five months, to traverse the greater portion of this province, comprising almost the whole Northern half of the North Island, and to accomplish my object on a larger scale according to a fixed plan.

For the first two months, January and February, the town of Auckland itself continued to be the centre of my excursions, the season not yet appearing favorable for longer journeys on foot into the interior of the country. First of all, therefore, I purposed to finish the researches on the brown coal in the vicinity of the capital, and on the remarkable Auckland-volcanoes, which I had begun during the stay of the Novara: which object I accomplished by making a detailed geological survey of the district of Auckland. Indeed, the nearest environs of the capital even, despite a few preparatory sketches of Mr. C. Heaphy, were in a geological respect almost as unexplored as the remote interior of the country. Sketches and outlines of a topographical map on a large scale (1 inch = 1 sea-mile) furnished me by the Provincial surveyor's office, served as a basis for a geological map of the Auckland District.

On a closer examination the country presented a far greater variety of geological formations than I had anticipated. The remarkable, extinct volcanoes on the Isthmus of Auckland engaged my attention the most, unique as they are in their kind both with respect to their number on so small a space, and the peculiar shape of their cones and craters and their streams of lava. In a circumference of only 10 miles from Auckland I had to note down no less than 61 extinct points of eruption. An excursion southward to the Manukau Harbour and the mouth of the Waikato River led to the discovery of very interesting fossils on the Southside of the mouth of the Waikato and along the West-coast, to the discovery of belemnites and beautifully preserved fossil ferns. By this circumstance the existence of *secondary strata* upon New Zealand was for the first time proved beyond a doubt. Subsequent excursions towards the Drury and Papakura Districts, as also to the Wairoa-River had for their aim the establishing of the extent of the brown coal-formation, and in a northern direction I advanced as far as the Waitakeri River and the peninsula Wangaparoa.

The geological map, as far as I had finished it by the end of February, comprised the whole environs of Auckland in a circuit of about 20 miles. My collections, on the other hand, contained quite a number of new fossils, and interesting specimens of rock; my cabinet of botanical and zoological curiosities had also greatly accumulated through liberal contributions from all quarters.

The question now arose, whether I should choose the North or the South of the province as the object of my further explorations, as my limited stay would not admit of my exploring both. I did not hesitate to decide in favour of the South from various reasons. The southern portions of the Auckland Province were almost exclusively inhabited by natives. None but missionaries, a few government-officers and tourists had traversed those remarkable, and as yet but little known regions; whereas the North of the island was far better known. Numerous European settlers inhabit the coasts of the bays of the northern peninsula. From verbal and written reports of the colonists I gathered much valuable information concerning the nature of that country. *Dieffenbach* [1] had visited all the important parts in the North, and published detailed accounts of them, which are very creditable to the writer. The celebrated American geologist *Dana*, in the grand expedition sent by the United States to the South Sea, had touched the Bay of Islands, the most important harbour of the North, and made a geological exploration of its environs. [2] Moreover, my friend the Rev. Mr. *A. G. Purchas* and Mr. *Ch. Heaphy* had during my stay at Auckland, visited the districts of the North, and brought me all sorts of collections and drawings, so that I was by no means a stranger to that part. On the other hand, the

[1] Travels in New Zealand by C. Dieffenbach. 2 Vols. London 1843.
[2] United States Explor. Exped. Vol. X. Geology by James Dana. 1849. Chap. VIII. p. 437.

remote interior of the southern portion of the Province, was, as yet almost unexplored. Since Dieffenbach's memorable travels in the year 1840, no naturalist had beheld the high volcanic cones in the interior of the island, or the beautiful inland-lakes, the boiling springs, the steaming solfataras and fumaroles of that country. Rumours about the existence of extensive coalfields, of lead, and copper ore were afloat. The geological information gleaned from Dieffenbach's account of that country, could not suffice, nor was the topography of the interior known to any extent. Hence a journey in that direction promised the most remunerative results; and towards the end of February I made the necessary preparations for it.

Captain (now Major) *Drummond Hay*, a gentleman thoroughly versed in the Maori language, was appointed by the Governor as my travelling-marshal and interpreter. The Provincial Government, on the other hand, was so kind as to invite my friend Dr. *Haast* to join our party; it moreover most readily complied with my request to furnish me a photographer, and also an assistant to make metereological observations and to aid me in my collections and drawings. In the latter capacity I engaged a willing and very useful young German, Mr. *Koch;* in the former the artist, Mr. *Bruno Hamel.* Besides these, two male attendants, a cook and fifteen natives to carry our luggage were hired.

Thus provided with letters of recommendation from the Governor to the most influential chiefs, and supplied with every article necessary for a long journey on foot through thinly populated countries, and for nightly bivouacs in the open air, I left Auckland on the 6th of March for Mangatawhiri on the Waikato. In the canoes of the natives I proceeded up Waikato, on the banks of which, near Kupakupa below Taupiri, extensive brown coal beds crop out; thence up the Waipa, a tributary of the former, for the purpose of visiting the harbours of Whaingaroa, Aotea and Kawhia on the West-coast, — localities of great interest on account of the fossils found there. On Kawhia Harbour I discovered besides belemnites the first specimens of ammonite in New Zealand. From Kawhia, I proceeded in an inland direction through the upper Waipa country as far as the

Mokau district. Hence across numerous bush ranges, I journeyed to the sources of the Wanganui River in the Tuhua district; and on the 14th of April I arrived at the extensive Lake Taupo, 1250 feet above the level of the sea, surrounded by majestic volcanic cones. Here I was in the very heart of the country, at the foot of the steaming volcano Tongariro, and its extinct, ever snow clad neighbour, the Ruapahu, 9200 feet high. At the South-end of the lake, there stands a missionary's house, where I met with a truly hospitable reception, my Maoris being hospitably entertained in the neighbouring Pah Pukawa by the famous Maori chief *te Heuheu*.

After having sketched the outlines for a map of the lake, and examined the hot springs on its shores, I followed, from where the Waikato leaves the lake, the remarkable line of boiling springs, solfataras and fumaroles, situated in the Bay of Plenty in a north-easterly direction between two active volcanoes: the Tongariro and the island-volcano Whakari or White Island on the East-coast. The *Ngawhas* and *Puias* of New Zealand, boiling fountains and geysers with silicious deposits as in Iceland, developing themselves on the grandest scale near the lakes Rotorua, Rotoiti, and Rotomahana, the lake-district of course demanded a prolonged stay. Iceland excepted, I consider this the most remarkable and most extensive hot-spring territory known.

In the beginning of May I reached the East-coast near Maketu, and proceeded along the coast as far as Tauranga Harbour; thence I turned inland towards the Waiho Valley, or the Valley of the New Zealand Thames, and struck again the Waikato near Maungatautari. I roamed through the fertile plains of the Middle Waikato Basin near Rangiawhia, paid a visit to the Maori King *Potatau te Wherowhero* in his residence Ngaruawahia at the junction of the Waikato and the Waipa; and towards the end of May, I passed down the Waikato, and returned to Auckland via Mangatawhiri.

The results of this nearly three months expedition were, in every respect, to my entire satisfaction. Greatly favoured by the weather, I did not meet with any insurmountable difficulties, although our way led through many a bog, river and almost inaccessible tracts of primeval forest. My travels happening to fall in the New Zealand autumn, after the gathering of the wheat and potatoe-crops, we experienced no want of provisions. We were always sure of being received with the heartiest welcome at the missionary stations scattered through the interior of the country; nor would the Maori chiefs be behind hand in receiving the Te Rata Hokiteta, as they called me in their native tongue, and my companions within their pahs with all due honour, and entertaining us with unfeigned hospitality. My Maoris behaved admirably; they were always ready and willing, and "merry as skylarks;" and, being moreover most zealously supported by my friends Haast, Hay, Koch and Hamel, the results were in every respect such as to gratify my most sanguine expectations. I had a considerable store of geographical, geological, botanical and zoological materials on hand; and for ethnographical studies, also, I had found ample opportunities.

However, the chief object I had always in view was the geology and geography of the country. In order to make geological surveys, I was obliged to work at the same time topographically; for the few existing maps of the interior were merely outlines traced on the evidence of the reports of missionaries and tourists. The sketch of a map I had brought with me from Auckland, presented nothing but standard-points for the coast; and at a distance of a few miles from Auckland it was but little better than a blank sheet of paper. From the very commencement of my travels, therefore, I had adopted by means of the Azimuth-compass, a system of triangulation which I based upon Captain Drury's nautical coast-survey, and carried through from the West-coast to the East-coast. The natives, who from an innate distrust of Government land-speculations were always certain to raise objections on seeing English engineers of the provincial corps upon ground not purchased by Government, with instruments for measuring and surveying, allowed me to proceed unmolested. They knew I was a stranger, and only on a short stay in the country; so they aided me in every way, that I might be able to give a fine and glorious account of their country on my return to my home. It was the chiefs themselves, who acted as guides; they would climb with me to the tops of mountains, whence I made my observations; they were ever ready to tell me the names of mountains, rivers, valleys and lakes, and in their own way instructed me in the geography of their country. I noted down carefully all the native names, and believe that I have rescued many a melodious and significant Maori-name from oblivion. [1] The peculiarities of the ground I always sketched on the spot; and thus procured a supply of reliable material, from which on my return to Auckland I compiled a topographical map on a large scale of the southern portions of the province; in fond anticipation of ampler leisure-time at some future period, when I should be enabled to subject the provisional sketch to a careful revision, and to draw from it some detailed maps of special districts. [2]

The observations of the Royal Engineer's Observatory in Auckland, synoptical tables of which my friend, Colonel *Mould*, had the kindness to furnish me, served as corresponding observations for the barometrical measurements made during the journey. I must also make mention of numerous drawings and photographs executed by the members of the expedition; likewise of valuable sketches of landscape scenery from the talented pencil of Mr. *Heaphy*. [3]

[1] Names in New Zealand are partly corresponding to certain mythical conceptions, partly derived from the qualities of the object named.

[2] One copy of my original map, on a scale of 2 miles = 1 inch, I left in Auckland for the use of the Government; Mr. J. Arrowsmith in London received a second copy, to aid him in the compilation of a large map of New Zealand in six sheets, which he had then in view; with the special intimation, however, that, the map being merely a copy of that left in Auckland, it should be considered merely as a temporary, rough sketch of my observations. To my much esteemed friend, Dr. A. Petermann in Gotha, I am indebted for the overhauling and revising of all my original sketches and surveys, and the map on a reduced scale, such as is found annexed to this work, is the final result of my observations. It is self evident, that a map, compiled within the short space of three months, with only the help of compasses, and comprising more than one fourth of the North Island, can make no pretensions to trigonometrical precision. However, it gives a correct view of the river and mountain systems of the country travelled through, and will be of service till something better shall have been substituted. A geological and topographical Atlas of New Zealand, lately published at the geographical institute of Justus Perthes in Gotha, contains all the rest of my surveys and sketches.

[3] Many of these drawings, photographs and other views serve as illustrations to this work.

A very interesting part of the country, in the vicinity of Auckland, still remained to be explored, viz. Cape Colville peninsula on the eastern coast of the Hauraki Gulf. The discovery of gold, in the neighbourhood of Coromandel Harbour on said peninsula, had created quite a sensation several years ago. Accordingly I availed myself of a few pleasant days in the month of June for the purpose of visiting the Auckland gold field, which, although bearing no comparison with the abundant and extensive gold fields of the South Island, yet offered much of geological interest. An intended visit to the copper mines of Great Barrier Island and of the Island Kawau, was, I am sorry to say, frustrated by inclement weather.

Thus my stay in the Province of Auckland was fast drawing to a close. Previous to my departure, at the request of the members of the *Mechanics' Institute*, I delivered in their hall, on the 24th of June, a lecture on the geology of the province, in which I presented a collective view of the chief results of my researches, illustrating the same by means of maps, plans drawings and photographs then ready, and exhibited at the occasion.[1] The arranging and packing of the collections, and the drawing of maps delayed my departure a few weeks longer; and that period of labour was followed by a pleasant time of social amusement and festive demonstrations up to my taking a final leave of the people of Auckland. Thousands of tokens of New Zealand were in my possession; my collections consisted of all sorts of lasting mementos of the forests and mountains of New Zealand; and it also became my pleasant duty to have to tender my warmest thanks the amiable inhabitants of Auckland for a special memorial of their kindness and generosity. On the 24th of July, on a festive occasion, a complimentary address couched in terms but too flattering to myself, was delivered to me in the name of the people of that province, accompanied by a valuable testimonial.[2]

[1] "Lecture on the Geology of the Province of Auckland, New Zealand, delivered to the Members of the Auckland Mechanics' Institute, June 24. 1859." The lecture appeared first in the Auckland papers, "New Zealander" and "Southern Cross"; afterwards somewhat enlarged in the "Province Govt. Gazette" of Auckland No. 14, July 8. 1859; and in the "New Zealand Govt. Gazette" No. 23, July 14. 1859.

[2] See Appendix to Ch. I.

May my friends on the other side of the globe permit me herewith to repeat to them my heartfelt thanks for the munificent present and for the honours and distinctions, with which they have been pleased to overwhelm me. May they also in my endeavours to present to them the full and detailed results of my various researches, in which they aided me so vigorously and extensively, recognize my earnest, eager desire to repay them a debt of gratitude.

I was very sorry, that from want of time I could not accept the kind invitations to make a stay also in Wellington, New Plymouth (Province Taranaki), and Ahuriri (Province Hawkes Bay), for the purpose of exploring those provinces, which according to the representations made to me by Messrs. *J. Crawford* in Wellington, *A. S. Atkinson* in New Plymouth, und *Triphook* in Ahuriri Bay, promised so many points of attraction. Nor could I avail myself of the Governor's kind and friendly invitation to accompany him on a trip to the southern parts on board the English man-of-war "Iris"; because, I had accepted a prior invitation of the Superintendent of the Province Nelson to visit the Southern Island, which, however short the stay, seemed to me of the utmost importance. It was not merely the fine name of "the garden of New Zealand," as Nelson is styled, that enticed me to that step, but rather the manifold mineral treasures, such as copper, gold and coal, which had obtained for Nelson the fame of being the principal mineral and metal district in New Zealand. Besides, how could I think of returning to Europe without having seen, even though only from afar, the magnificent Southern Alps with their dazzling summits of perpetual snow?

Consequently, on the 28th of July, I went in company of my friend Haast on board the steamer "Lord Ashly" bound for Cook-Strait. As the steamer lay to near New Plymouth, and moreover, before entering Blind Bay, anchored off Wellington, the voyage to Nelson afforded me at least an opportunity of hasty visiting those places. Thus on the 30th July I enjoyed the magnificent view of the Taranaki Mountain (Mt. Egmont), 8270 feet high; and from the sugar-loaf cliffs of the Taranaki-coast I had an opportunity of studying the trachyte-lava of that volcanic cone, the most regular of all its kind upon New Zealand. After a stormy passage through Cook Strait we landed in Wellington on the 1st of August, and on the 3[d] we arrived in Nelson.

Mount Egmont, the Taranaki Mountain seen North-East from Otamatua.

The inhabitants of Nelson, who, already at the time of the Novara's stay in Auckland Harbour had tendered such a cordial invitation to the members of the expedition, gave me a most hearty reception in their city, where I found pleasant and commodious quarters already awaiting me in Mr. Luck's Trafalgar Hotel. The Provincial Government, with the amiable and excellent Superintendent, Mr. *J. P. Robinson* [1] at its head, had left nothing undone to enable me to make the best of my time in geological explorations. They placed the steamer "Tasmanian Maid" at my disposal, in order that I might examine the most important points on the shores of Blind Bay and Golden Bay in rapid succession.

The most beautiful weather fully confirmed the report of the far-famed climate of Nelson; and my first excursions opened to me so new and important a field of exploration, that I gladly decided on prolonging my stay, originally intended to last but one month, to the end of September. Thus I was enabled to subject the gold and coal fields in the vicinity of Nelson, and likewise the

---

[1] Unfortunately the most respected Superintendant met with an untimely end in 1864. He was drowned on the West coast of the province.

Dun Mountain copper-mines to a closer examination, and to trace out a geological map of at least the northern portions of the province.

The result of my researches with regard to the vaunted mineral wealth of that province was on the whole quite favourable. It is true, I could not confirm the sanguine belief of some enterprising miners as to the existence of inexhaustible though still undiscovered copper-treasures in the Dun Mountain; there were, however, beside some copper-ore, distinct evidences of a quantity of chromate of iron. But above all I was convinced, that if properly worked, the goldfields of the Aorere and Takaka Valleys near Golden Bay would prove very productive and that the discovery and working of those first goldfields of New Zealand would be followed by that of new goldfields extending all along the mountain-range of the Southern Island. The discovery of such fields, I was certain, would in the course of few years secure for New Zealand an increasing importance amongst the gold-countries of the world. Finally, besides beds of brown coal similar to those found in the North Island, there appeared in the Province of Nelson also traces of an older coal of better quality. The excellent but scanty coal-seams of Pakawau gave reason to hope that in other parts probably coal deposits may be found of larger dimensions and easier to work. Happily my friend Haast has since actually discovered such on the Buller and Grey Rivers, on the West-coast of the Province of Nelson.[2]

My collections increased during my stay in the Province of Nelson to an extent quite unexpected. Upon the North Island I had searched in vain for remains of the extinct gigantic birds of New Zealand, for the bones of Dinornis and Palapteryx or for the Moas of the natives. Those researches were all the more successful on the Southern Island. The lime stone caves of the Aorere Valley opened to us rich stores of Moa bones. Through the exertions of my companion Dr. Haast not only single bones, but more or less complete skeletons were brought to light. To these was added

---

[1] The rich goldfields in the Province of Otago, were discovered in 1861, the goldfields on the west coast of the Province of Canterbury in 1864. See Ch. V.

[2] See Ch. IV.

an almost complete skeleton of *Palapteryx ingens*, a very valuable present given by the trustees of the Nelson Museum to the Imp. Geological Institution of Vienna. Two Kiwis *(Apteryx Owenii Gould)* living representatives of the wingless birds, which had been caught for me by natives in the mountains on the Aorere River, formed an interesting counterpart to those remnants of extinct wingless birds of New Zealand. I may be allowed also to mention the kindness and attention of my friends, Dr. *D. Monro, W. T. L. Travers*, Capt. *Rough, N. Adams, H. Curtis, M. Mackay, Th. R. Hacket, Wrey, Wells*, and many others, who by their liberal contributions of minerals, plants, and zoological specimens have helped me to complete my collections. To Messrs. *A. L. C. Campbell* and *Burnett* I am indebted for pretty landscapes and other sketches; and to the Provincial Government for numerous photographs of the environs of Nelson.

I was quite reluctant to leave a country, where so many discoveries and explorations remained still to be made. The pleasure of ascending to and penetrating the higher, remoted regions of the New Zealand Alps, at that time almost untrodden by the foot of man, was not in store for me. From Lake Arthur (Rotoiti), the southernmost point I reached in the southern Island, I saw from afar the stupendous peaks of the southern mountain-ranges with their summits of perpetual ice and snow glistering towards me. I was allowed but a distant glimpse of the grandeur and majesty of those mountains, which my friend and fellow-traveller, Dr. *J. Haast*, explored so successfully during the last years, under many difficulties and privations, but to the lasting honour of German perseverance and science.

My time had now almost expired, and I had to think of returning to Europe. In a lecture on the geology of the Province of Nelson, delivered on the 29th of September in the Wesleyan church, which had been most readily assigned me for the occasion, I presented a summary report of the results of my observations. A duplicate of this lecture illustrated by a geological map I transmitted to the Provincial Government of Nelson and to the

Colonial Government of Auckland.[1] Amid the joyful recollections of those happy days, I can not pass over in silence the agreeable surprise made me by the inhabitants of Nelson by a most flattering address,[2] presented to me at the conclusion of my lecture, and accompanied by a beautiful and significant present of valuable specimens of the gold-fields of Nelson, which were enclosed in a box ingeniously composed of various kinds of New Zealand fancy woods.

On the 2d of October I embarked on board the steamer "Prince Alfred" bound for Sydney. Feeling at heart as though I parted from my native home, I waved a last farewell to my numerous friends assembled on shore, and bade adieu to the coast of New Zealand. A perfect stranger, I had met with a truly hospitable welcome and reception at the hands of the brave and generous colonists on those distant shores. As a member of a Government Expedition, promoted by a magnanimous prince of an Imperial house for the noble ends of science, I was zealously supported in New Zealand by the representatives of a friendly Government. As a naturalist, I was most disinterestedly aided by men, who may justly be proud of belonging to a nation, whose banners wave in every quarter of the globe; a nation, that with equal energy pursues both the practical interests of life and the nobler ends of science and civilization. I was deeply impressed by the fact that the man of science, of whatever nation or country, is at home wherever he labours; and that the field of his researches, even though it were the remotest end of the earth, will become to him a second home.

After a short visit to the gold-fields of the Colony Victoria, I proceeded on board the steamer "Benares" via Mauritius and Aden to Suez, and on the 9th of January 1860 arrived in Triest, where, for the first time after an absence of almost three years I hailed again my native soil.

---

[1] "Lecture on the Geology of the Province of Nelson" appeared first in the "Colonist" and "New Zealand Examiner" of October 1859, then in the "New Zealand Government Gazette" No. 39 of December 6. 1859, and in other New Zealand papers.

[2] See Appendix to Ch. I.

Appendix.

A. Official documents.

Government House, Auckland, New Zealand,
January 4th 1859.

Sir,

I do myself the honour to express to you the gratification which the visit of His Imperial Majesty's Frigate "Novara" has afforded to the inhabitants of Auckland and to myself.

I beg also to convey to you and to the Officers of the scientific department of your Expedition my best thanks for the valuable information supplied by the investigations of these gentlemen.

It will be my agreeable duty to report to Her Majesty's Government on the subject and I am satisfied, that Her Majesty will receive the communication with pleasure and will recognise the importance of the services rendered to one of Her Dependencies.

Wishing you a prosperous voyage and success in the interesting objects of your pursuit, I beg to subscribe myself

Your faithful servant

THOMAS GORE BROWNE,
Col. H. M. S.
Governor of New Zealand.

Government House, Auckland, New Zealand,
January 5th 1859.

Sir,

Having already endeavoured to express my thanks to yourself and the Officers of the scientific department of your Expedition for the valuable aid afforded to this Colony, I now venture to ask you to confer a still greater favour, by giving permission to Dr. Hochstetter to extend his researches for a few months longer.

In the event of your granting this permission, the means necessary to enable him to explore effectually, will be provided at the expense of the Colony of New Zealand.

I feel less diffidence in making this request to you, as Representative of the Imperial Government, because Dr. Hochstetter's labours in this Colony may be made the means of furthering the objects, which His Imperial Majesty the Emperor of Austria had in view, when He despatched the Expedition under your command.

I beg to add, that, should you feel it compatible with your duty to accede to the application I have now the honour to make, every assistance shall be afforded to Dr. Hochstetter, whilst engaged in this Colony, to enable him to make his scientific researches as valuable as possible to the Expedition of which he will remain a member, and care shall be taken to facilitate his return to Europe at the expense of this Colony by such route as he shall prefer.

I habe the honour to be

Sir

Your most faithful servant

THOMAS GORE BROWNE,
Col. H. M. S.
Governor of New Zealand.

On Board H. I. R. M. Frigate Novara, Auckland-Harbour,
January 5th 1859.

Sir,

I reply to your official Note, dated Government House, Auckland, January 5th a. c. in which, as the Representative of the Imperial Government, you prefer the request, that I would give Dr. Hochstetter permission to extend his geological researches in this colony for a few months longer, I am most happy to accede to your application, and to give Dr. Hochstetter, in his capacity as geologist of the Imperial Expedition, leave for that purpose, under the following conditions, which are nearly the same as those stated in your kind note.

1. That Dr. Hochstetter's sojourn in New Zealand may not exceed six months, and thus enable him to return to Europe nearly at the same period as the I. R. Frigate is most likely to arrive there, namely in November or December next.

2. That the Novara-Expedition, of which Dr. Hochstetter still remains a member, may likewise enjoy the benefit of the observations, collections and publications made by Dr. Hochstetter during his stay in New Zealand.

3. That the means necessary to enable Dr. Hochstetter to explore the Country effectually, shall be provided at the expense of the Government of New Zealand; that every assistance shall be afforded to this gentleman, whilst engaged in these geological explorations, and that care shall be taken to facilitate his return to Europe (viz. Trieste) at the expense of the Government of New Zealand by such route as he shall prefer.

Upon this understanding I shall not only consider it compatible with my duty, to accede to Your Excellency's application and give Dr. Hochstetter permission to remain for the time stated in the Province of Auckland, but shall also feel quite certain, that the Imp. Austrian Government, as well as the Academy of Sciences whose delegate Dr. Hochstetter must be considered, will be highly gratified to learn, that it was in the power of the first Austrian Exploring Expedition to become serviceable to a nation, which has done so much for the advancement of science and the development of natural resources in almost all parts of the world.

With hope that the friendly arrangement thus entered into on this subject may create a lasting bound of union and communication between the scientific men of both countries,

I have the honour to subscribe

Your faithful servant

B. v. WÜLLERSTORF.

B. Address presented by the inhabitants of Auckland.

Dr. Hochstetter,

On the conclusion of your Geological Examination of a large and most interesting portion of this Province of New Zealand, we, the assembled inhabitants of Auckland representing every section of the community, and for the most part intimately connected with the Agriculture and Commerce of the Province — desire to express our admiration of the eminently scientific manner, and unwearied activity, with which you have conducted your researches into the Geological Formations and Mineral Resources of Auckland. We have also to thank you for the valuable information upon these objects, which you have already placed in our possession in the public lecture delivered by you in this hall on the 24th of June, and in the reports, you have forwarded to the General and Provincial Governments.

The report of a member of the "Novara" Expedition on the physical characteristics of this portion of New Zealand — of which so little has hitherto been known — will be acknowledged in Europe as both impartial and authentic.

To us, as a community, the information contained in that Report and the maps you have constructed, together with those additional details we hope to receive from you after your return to Europe, will be of essential service in a material point of view. We also desire to convey to you our sense of the impartiality of your reports — which, whilst they lay open to our view those resources of the country that will eventually aid to its wealth and its general prosperity, in no way exaggerate their value or tend to lead to extravagant ideas or speculations that might only result in disappointment.

Arriving in Auckland a stranger, upon whose sympathies we had no claim, you have exerted all your energies to condense the results of your scientific exploration into practicals forms, for the benefit of the people of the foreign country you visited for purely scientific purposes, or for the special advantage of your own country.

On all these accounts we feel, that our warmest thanks are due to you for your disinterested exertions for the promotion of our welfare. As an enduring testimony thereof, we request the acceptance of this Purse, — the contents of which we beg you will devote to the purchase of some piece of plate, that we trust may be regarded by your family and your countrymen not only as a tribute of respect to your varied talents, but as a well-merited memento of the grateful acknowledgement by the people of the Province of Auckland of the eminent scientific and practical services rendered to them by you.

We are desirous that the plate should bear the following inscription:

"Presented to Dr. Hochstetter, Geologist attached to the Imperial Royal Austrian Scientific Expedition in the Frigate "Novara", by the inhabitants of the Province of Auckland, New Zealand, in' testimony of the eminent services rendered to them by his researches into the Mineral and Agricultural resources of the Province."

Signed on behalf of the subscribers:

R. MOULD,
Col. Com. R. Engineers,
Chairman of Committee.

JOHN WILLIAMSON,
Superintendent,
Province of Auckland.

C. Address presented by the inhabitants of Nelson.[1]

Dr. Ferdinand Hochstetter,

Before your departure from among us, we, the inhabitants of the Province and City of Nelson beg to express to you our great obligations for the benefits which you have conferred upon us as a community.

Though we cannot but congratulate you upon your approaching return to your country and your family, we have strong personal reasons for looking upon it with regret. We feel, that it has been no light or trifling advantage to have had among us one of that small class of men, who conduct the great national Expedition by which the benefits of science are distributed over the world.

We know, that such an one comes invested with the highest possible authority to speak decidedly on the subjects of his investigations, and are sure that we may place the most implicit confidence in his statements. It

[1] Presented in Nelson on the 29th September 1859.

is the great characteristic of such scientific pursuits, as you are engaged in, that, though on the one hand they are joined to the deepest and inmost principles of nature, on the other they are linked to the daily wants and commonest necessities of life. We believe therefore that your visit here will not be barren of practical results. We believe, that it will give us both a desire to develop, as far as possible, our share of the gifts of nature, and a knowledge how we may best do this.

We know, that we have had no special claims on you for the interest you have taken in our welfare. The advantage, which we have derived from it are, however, of such a kind, that both those who give and those who receive, may be proud of. We have had many opportunities of noticing how earnestly you pursue knowledge for its own sake, and are glad to find that those who do so, are the most ready to employ for the benefit of others what they have acquired themselves. You have done this in our case with considerable personal exertion and discomfort, which have been cheerfully encountered by your diligence and activity.

We do not wish to do more than allude to considerations of a personal kind. But we must express our appreciation of your courteous and kind behaviour towards us and assure you that few men could habe been among us for so short a time and have acquired so much of the character of a personal friend.

We beg your acceptance of the accompanying Testimonial, the product of our Goldfields, and we ask you to apply it to the purchase of a piece of plate, which may help to keep us in your remembrance and on which we ask you to place the following inscription:

"Presented to Dr. Ferdinand Hochstetter, Geologist to the Imperial Royal Austrian Scientific Expedition in the Frigate "Novara", by the inhabitants of the Province of Nelson, New Zealand, as a record of their appreciation of the great benefits conferred upon them and the Colony by his frank communication of the results of his zealous and able researches into the geological character and mineral resources of the Province."

We earnestly hope, that all good may go with you on your return to Europe, and that after a pleasant and speedy voyage you may reach in safety your home and friends. And with this wish we bid you heartily "Farewell".

Signed on behalf of the inhabitants of Nelson.

J. P. ROBINSON,
Superintendent of the Province of Nelson,
New Zealand.

CHAPTER II.

New Zealand. Sketch of its physical Structure.

Position. — Area and shape. — Name of Islands. — The South Island. — The Southern Alps. — Physical features and natural scenery. — West coast. — East coast. — The North Island. — Continuation of the Southern Alps. — Volcanic zones. — Taupo zone. — Auckland zone. — Bay of Islands zone. — Harbours.

On the opposite part of our globe, and just below our feet — nearly 180° long. distant from us, and as far to the South of the Equator as Italy is to the North of it — there lies in the South Pacific Ocean, between the Australian and S. American Continents, a country, of which *Tasman*, in the middle of the 17 century, brought the first tidings to Europe. The Dutch named it *New Zealand*. [1] A marvellous country it is, that land of our Antipodes, where people — as we used to think when children — walk with their feet towards us, and their heads downwards: "all upside down," because,

[1] English writers on New Zealand protest strongly against this name. *Hurst-house's* opinion is, that no one has even done less for the land of his discovery, than Tasman, the Dutchman; he came, saw and went off again, without even setting foot on the land. He named it after a small, flat province in Holland, that resembles it as much as a herring a whale. '*South Britain*' or *Britain of the South* would be much more appropriate, and '*King of South Britain*' would sound much better than '*King of New Zealand*', a title that might induce some mischievous court-lady, to sing His Majesty the song of the "*King of the Cannibal Islands*". *Taylor* says, that, whereas the more euphonious names of *Tasmania* and *Australia* have been substituted for *Van Diemensland* and *New Holland*, so also New Zealand ought to be re-christened. To this end he proposes names such as "*Austral Britain*", or "*Austral Albion.*" *Zealandia* was likewise proposed.

as we had heard at school, it is winter there, when we have summer; and day there, when we have night. Of course, these facts appearing so very contradictory within the limited range of our childish comprehension, have long since lost every trace of strangeness with the advance of a maturer age; but nevertheless New Zealand still remains to us a most wonderful country.

Far from all continental shores, and out of the limits encircling the numerous clusters of islands in the equatorial zone of the Pacific Ocean, it towers amid the greatest mass of waters on the face of the earth, washed by the ever restless waves of the vast ocean; more isolated than any other land of equal extent.

Not inhabited probably till within late centuries of the history of man, and then but thinly populated, and only along the coasts and along the banks of navigable rivers, — New Zealand has fully preserved within its interior, the originality and peculiarity of its remarkable animal and vegetable kingdoms up to our present time. No monuments of any kind; no tombs of kings, no ruins of cities, no time-honoured fragments of shattered palace-domes and temples, are there to tell of the deeds of ages or nations past and gone. But Nature, through her mightiest agencies, through fire and water, has stamped her history in indelible characters on the virgin-soil of the island. The wild Alpine heights of the South, towering in silent grandeur to the sky, their lofty summits crested with fields of ice and decked with glacier-robes; the Volcanoes of the North looming up into the regions of perpetual snow, glisten from afar, dazzling the wondering eyes of the mariner, as he approaches the coast. Fertile and well-watered alluvial plains are there awaiting the enterprising settler, — the virgin-soil, on which he founds a new home; a land, blessed with the most genial climate, where he has but to battle with and subdue the wilderness to reap the never failing fruits of his labours.

New Zealand consists of two large and several small islands, forming a broad stripe of land extending from South-west to North-east, the North-end of which is prolonged by a narrow peninsula running in a north-westerly direction. Its outlines are very similar to those of Italy in a reversed position. It lies between the parallels $34\frac{1}{2}^\circ$ and $47\frac{1}{2}^\circ$ South-latitude, and the meridians $166\frac{1}{2}^\circ$ and $178\frac{3}{4}^\circ$ long. East of Greenwich. Its length is 800 sea-miles from North to South; which extent, if measured on European soil, would reach from the Southern extremity of Italy far over the Alps to the vicinity of Munich. The mean width from East to West is 120 sea-miles, and the area of the whole group of islands has been estimated by Dr. Petermann at 99,969 Engl. sq. miles (or 4703 German sq. miles). [1]

Hence New Zealand is about as large as Great Britain and Ireland, or four times as large as the ancestral estates of the Roman Empire in Italy; an area extensive enough to fit it in future for a maritime state destined to extend its sway around — a "Great Britain of the South-sea." [2]

If we moreover contemplate the situation of New Zealand, as it appears upon a planiglobe on Mercator's projection, we find it almost in the centre of an immense semi-circle dividing the globe,

---

[1] English accounts state the area and circumference of New Zealand to be as follows:

	Area. Acres.				Coast Extent (in sea-miles).
North Island . . .	31,174,400	=	48,710	sq. mil.	1,500
South Island . . .	46,126,880	=	72,072	" "	1,500
Stewart's Island . .	1,152,000	=	1,800	" "	120
Total =	78,452,480	=	122,582	" "	3,120.

The whole of New Zealand has only 50,000 acres less than Great Britain and Ireland together; the Northern Island is $\frac{1}{32}$ less than England, Wales and Scottland excluded; the Southern Island is $\frac{1}{9}$ less than England and Scottland together; $\frac{2}{3}$ of the whole or 52,000,000 acres are calculated to be land fit for agricultural or pastoral purposes; the rest comprises inaccessible mountains, coast-sand-hills, swamps, lakes and rivers.

E. v. *Sydow* has computed the area of New Zealand from Arrowsmith's map (of 1851), as follows:

North Island	2207	German sq. miles
South Island	2640	" " "
Stewart's Island	28	" " "
Smaller adjacent islands	30	" " "
The whole of New Zealand	4905	" " "

Consequently New Zealand is nearly as large as the Kingdom of Prussia before the war of 1866; or as Hungary, Bohemia, Moravia and Silesia together.

[2] Ritter: The Colonization of New Zealand p. 11.

from the Cape of Good Hope to Behring Strait in the Old World, and from Behring Strait to Cape Horn in the New World. Such is the situation of "Young Albion of the Antipodal World."

Two arms of the sea, Cook Strait in the North (41⁰ N. lat.) and Foveaux Strait in the South (46⁰ 40′ N. lat.) divide New Zealand into three parts of different sizes; two larger islands, which for want of better names have been designated in geography as *North Island* and *South Island;* and a third, small island, favoured with the special name of *Stewart's Island.* [1]

The three islands constitute only parts of one and the same system of mountains, which running from South-West to North-East forms a distinctly marked *line of elevation* in the Pacific Ocean. This longitudinal line is crossed by a second almost at right angles, which is indicated by the direction of Foveaux and Cook Straits, but still more so by the direction of the long-stretched N. W. peninsula of North Island; [2] this N. W. peninsula corresponds to the line, striking to N. 52⁰ W., and designated by *Dana* [3] as the *axis of the greatest depression* in the Pacific Ocean.

Dana observes that a line drawn from Pitcairn's Island

---

[1] The usual denominations of the three islands were *North, Middle* and *South Island.* Hereby, however, the last named island was unjustly classed with the two larger islands; and it is certainly more proper to distinguish only the two principal islands as *North Island* and *South Island,* and to allow the third, small island to pass by the exclusive name of *Stewart's Island.* The latter denomination has of late been adopted as the more appropriate.

The names *New Ulster, New Munster, New Leinster* (after the three Irish provinces), officially introduced by the first governor, Captain Hobson, which are still occasionally seen on maps, are now scarcely known to the colonists even as antiquated reminiscences. The original names *Te Ika-a-Maui, Te Wahi Punamu* (Cook wrote: *Ea heino mauwe* and *Tavai Poenammoo;* Dumont d'Urville: *Ika Na Mavi*) and *Rakiura* were not made for European ears and tongues. On the map appended to Polack's work on New Zealand the name *Victoria Island* is suggested for *South Island,* and at the time the French thought of extending the dominion of their flag from Bank's peninsula over the whole of New Zealand, they named South Island "La Nouvelle France". The *Chatham Islands,* situated 400 sea-miles East of South Island, and the "Snares", 60 sea-miles South of Stewart's Island are usually likewise numbered with New Zealand.

[2] The names of "North Island" and "South Island" being thus taken as proper names, will be used in this work without the pronoun.

[3] Dana, United States Expl. Exped. Vol. X. pp. 394—395.

(Paumotugroup) in a westerly direction, to the North of Society Islands, past Samoa and Salomon Islands as far as the Pelew Islands (East of the Philippine Islands), is pretty nearly the boundary-line between the *Low Islands* or Atolls to the North, and the *High Islands* to the South. He designates the extensive sheet of water between said line and the Sandwich Islands, — an area, nearly 2000 sea-miles wide and 6000 long, and studded with nearly 200 low Coral Islands, — as one vast *area of depression.* Nearly all the groups of islands in this area have a north-westerly course, and an imaginary line drawn from Pitcairn's Island in a north-westerly direction to N. 52^0 W. as far as the Japanese Islands, would be the mean-line of that field of depression, or the axis of the greatest depression. But on actually drawing this line upon a map on Mercator's projection, we shall find that a line drawn from Pitcairn's Island to N. 45^0 W., — its north-westerly prolongation touching precisely the North-coast of the Japanese Island Yezo, — would constitute perhaps with still more correctness the mean-line of that area of depression. If moreover according to the mean direction of the Southern Alps, the highest mountain-range upon the Isles of the South-sea, we imagine a line running in the direction of N. 45^0 E. as the mean longitudinal direction of New Zealand, the two lines, viz. the above mentioned line of depression and this line of elevation intersect each other at right angles. Noteworthy is the circumstance that the general geological importance of those two directions in the Pacific Ocean displays itself also in the direction of the eastern coast-line of the Australian continent. The East-coast of Australia and the Westcoast of New Zealand form lines very nearly parallel; their distance about 1000 sea-miles. To the north-easterly direction of the line of elevation there correspond upon New Zealand also the principal eruptions, plutonic as well as volcanic; to the north-western line of depression, on the other hand, the transverse ruptures, by which Foveaux and Cook Straits were formed, and a third dislocation, to which the N. E. coast-line of North Island owes its origin.

The most striking and important feature of New Zealand is

an extensive, longitudinal mountain-range, which, interrupted by Cook Strait, runs through the whole length of the two larger islands in the direction of S. W. to N. E., from the South Cape to the East Cape. This range, consisting of upheaved zones of stratified and massive rocks of different ages, constitutes the powerful back-bone of the islands. At its foot, — on South Island at the eastern foot, on North Island at the western (perhaps partly also at the eastern), — it is accompanied by a volcanic zone, on which abysso-dynamical forces have acted powerfully till within the most recent periods of the earth. The volcanic ·outbursts, forming an extensive table-land of volcanic tuff, from which isolated cones arise, together with sedimentary formations of the tertiary and quartary period, gave the islands but very recently their present form; which, however, still continues to be subject to various sudden and periodical changes, in consequence of earthquakes, as well as secular elevations and depressions.

The main range attains its grandest development on South Island. Here its numerous peaks capped with perpetual snow and decked with glacial shrouds impart to it a truly Alpine character, justly entitling it to the name of the "Southern Alps". High, precipitous, craggy mountain-ranges, intersected by narrow longitudinal valleys, run parallel to each other from Foveaux Strait to Cook Strait; they are connected by transverse ridges, and intersected by the deep transverse valleys of the various rivers. In the centre of this range are seen towering up in majestic grandeur the peaks of Mount Cook, Mount Tasman and the adjacent mountain giants, glistering with perpetual snow and ice, to a height of 13,000 feet above the level of the sea, almost as high of Mont Blanc. Splendid glacier-streams, lovely mountain-lakes, magnificent cataracts, mountain-passes and gloomy ravines with roaring mountain-streams rushing through, — such are the charms of a wild and uninhabited alpine region, but seldom trodden by human foot; a scenery so grand, that according to the enthusiastic reports of the travellers, who ventured to penetrate into this wilderness, it scarcely has its equal anywhere.

Towards the West this range of mountains sinks abruptly, leaving

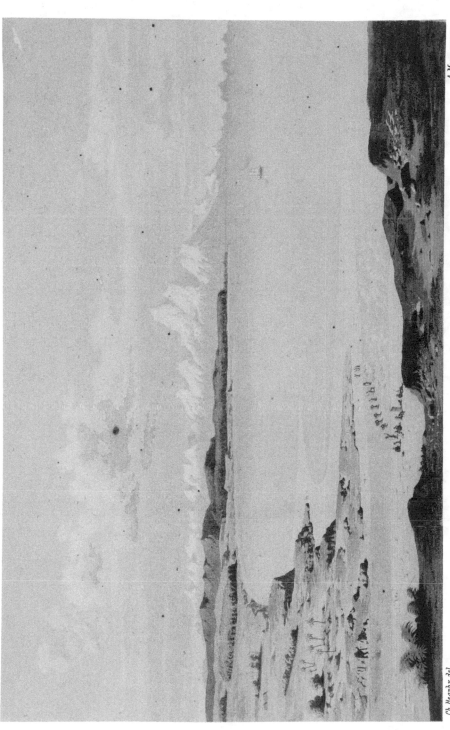

New Zealand, South Island.

Ch. Heaphy del.

(Mt. Cook 13,200 feet.)

A. Meermann sc.

The Southern Alps.

View from the mouth of the Teakawa or Brunner River on the Westcoast.

only a narrow slip of fertile land between its base and the sea, or forming vertical precipices, in some places from 3000 to 4000 feet high.
Only where the outlines of the coast cease to run parallel with the
mountains, taking a course across the axis of the range, the coast-
line is indented with deep inlets and sounds, which extend far
into the land between the high mountain ridges. Such is the case
in the South from Milford Haven as far as Foveaux Strait, and
in the North on Cook Strait from Cape Farewell to Cape Campbell. At these two extremities of South Island are those capital
harbours situated, such as Dusky Bay in the South, and Queen
Charlotte Sound in the North, which have been the safe places of
shelter for the bold seafarers, who first ventured out to these distant
shores.

Very different are the features on the East coast. The attention of the scientific observer has been quite correctly drawn to
the close resemblance between the formation of the coasts of New
Zealand and of South America (Patagonia and Tierra del Fuego);
as also to the general fact, so very obvious in New Zealand, that
the destructive tendency of the sea displays itself principally on
the West and South-West-coasts of islands and continents, continuing
to wear away the land until it reaches a powerful mountain-range,
which then serves as a bulwark for the low lands and plains at
its eastern foot.

At the foot of the range on the East-side runs a long row of
trachytic cones varying in height from 3000 to 6000 feet above the
level of the sea (Mt. Sommers 5240 feet; Mt. Hutt 6800 feet; Mt.
Grey 3000 feet etc.); the inland Kaikoras on the North East-side
of the island reaching the considerable height of 8000 to 9000 feet.
These caps and cones consisting of trachyte, andesite and phonolite,
surrounded with extensive deposits of tuff and amygdaloid, [1] and without the least trace of crater formation or of lava streams, present a
remarkable line, — parallel with the axis of the mountain, — of
eruptions belonging probably to the tertiary era.

---

[1] These amygdaloidal rocks and tuff abound in amethysts, chalcedony, agate,
opal, and similar secretions of silica.

Parallel to this line of eruption, and likewise in the direction from S. S. W. to N. N. E., but a little more Eastward, there runs a second zone of newer volcanic eruptions. To it belong the groups of volcanic mountains built up of doleritic and basaltic lava streams, and varying in height from 2000 feet to 3000 feet above the level of the sea, which project peninsula-like from the East coast far into the sea, their deep craters open towards the sea, constituting the only safe harbours on the East-side of the South Island. This is Banks' Peninsula with the four harbours, Port Cooper, Levi Bay, Pigeon Bay, Akaroa, and Otago Peninsula with Otago Harbour. On South Island, however, volcanic action has entirely ceased according to observations hitherto made. Some hot springs recently discovered in the Kaikoras may perhaps be considered as its last struggling efforts.

Between the trachytic and basaltic zones, there are extensive plains and alluvial tracts most admirably adapted to agriculture and pasturage. They attain near the base of the mountains a height of from 1500 to 2000 feet above the level of the sea, are in some parts 40 miles wide, and slope gently towards the sea, where they are bordered by a long line of sand-hills. They are crossed by numerous mountain-streams, which bring large quantities of detritus from the mountains, rolling them in broad shingle-beds into the sea. In these plains, especially in the vicinity of Christchurch, *Fata Morgana*, — mirages, — are of no rare occurence. The sea on one side, and the mountains on the other, seem to approach together, and ti-trees *(cordyline australis)* on the plains are seen inverted.

On North Island, on the other side of Cook Strait, the mountain-range extending along the East-coast from Cape Palliser to East Cape, is to be considered as a continuation of the Southern Alps, — at least of their eastern ranges. The various parts of it, such as the rugged forest ranges of Tararua, Ruahine, Tehawera, Kaimanawa, Tewaiti, fall far short of the height of the Southern Alps, — their highest peaks attaining a height of only 5000 to 6000 feet — and are almost a *terra incognita*, that may conceal yet many an unexpected treasure. [1] Whether the zone of trachytic eruptions on South Island has likewise a northerly continuation on the East-side of this mountain-range, future researches must determine. Perhaps the conical peak Mount Hikurangi (5500 feet) near East Cape, so celebrated in the legends of the natives, is such a trachyte or andesite-dome.

Nevertheless the northern Island abounds in volcanic phenomena of every kind. The table-land on the Westside of the mountain-range, sloping off gradually towards North and South and constituting the remaining portion of North Island, is perforated in more than a hundred places by the volcanic forces from below, which have continued their operations to this very day.

High trachytic cones of volcanic character; a large number of smaller basaltic cones of quite recent geological date; a long row of hot springs and steaming fountains, which, intermittent like the geysers upon Iceland, eject at shorter or longer intervals masses of seething water high into the air; fumaroles and solfataras of the grandest variety, afford the geologist a rich field of observation, and the traveller a series of sights the most remarkable in nature. No wonder then, that these striking, far-famed phenomena, — the peculiar nature of which was as well comprehended by the very first settlers even without a minute geological knowledge, as by the natives, — led to the general supposition that the soil of New Zealand is principally of volcanic origin.

In reality, however, the volcanic formations on North Island are limited to three separate districts, or to three zones, which, — different from the volcanic zone of South Island situated on the East-side of the Southern Alps, — lie all west of that range of mountains, which may be considered as a continuation of the Alps on North Island.

Immediately contiguous to that northern range of mountains is the great central zone; I call it the *Taupo Zone*. It contains the grandest and rarest volcanic phenomena that New Zealand can boast

[1] This country is now being explored by Mr. Crawford, the Provincial Geologist of Wellington.

of. Quite close to the centre of the island, on the southern shore of the large inland-lake Taupo, the waters of which fill a deep reservoir upon a sterile pumice-stone plateau of about 2000 feet above the level of the sea, the two giants among the volcanic cones of New Zealand, *Tongariro* and *Ruapahu* rear their colossal heads. The Tongariro volcano, 6000 feet high, with two powerful, ever steaming craters, is still active at least as solfatara; the Ruapahu on the other hand, which is over 9000 feet high and covered with perpetual snow, appears to be totally extinct. These two mountains are surrounded by a number of smaller cones, likewise extinct, such as Pihanga, Kakaramea, Hauhanga and others, which the natives call the wives and children of the two giants. A third giant, — so says tradition, — named *Taranaki*, stood formerly beside Tongariro and Ruapahu; but having quarrelled with the latter he was obliged to flee to the West-coast, where, a lonely exile on the coast, he now rears his hoary head amongst the clouds; this is the snow-capped Taranaki or Mount Egmont, 8270 feet high.

The two principal rivers of the North Island, rising near the Tongariro and Ruapahu, are the Waikato flowing North, and the Wanganui running South into the Cook-Strait. The land, consisting wholly of quartzose trachytic rocks, — recently termed *rhyolite* to distinguish them from the common trachytes, — and of pumice-stone, slopes from the foot of those mountains gradually to the north-eastern coast at the Bay of Plenty. A few miles from this coast lies the small island Whakari or White Island, 863 feet high, the conical peak of which, visible at a great distance by its continually ascending white clouds of steam, contains the second active crater of New Zealand. The distance from Tongariro to the Whakari Volcano is 120 nautical miles. Over this whole distance, almost on the very line between these two active craters, it seethes and bubbles and steams from more than a thousand crevices and fissures, that channel the lava-beds, of which the soil consists, — a sure prognostic of the still smouldering fire in the depths below — while numerous freshwater lakes, of which Lake Taupo, twenty miles in diameter, is the largest, fill up the larger depressions of the ground.

This is the *"Lake District"* in the south-eastern portion of the Province Auckland, so famous for its boiling springs, its steaming fumaroles, solfataras, and bubbling mud-basins, or as the natives call them, the *Ngawhas* and *Puias*.[1] Till now, none but missionaries, government officers and some few tourists have ventured by the narrow Maori paths through bush and swamps to visit this marvellous region; but all who have witnessed with their own eyes the wonders of nature displayed here, were transported with amazement and delight.

Only the natives have hitherto made practical use of those hot springs, which are the grandest in the world, and sought relief in them for their various complaints and diseases. But when once, with the progressive colonization of New Zealand, these parts have become more accessible, then thousands dwelling in the various countries of the Southern Hemisphere, in Australia, Tasmania or New Zealand, will flock to these parts where nature not only exhibits such remarkable phenomena in the loveliest district with the best and most genial climate, but has also created such an extraordinary number of healing springs.

The other two volcanic zones, — although now entirely extinct, yet decidedly of a later period than the Taupo zone, — belong to the long-stretched and singularly shaped north-western peninsula of North Island. I have designated them as the *Bay of Islands zone* and *Auckland zone*; the latter with no less than 63 points of eruption, rendering the narrow isthmus, that joins this peninsula to the main-body of North Island, a truly classical soil for the study of volcanic formations.

The *Bay of Island zone* comprises a number of smaller, extinct volcanic cones, situated on the northern half of the peninsula,

---

[1] The mythical notions of the natives have not failed to be influenced by these phenomena; fire-spreading goblins and spirits act a great part in the mythology of the tribes on the Bay of Plenty, while the Thames and Waikato-tribes by preference assign to themselves the protection of the deities of water and fogs.

Pohutu, Solfatara and intermittend fountain at Whakarewarewa on the Rotorua.

between the Hokianga River West and the Bay of Islands East, and fully corresponding as to their geological character with the Auckland volcanoes. Some few hot springs and solfataras south of Waimate in the Otoua district are the final effects of volcanic action in this zone.

North Island, like its southern mate, has its best harbours at its two opposite extremities; Port Nicholson, the harbour of Wellington on Cook Strait in the South, and the numerous harbour bays on the East coast of the northern peninsula from North Cape to Cape Colville, among which the Bay of Islands, and the Waitemata or Auckland Harbour are the most important. All the harbours on the West coast of North Island have sandbars at their entrances, and Manukau Harbour is the only one accessible to larger vessels.

On looking over the numerous colonies of great and glorious Albion, — that vigorous, parent tree, the branches and saplings of which are taking root and thriving in all parts of the world, — and comparing them with New Zealand, it is at once evident, that of all the colonial provinces of the British crown, New Zealand bears the most resemblance to the mother-country by virtue of its insular position, its climate, its soil, and the whole form and structure of the country. It is an empire of islands, a double island, which, — thanks to the power of steam, that now-a-days shortens every distance, — lies towards the neighbouring Australian Continent,[1] like Great Britain towards Europe. Blessed with a genial oceanic climate, so admirable suited to the Anglo-saxon race; with a fertile soil well watered and splendidly adapted to agriculture and farming; with a manifold coast-line, suiting perfectly to the notions and habits of the first maritime nation of the world: it is a country without dangerous animals, without poisonous plants, but rich in mineral treasures; a country, where horses, cattle and sheep thrive; where fruit, grain and potatoes grow most abundantly; a country adorned with all the charms and beauties of grand natural scenery; a country which can easily support a population of 12 millions; which promises the bold and persevering immigrant a lucrative and brilliant future. Such a country appears indeed destined before all others to become the mother of civilized nations.

[1] The distance from the coast of New South Wales is 1200 sea-miles. Sydney may be reached from Auckland by steam in 5 or 6 days; sailing-vessels take from 10 to 14 days.

CHAPTER III.

Geology and Palæontology.

History of geological and palæontological explorations. — Present state of knowledge. — Synoptical view of the formations and strata at present known. — Secular elevations and depressions. — Former connection with other bodies of land. — Earthquakes.

The discovery and first exploration of New Zealand by Captain Cook and his companions towards the close of the last century took place at a time when geology as a science had scarcely taken its first start. Rich in results as these earliest voyages were for zoology and botany, they yielded scarcely any thing noteworthy with regard to the geology and palæontology. The scientific expeditions of the French, English and North Americans, that touched New Zealand after Cook, also found but little of geological interest on the coasts and the much frequented harbours North and South.

White Island, Whakari of the natives, on the East coast of North Island was the first volcano noticed upon New Zealand; and in 1839 Mr. *Rule* brought the first fragment of a fossil bone found upon the North Island to London, from the structure of which Prof. *Richard Owen* was able to prove, that it had once formed part of a huge bird.

These are the first facts, that have become known with regard to the Geology and Palæontology of New Zealand, and till within the last years, the communications from missionaries, colonists and tourists have treated almost exclusively of the *volcanoes* and *volcanic phenomena* of North Island or of new specimens of *"Moa-bones"*, the remains of the extinct gigantic birds of New Zealand.

In the first instance we are indebted for valuable informations to the enterprising German traveller, Dr. *Ernst Dieffenbach*, who as naturalist accompanied in 1839 the expedition sent out by the New Zealand Company for the purpose of establishing a colony on Cook Strait. *Dieffenbach* examined the shores of Cook Strait, he was the first to ascend in December 1839 Mount Egmont or Taranaki (8000 feet high), and traversed in 1840 North Island from Cape Reinga to the volcanic regions of Lake Taupo. In his instructive and interesting work correct views were given of two of the volcanic zones extending through North Island. Dieffenbach moreover mentions in various passages the tertiary strata, so extensive upon North Island, with their numerous fossils.

Prof. *Owen's* discovery confirmed the, at first discredited, accounts of the natives about the giant birds *"Moa"*, which were said to have populated those islands in times long past, and gave an impulse for new investigations. By the zeal and energy of missionaries, colonists and natives on both North and South Islands thousands of individual bones, and also whole skeletons in a more or less perfect state of preservation were soon collected, furnishing Prof. *R. Owen* with abundant material for his celebrated treatises on the Genera *Dinornis* and *Palapteryx*, which have become extinct within the very latest period of the earth (see transactions of the Zoological Society, London 1843—1856).

In the person of Mr. *Walter Mantell*, eldest son of the celebrated author of the *"Coins of Creation"*, a settler came to New Zealand, who besides an untiring industry for making collections was moreover possessed of valuable geological acquirements, and contributed, in his communications to his father Dr. *G. A. Mantell* very interesting items to the Geology and Palæontology of New Zealand. In 1848 Dr. *Mantell* (in the Quarterly Journal of the Geolog. Soc. in London), gives an account of the extensive collection of Moa bones, which his son had made up, and of the probably very late, if not quite recent strata, in which those remains were found. He took occasion to add some general remarks on the natural character of New Zealand, comparing this remarkable insular province, on account of the prevalence of fernweeds, lycopodiaceæ and other cryptogamians, on account of its gigantic birds and the absence of every kind of mammalia, with the state of European countries during the coal and trias-period; in a similar manner as Australia with its *Cykadeæ*, *Araukariæ* and marsupial mammalia reminds us of the oolite-period; the Galapagos Islands with their graminiverous land and sea-saurians, with their reptiles and tortoises, of the age of the *Iguanodon* or the wealden-period.

James Dana (in 1849), in the admirable volume comprising the Geology of the United States Exploring Expedition under Ch. Wilkes 1839—1842, gave a brief sketch of the geological features of the country adjacent to the Bay of Islands, thus acquainting us with the third volcanic zone of New Zealand, the Bay-of-Islands zone.

Dr. *G. A. Mantell* (in 1850) published in the Quarterly Journal a geological sketch of the East coast of South Island from the volcanic Bank's Peninsula to the Molyneux River, in which he distinguishes volcanic formations, clay-slate, quartz-conglomerates and various other fossiliferous sedimentary formations. It is in this treatise also, that for the first time fossils were described and represented, and different groups of strata were distinguished according to the different fossils found in them. The *"Otatara limestone"*, resembling the cretaceous rocks of Faxoe and Mastricht, with *Terebratula Gualteri Mant.*, with a body resembling a belemnite (without however being a belemnite), and a series of foraminifera, which by Mr. R. Jones were partly identified with species of the chalk formation, corresponds, according to Mr. *Mantell's* views, to the upper cretaceous or to the eocene formation. The clayey strata of Onekakara and Wanganui (North Island), on the other hand, containing chiefly living species, such as *Turitella rosea Quoy*, *Struthiolaria straminea Low.*, *Fusus australis Quoy*, *Murex Zealandicus Quoy*, *Venus mesodesma Gray*, *Pecten asperrimus Lam.* etc., were numbered with the pleistocene period. As deposits of the most recent age, Mantell describes alluvial sediments and magnetic iron-sands of the coast with remains of *Dinornis*, *Palapteryx*, *Notornis* etc. occurring here and there. Mention is made also of the infusorial earths from Taranaki and from Lake Waihora near Bank's Peninsula, entirely composed of diatomaceæ and polycistines.

To Dr. *Mantell's* treatise a short note is added, in which Prof. *E. Forbes* mentions two localities of South Island, Banks River and the cliffs about Nelson, remarking on the fossil specimens from the localities named, which were presented by Mr. Cumming to the Museum for Practical Geology, that they cannot be identified with living species, but that their general habitus reminds the observer very much of eocene fossils from the Bognor beds.

In 1854 and 1855, Mr. *Heaphy* of Auckland published geological notes on the Coromandel District near Auckland and on the gold-diggings on Coromandel Harbour. The trachytic rocks found there were, however, mistaken for granite.

In the XII. Vol. of the Quarterly Journal, 1855, *C. Forbes*, Surgeon on board H. M. Ship "Acheron", gives an interesting description of the geological features along the coast of North and South Islands, attaching thereto remarks on the coal seams upon Preservation Island, at Motupipi, on the West coast of North Island not far from the mouth of the Waikato, and on Saddle Hill near Dunedin.

In addition to this we find in the same volume a short note by Mr. *J. Crawford* on the geological structure of the vicinity of Port Nicholson, where more or less vertical strata of clayslate constitute the mountain ranges, and the plains consist of tertiary strata and alluvial deposits. Mr. Crawford remarks also, that about Port Nicholson and likewise near Whakapuaka there are indications of the coast having been raised within a very recent period.

In 1859, Prof. *T. H. Huxley* gave an account concerning fossil bones of *Paleudyptes antarcticus*, a species belonging to the family of Pinguins, and cetaceous remains of *Phocænopsis Mantelli* from apparently tertiary strata.

In 1861, Prof. *Owen* surprised the geological section of the British

Association in Manchester with the news, that Mr. Hood in Sydney had forwarded to him fossil bones from the Waipara River in the Province Canterbury, South Island, which belong to a plesiosauric reptile, *Plesiosaurus australis*, and seem to indicate the existence of jurassic strata.

It was accident rather than design or scientific researches, that led to the discovery of the mineral treasures, which have continued to be developed during the last years: of coal, gold, copper, iron, chrome-ore and graphite. Where such a variety of useful minerals was presented to the view almost without any intentional effort on the part of the discoverers, what remunerative harvests could be expected from a systematical exploration! The well-educated class of colonists, for which New Zealand is noted, were fully aware of the importance of explorations to be made by scientific men in behalf of physical geography and geology, and that scientific knowledge aids in the extension and improvement of the industrial arts. The provincial Governments shunned no expenses to obtain the services of men, by whose aid the geological and mineralogical exploration of the country could be carried through.

Thus by a lucky juncture of circumstances, I had, in 1859, the pleasure of beginning the explorations in the Provinces of Auckland and Nelson, and of sketching the first geological maps of parts of New Zealand; and my friend and fellow-traveller, Dr. *Julius Haast*, had the honour of being the first Government Geologist in New Zealand. After having finished, in 1860, some geographical and geological explorations in the western Districts of the Province Nelson with the best success, he was appointed Geologist by the Provincial Government of Canterbury. This laudable example was soon followed by other provinces. Towards the close of 1861, Dr. *James Hector*, the travelling companion of Captain *Palliser* during his expedition through the Rocky Mountains (1857—1859), was called as Geologist to Otago, and in 1862 Mr. *J. C. Crawford* was appointed Provincial Geologist of Wellington. This was the beginning of a new era, in which the geological survey of New Zealand is progressing rapidly and systematically.

Concerning the history and development of the geography and chartography of NewZealand, I may well be allowed to direct the attention of the reader to the remarks, which Dr. *A. Petermann* has attached to the "*Geological and Topographical Atlas of New Zealand.*" [1]

The extraordinary variety of the surface, — as displayed, in a vertical direction, in the long and rugged mountain chains, in isolated mountain groups or in extensive plateaus and broad flats, and, in a horizontal direction, in the numerous indentations of the coast-line, — leads one to infer a very manifold geological composition of the soil. The first geological explorations of North and South Islands proved this very satisfactorily.

The geological maps of districts of the Provinces of Auckland and Nelson, as published according to my own and my friend Dr. Haast's observations, — although these maps in comparison to the geological surveys of West European countries may be designated as only the first outlines of the geological constitution of those districts, — nevertheless indicate an extremely manifold series of formations and rocks. And although it has been impossible hitherto, to parallelize the New Zealand strata according to their palæontological character with the European series of strata; yet this much appears from the facts gathered from observations hitherto made, that the stratified rocks from the oldest metamorphic to the newest sedimentary formations, and the eruptive formations from the oldest plutonic rocks to the youngest volcanic lavas are all represented in this country.

In the tabular view of the formations and strata occurring upon New Zealand, which I shall present in this chapter, I have first laid down the results of my own observations upon North and South Islands, without omitting, however, to add to them the most important facts ascertained up to the conclusion of this work by the investigations of Dr. *Haast* in the Provinces of Nelson and Canterbury, and by the researches of Dr. *Hector* in the Province of Otago

---

[1] See Dr. Fischer, the Geology of New Zealand, Auckland 1864. p. 3.

and in other parts of New Zealand. In the chronological arrangement of the strata and formations it was chiefly observations of the stratigraphical order of the different beds that directed my judgement. Palæontology, which in Europe, in the case of districts laying close together and thoroughly explored, is the surest guide for determining the relative age of the strata, affords upon New Zealand but few aids to judgment. The fossils found in favourable localities upon the extensive area, — where geological explorations have but commenced, and where the soil is laid bare merely by natural openings, such as we find them on sea-coasts, river banks, mountain slips etc. — admit, according to the present standard of science, of no other conclusions than such as have reference to the grand epochs in the history of the earths development. The researches upon New Zealand and upon the Australian and South American Continents have not as yet progressed so far, that according to palæontological criteria a more detailed comparison might be drawn between the range of the various formations of the extensive districts upon the Southern Hemisphere; nor does palæontology, as yet, afford a method, according to which for two continents so far apart as Australia and Europe, the synchronism of two strata might be proven. The first requisite is the establishment of at least the main features of a geography of plants and animals for the older geological periods also; we must first become acquainted with the palæozoic provinces in the same measure as we have gradually become acquainted with the neozoic districts of their distribution; — then, and not till then, will it be possible to decide, how far a comparison according to mere palæontological principles is at all feasible. Prof. *Agassiz* [1] is of opinion, that the mere comparison of the fossils of America with those of Europe justifies him in inferring, that between animals which lived at a great distance from each other, no specific identity can be traced, even though they be cöeval; that, on the contrary, species of the same family, belonging to different geological epochs, are more closely related to

---

[1] Agassiz: Ann. Rep. of the Museum of comparative Zoology, Boston 1862.

each other, — if they only originate from the same degree of latitude, — than species of the same geological age from different geographical zones. If such be the case, we may well assert, that the identity or close relationship of the fossil remains of one or the same geological period, such as they are laid down in the geology of to-day, is principally to be ascribed to the fact, that these remains were collected in *the same geographical zones*. But as the faunas of the present period, in continents at great distance from each other, are essentially different, so Prof. Agassiz believes himself justified in supposing that the same was the case also with the faunas of the older periods. The geology and palæontology of the countries in the Southern Hemisphere are, no doubt, the chief means to prove or disprove this view of the celebrated zoologist and palæontologist, and to reduce it to its proper standard. And New Zealand with its very peculiar flora and fauna of the present day, which show so very little analogy to the countries nearest to it, to Australia, the South Sea Islands and South America, — New Zealand, I say, is the very country, that might afford the necessary favourable or adverse proofs by means of its fossil plants and animal remains.

Although without the intention of drawing general conclusions from the little that is now known, I cannot refrain from simply enumerating the facts.

Most striking to the observer, on comparing the living land fauna of New Zealand with those of the countries nearest to it, is the nearly total absence of quadrupeds. New Zealand with its two, perhaps three, endemial species [1] is far surpassed in the number of mammalia by many much smaller islands of the South Sea. As substitutes for mammalia we have there most marvellous forms of wingless birds, the *Apteryx* species, which have been found to exist nowhere else. We might suppose, that perhaps the defunct land fauna of older periods displays more relationship with the faunas of the neighbouring continents of Australia and South

---

[1] A bat, a rat, and another animal, not yet described, resembling an otter, and living on the lakes of the South Island, are the only quadrupeds of New Zealand.

America. But, what is as yet known of the remains of extinct animals, is by no means in favour of such a supposition.

It is true, we have hitherto become acquainted with nothing but the remains of giant forms of wingless birds: *Dinornis* and *Palapteryx*. With these the present small representatives, the *Apteryx* species, are connected in the same manner as the now living Kangaroos of Australia with the gigantic forms of the extinct Marsupialia *Nothotherium (Zygomaturus Macleay)* and *Diprotodon*, which were found in the bone-caves and post-tertiary freshwater beds of Australia; or as the present Edentata of South America with the extinct giant-sloths *Megatherium* and *Mylodon*, the remains of which are dug from the diluvial deposits of the Pampas. From older than post-tertiary formations there are as yet no remains of warm-blooded vertebral animals known upon New Zealand. The fossil land fauna in New Zealand, therefore, as far as known, is as different from the fossil land fauna of the countries nearest to it, vz. of Australia and South America, as the living land fauna.

With regard to the marine fauna, the results which my friend Dr. *Zittel* has obtained from examining the fossils brought by me, prove that the mollusca of the upper tertiary strata stand in close connection with the living shells, bearing to the latter about the same relation as the fauna of the Sub-Apennine formation in Italy to the fauna of the Mediterranean. The same genera occur both fossil and living; and even the species are not seldom found to be identical. But at the same time there appears a striking resemblance to the tertiary fossils of Chili and Patagonia, as described by *Sowerby* and *d'Orbigny*, i. e. to a coëval fossil fauna of the same latitudinal zone.

On examining the fossils belonging to older formations, we observe even in the Ammonites, Belemnites, Inocerami etc. of North Island, which belong to strata of the mesozoic period (Jurassic or Cretaceous strata), such a striking resemblance to European forms of the same period, that we are tempted to place them on a level with European species. It is especially the belemnite of the group of *Canaliculati d'Orb.*, that shows such an analogy to the *Belemn. canaliculatus Schloth.*, that it is quite difficult to trace differences sufficient to justify a particular name for it.

The oldest fossiliferous strata, which I found on South Island at Richmond, not far from Nelson, contain species of Monotis and Halobia, which cannot be distinguished from the European forms *Monotis salinaria* Br. and *Halobia Lommeli* Wissm. from the Trias of the Alps.[1]

If it were allowed to consider these two facts already as sufficient proofs, we might at once infer as a natural consequence, that the faunas of former periods upon the Northern and Southern Hemispheres present an analogy and intimate relation, such as is no longer found in the living fauna; which inference seems by no means adapted to confirm the above opinion pronounced by Prof. Agassiz; but is wholly coinciding with the prevalent opinion, that the older the formations, the more analogy is exhibited in their fossils also in countries at great distance from each other.

Synoptical view of the geological formations and strata, presented upon New Zealand, in chronological succession.

I. Metamorphic Strata. (Foliated schists.)

None as yet proven upon the *North Island*.
Developed on a grand scale on the *South Island:*
a) in the Western ranges of the Province of Nelson as Gneiss, Micaslate, Quartzite and Phyllite with more or less vertical, partly fan-shaped stratification (Mt. Olympus); granite and syenite also appears in zones of considerable longitudinal extent. — These foliated schists are more or less impregnated with quartz and constitute the original beds of gold in the Province of Nelson.
b) On the West coast of the Province of Canterbury as a narrow belt consisting of the most different crystalline schists with more or less vertical stratification, and of granite. The gold-fields of the Province of Canterbury belong to this series. (Dr. Haast.)

[1] Recently strata with the same fossils have been found in New Caledonia.

c) In the southernmost part of South Island, in the Province of Otago, Gneiss-, Mica-, Chlorite-, Talc-, Quartz- and Clay-slates compose the larger portion of the province, especially the vast central mountain-ranges of 5000 to 9000 feet height; the Quartz interbedded in or associated with these slates is regarded as the matrix of gold. Dr. Lindsay compares these schists with the metamorphic slates of the Grampian Hills in Scotland. On Preservation Inlet granite is predominant. (Dr. Hector.)

d) Stewart's Island according to Dr. Hector is wholly composed of granite.

II. Palæozoic (primary) Formations.

Upon *North Island* dark coloured, quartzy clay-slate, gray sandstone, silicious schist (chert), and jasper with dioritic rocks (aphanite) intervening. Fossils hitherto none found, hence their age not yet determined.

a) On Bay of Islands (Dana).

b) On the Islands of Hauraki Gulf; on Great Barrier and Kawau Island containing copper-ores (copper-pyrites, copper-black and small quantities of red copper-ore), which have been worked for several years past. Upon Whaiheki near Auckland with beds of jasper and with psilomelan (Manganese).

c) On Cape Colville Peninsula (Province Auckland) with gold-bearing "Quartz-reefs," which since 1862 have given rise to considerable mining enterprises. (Coromandel gold-field.)

d) In the mountain-ranges on the West side of the Firth of Thames (Wairoa ranges), and South of them in the Taupiri and Hakarimata ranges.

e) In the mountain-ranges between Port Nicholson and East Cape.

Upon *South Island* in the Alpine ranges; only a few points admit of an approximative determination of their age.

a) In the Western ranges of the Province Nelson, on Mount Arthur slates with *Trilobites, Leptaena, Orthis* and corals of probably Silurian age. (Dr. Haast.)

b) In the Eastern ranges of the Province Nelson gray sandstone and clay-slate crossed by quartz-veins, as yet without fossils.

c) In the Southern Alps conglomerate, gray sandstone and clay-slate with more or less vertical stratification, composing the main-body of the mountains and also the highest peaks. In a northern side-valley of the Clyde, on the Upper Rangitata, Dr. Haast has discovered fossils indicating a Devonian age.

d) in the Otago Province according to Dr. Hector:

Kahiku series: Quartz, clay-shales, sandstone, diorite slate, black cross-cleaved slate, siliceous and true clay-slate.

Anau series: Porphyritic conglomerate, wacke, claystones, glossy slates and diabase, and porcellanite.

Oldest coal-formation in New Zealand in the eastern portions of the Southern Alps at the sources of the river Hinds, on Mount Harper, in the Malvern Hills, on the upper Ashburton River (Province Canterbury); the fossil plants, species of *Glossopteris*, indicate an age co-equal with the coal-fields of New Castle and Hunter River in *New South Wales.* (Dr. Haast.) See pag. 77.

Eruptive rocks in dykes of long extent: diorite and diabase.

III. Mesozoic (secondary) Formations.

These I have arranged according to their probable age and the various localities as follows:

1. Triassic group.

Upon *South Island.*

a) *Maitai* series in the Eastern ranges near Nelson; red and green clay-slates with highly inclined strata, and limestone at the base. (For example, on Wooded Peak, in Croixelles Harbour and Current Basin near Nelson.) As yet no fossils found.

b) *Richmond sandstone* about Richmond not far from Nelson; a ferruginous sandstone with *Monotis salinaria* var. *Richmondiana* Zitt. filling whole banks., *Halobia Lommeli* Wissm., *Mytilus problematicus* Zitt., *Spirigera Wreyi* Suess; Astarte, Turbo etc., fossil woods (*Dammara fossilis* Ung.)

2. Jurassic group.

Upon *South Island*:

a) *Waipara* beds, clay-marl with fossil remains of *Plesiosaurus Australis* Owen.

b) In the *Amuri District* (south-eastern portion of the Prov. Nelson) a complex of strata with saurian, fish and numerous shells (Dr. *Haast*).

c) *Shaw's Bay series* at the mouth of the Clutha River in the Prov. Otago with *Spirifer, Ammonites,* shells resembling mytilus etc. (according to Dr. *Lindsay*).

3. Cretaceous group.

A. Strata containing Ammonites and Belemnites.

Upon *North Island:*

a) On Waikato Southhead inclined beds of gray marl and sandstone with *Belemnites Aucklandicus* v. Hauer, *Aucella plicata* Zitt. etc.

b) On the Kawhia Harbour, similar beds with *Belemnites Aucklandicus var. minor, Ammonites Novo-Seelandicus* v. Hauer, *Inoceramus Haasti* Hochst. etc.

B. Carboniferous Strata.

Upon *North Island:*

a) On the West-coast of the Prov. Auckland, South of the mouth of the Waikato River, sandstone, marl and shale with small coal-seams and fossil ferns: *Polypodium Hochstetteri* Ung., *Asplenium palæopteris* Ung.

b) On the North side of the Harbour of Parengarenga, in the vicinity of the Harbour of Wangarei and on the Kawa-kawa-River, Bay of Islands. Mines very recently opened (Dr. Hector).

Upon *South Island:*

c) The *Pakawau coal-field* on Golden Bay, Prov. Nelson, with workable seams of very bituminous coal and indistinct fossil plants *(Neuropteris, Equisetites, Phoenicites)* in a coarse-grained sandstone.

Under this head I will also mention the coal-fields on the West coast of the Prov. Nelson, examined by Dr. J. Haast:

d) The *Buller coal-field*, on the Buller (Kawatiri) River, ten miles above its mouth, with seams 8 feet thick.

e) The *Grey coal-field*, on the Grey (Mawhera) River, seven miles above its mouth, with beds 12 to 17 feet thick. Among the fossil plants Dr. Haast mentions: *Zamites, Pecopteris, Equisetum* and dicotyledon leaves.

f) The existence of good bituminous coal in thin beds on Preservation Harbour (Province Otago), upon Chalky Island, is mentioned by *Ch. Forbes* (Quart. Journ. Vol. XI. p. 528). According to Dr. *Hector* this coal belongs to the same class as those found on *Patterson's* Point in Australia.

4. Eruptive Formations of the Mesozoic Period.

Upon *South Island:*

a) The Serpentine dyke of Dun Mountain near Nelson, with lodes of copper-ore and chrome-ore, with Dunite (Olivine rock) and Diallage.

b) The Syenite of Wakapuaka and the Pyroxene porphyry of the Brook-Street Valley near Nelson.

c) Felsite porphyry and Melaphyre of the Southern Alps (Prov. Canterbury). Dr. Haast.

d) The Hyperites of Mt. Torlesse, Prov. Canterbury. Dr. Haast.

IV. Cainozoic (tertiary) Formations.

1. Older tertiary Strata.

A. Brown-coal-bearing, lower series.

Upon *North Island:*

a) The *Hunua coal-field* near Drury and Papakura, south of Auckland, discovered in 1858 by the Rev. Mr. Purchas, and worked since

1859 by the Waihoihoi Company. The coal is brown-coal, and contains a fossil resin, Ambrite *(Haidinger)*, which has often been mistaken for Kauri gum. The Price of this brown coal in Auckland is 30 to 32 s. per ton. The shales and sandstones accompanying the coal contain bivalves and leaves of dicotyledones: *Fagus Ninnisiana* Ung., *Fagus dubium* Ung., *Myrtifolium lingua* Ung. etc.

b) The *coal-fields* of the *Lower Waikato Basin* (Province Auckland). A seam of great thickness has been laid open near Kupakupa on the northern slope of the Hakarimata range.

c) The coal-beds on the western and southern borders of the *Middle Waikato Basin*, as yet wholly untouched.

I also enumerate under this head:

d) Various, hitherto little noticed, localities of brown coal in the North of the Province Auckland, on Manganui Harbour (Doubtless Bay), near Rodney Point, on Cape Colville Peninsula close by Coromandel Harbour etc.; in the South, on the Mokau and Wanganui Rivers.

Upon *South Island:*

e) *Jenkins' coal-mine* at Ennerglyn near Nelson. The strata have been greatly disturbed here; the coal-seam is crusted; in a ferruginous sandstone are imbedded leaves of dicotyledon plants: *Phyllites Nelsonianus* Ung., *Phyllites Brosinoides* Ung., *Phyllites quercioides* Ung., *Phyllites eucalyptroides* Ung., *Phyllites leguminosites* Ung.

f) *Motupipi coal-field* on Golden Bay (Prov. Nelson), opened since 1854; and the brown-coal beds on Rangiheta Point West of Motupipi. In the coal Ambrite is contained as at Drury.

g) Throughout the provinces of Canterbury and Otago similar coal-deposits occur, and more especially in the valleys or beds of the Selvyn, Rakaia, Rangitata, Ashburton and other rivers in Canterbury; and near Fairfield, on Saddle Hill, on the Tokomairiro, Clutha river, in the Waitahuna and Wetherstone-flats etc. in Otago. Marketprice of this coal in Dunedin £ 2 per ton (according to Dr. *Lindsay*).

B. Marine Strata. Upper series.

At the base frequently foraminiferous clayey strata alternating with sandy banks, — these strata perhaps coëval with the brown-coal beds — towards the top tabular limestones and sandstones, abounding in fossils. *Echinodermes: Brissus, Schizaster, Hemipatagus, Nucleolites* etc. *Brachiopodes: Waldheimia, Terebratula, Terebratulina. Conchiferae: Ostrea, Lima, Pecten, Cucullaea. Gastropodes: Neritopsis, Scalaria. Shark's teeths*, foraminiferæ, and bryozoes.

Upon *North Island:*

a) *Waitemata* beds: sandstone and shale on the Isthmus of Auckland and on North shore, generally scarce of fossils. — On the Orakei Bay near Auckland sandy glauconite-bearing strata, abounding in foraminiferæ and bryozoes, together with small specimens of pectens (*P. Aucklandicus; P. Fisheri* Zitt.), bivalves, and forms resembling belemnites (probably Vaginella-shells). On North shore and on St. George's Bay with pieces of drift-wood turned into brown-coal.

b) *Limestone* near Papakura: marls abounding in foraminiferæ and bryozoes, and tabular limestone opened in quarries, with *Turbinolia, Schizaster, Waldheimia gravida, Pecten Fisheri, Neritopsis* etc.

c) Finely granulated sandstone (resembling the Quader--Sandstein of Saxony): on *Waikato Southhead* resting unconformable above the belemnite-beds, and on the West coast south of the mouth of the Waikato, containing *Cidaris, Nucleolites, Schizaster, Fasciculipora, Retepora, Cellepora, Waldheimia, Pecten* etc.

d) The shales and tabular limestones on the West coast of the Prov. Auckland (Whaingaroa, Aotea, and Kawhia Harbours) with numerous foraminiferæ and other fossils.

e) The *cave-limestone* of the Upper Waipa and Mokau Country with caves, funnel-shaped holes in the ground and subterraneous water-courses.

Upon *South Island:*

f) *Motupipi* and *Rangiheta* Limestone on Golden Bay (Prov. Nelson), tabular limestone deposited upon the brown-coal bearing strata, with
  Brissus eximius Zitt.
  Pecten athleta Zitt.
   „ *Burnetti* Zitt.
  Waldheimia lenticularis Desh.

g) The sandy cave-limestones of the Aorere Valley, and the sandy limestones on Cape Farewell (Prov. Nelson). On Cape Farewell with a great abundance of fossils: *Hemipatagus formosus* and *tuberculatus* Zitt., very frequent; *Pecten Hochstetteri* Zitt.

h) The *auriferous conglomerates* of the Aorere Valley, developed especially on the "Quartz ranges", (perhaps belonging to the driftformation). Commencement of the gold-digging on the Aorere goldfield in 1857.

Here I enumerate also:

i) White and yellow, foraminiferous sandstones and green sands, compared by english geologists with the chalk-tuffs of Mastricht and Faxö:
  Mataura and *Shag Valley Series* (according to Dr. *Lindsay*).
  Ototora (Oamaru) Series (Prov. Otago) calcareo-arenaceous, greatly

abounding in foraminiferæ and bryozoes; with *Cythereis, Terebratula, Cereopora, Textularia, Bairdia, Eschara* etc. (according to *W. Mantell*, Quart. Journ. Vol. VI. p. 329).

Woodburn Series (on Saddle Hill near Dunedin) with *Ostrea*, and echinite quills (according to Dr. *Lindsay*).

Green Island Series (Dunedin), sandy glauconite-bearing strata full of foraminiferæ, together with *Terebratula*, Echinodermes and Shark's teeths (according to Dr. *Lindsay*).

2. Younger tertiary Strata.

Conglomerates, sandstones, limestones and clays upon North and South Islands with a fauna, very nearly related to the living Mollusca of New Zealand. The strata raised partly to a height of 2000 feet above the level of the sea and in some places (on the cliffs about Nelson) in an almost vertical position.

Upon *North Island:*

a) Kohuroa near Rodney Point, North of Auckland; dark clay-slate breccia with living and extinct species: *Terebratella dorsata* Gmel., *Rhynchonella nigricans* Sow., *Purpura textiliosa* Lam., *Turitella rosea* Quoy., *Turbo superbus* Zitt., *Crasatella ampla* Zitt.

b) *Hawkes Bay Series;* limestones, sandstones, and clay-marls replete with fossils: *Pecten Triphooki* Zitt., *Venus, Mytilus, Pectunculus, Trochita* etc.

c) *Wanganui River beds;* blue clay with sea-shells, covered by a volcanic conglomerate; in the clay are imbedded numerous recent species, — as in the Awatere Valley, South Island, — *Fusus nodosus* Quoy, *Murex Zealandicus* Quoy, *Venus mesodesma* Gray, *Venericardia Quoyii* Lam.; *Pecten asperrimus* Lam. (W. Mantell, Quart. Journ. Vol. IV. p. 239, and Quart. Journ. Vol. VI. p. 332); sands with large oysters: *Ostrea ingens* Zitt.

Upon *South Island:*

d) Highly inclined strata on the cliffs about Nelson with *Cardium, Pectunculus, Trochosmilia, Bulla, Cerithium, Buccinum* etc.

e) Blue clays in the *Awatere Valley* (Prov. Marlborough), raised to a height of 2000 feet above the sea level, with shells in an admirable state of preservation: *Arca* (a very large species), *Pectunculus, Voluta, Struthiolaria, Trochita* etc.
   With these I class also:

f) The *Waitaki Strata*, arenaceous, on the Waitaki River bounding the Provinces Otago and Canterbury, with bones of *Cetaceæ.* (*Dr. Haast.*)

g) The *Moeraki Series* (Onekakara), argillaceous, on the East coast of the Province Otago, described by Mr. *Mantell* as pleistocene (Quart. Journ.

VI. p. 330), with *Pustulopora, Struthiolaria, Ancillaria, Fusus* etc. The shells in an admirable state of preservation resemble those of living species that have lost the colour; at Moeraki with Septaria.

V. Post-tertiary and recent Formations.

1. Lignite-bearing Strata.

Plastic clay and sand with lignite, containing the fossil remains of living species of plants.

Upon *North Island:*

  a) The lignite beds *of the Manukau flats* with clays of different colours and extensive deposits of pumice-stone-dust.

  b) The lignite beds *of the lower Waikato Basin.*

Upon *South Island:*

  Lignite-beds of the Provinces Canterbury and Otago in various localities.

2. Glacial-Drift.

Upon *South Island:* Moraine deposits and loess. The moraines of the *glacier period* in the Southern Alps, especially on the Alpine lakes Rotoiti, Rotoroa (Prov. Nelson), Tekapo, Pukaki (Prov. Canterbury and Otago) etc. (Dr. J. Haast, Quart. Journ. 1865. p. 133.)

3. Marine, lacustrine and fluviatile drift-deposits.

Boulder, gravel, sand, and loam-deposits with very distinctly marked terraces, upon the table-lands, in river valleys, and in coastward deposits. Upon *North Island* volcanic and tufaceous deposits, to which especially pumice stones have furnished the material.

  a) The *pumice stone diluvium* of the Middle Waikato Basin (Province Auckland).

  b) The terraced pumice stone plains of the Upper Waikato Basin or of the Taupo plateau.

  c) The terraced pumice stone diluvium of the Wanganui district.

Upon *South Island* fluviatile deposits of the *drift period* and *terrace period,* for which the rocks of the Southern Alps have furnished the material.

  a) The Mutere Hills, Buller, Grey plains etc. in the Province Nelson, partly auriferous (on the Aorere, Wangapeka, Buller).

  b) The Canterbury plains and the drift in the Alpine valleys, Prov. Canterbury.

  c) The *great gold-drift of the Prov. Otago* (Otago gold-fields):

    α) an upper one consisting of conglomerate clays (boulder clays) and shingle, and

    β) a lower one characterized by lignite layers.

The principal gold district of the Prov. Otago is the country watered by the Lakes Hawea, Wanaka and Wakatip, and the Clutha River with its various branches (Tuapeka and Dunstan gold fields, Lindis and Arrow diggings). Auriferous deposits are found also along the tributaries of the Mataura (Nokomai gold-field), Tokamairiro (Woolshed gold-field), Shag and Taeri (Mount Highlay Diggings), Waikouaiti and on other rivers and creeks in various portions of the province, likewise on the coast (Moeraki Beach), and in and around the Capital Dunedin (Saddle Hill); consequently the larger portion of the Province Otago is auriferous. Together with gold are found Iserine, Cassiterite, Aquamarine (Beryll), Aventurine, Topaz, Garnets and other minerals. — According to Dr. *W. L. Lindsay.*

Commencement of the gold-diggings upon the Tuapeka gold-field in 1861; first escort of gold 12. July 1861; amount of gold obtained up to 31. March 1862: 359,639 oz. or £ 1,393,600. The Tuapeka gold-field yielded up to Sept. 1864 gold valued at about £ 6,000,000.

4. Recent coastward Deposits, estuarine or littoral.

Upon *North Island:*
  a) Sand-hills; most extensively developed along the West coast and on the coast of the Bay of Plenty.
  b) Deposits of titaniferous iron-sand along the West coast.
  c) Estuary mud with brackish shells in the estuaries of the East and West coasts.
  d) *"Submarine Woods"* on the coast of the Prov. Taranaki (Dieffenbach).
Upon *South Island:*
  a) Sand-hills on a very grand scale upon the Cape Farewell Sandspit, 20 miles long (Prov. Nelson).
  b) Boulder deposits in the Sounds and Fiords of the N. E. and S. W. coasts. A grand illustration of it, the "Boulder Bank", forming the Harbour of Nelson.
  c) Boulders of the West coast containing Nephrite (Punamu of the natives).

5. Recent inland-formations, lacustrine and alluvial (river-silts, shingles and deltas).

Upon *North Island:*
  a) Extensive swamps and turf-moors along the East coast in the Middle and Lower Waikato Basins and in front of the mouth of the Waikato.
  b) Deposits of Kauri gum in the northern portions of the Province Auckland; found everywhere on the surface in places where formerly forests of the Kauri pine *(Dammara Australis)* had stood.
  c) Gold-bearing gravel in some creeks of the Cape Colville Peninsula,

especially in the environs of Coromandel Harbour. In 1852 first washing experiments; amount of gold obtained scanty; lately a little better.

d) Pumice-stone gravel on Lake Taupo and on the Waikato River.

e) Deposits of *silicious earths* ("Infusorial earths") on the Bay of Islands, near Auckland in the Cabbage Tree Swamp; near Onehunga; near New Plymouth (W. Mantell, Quart. Journ. Vol. VI. p. 332).

Upon *South Island:*

a) Auriferous gravel of the rivers and creeks of the Nelson and Otago gold-fields.

b) Till deposits (glacier mud) in the Alpine lakes.

c) *Silicious earth* on Lake Waihora near Banks Peninsula (W. Mantell, Quart. Journ. VI. p. 333).

6. Recent (partly perhaps diluvial) deposits with Moa-remains.

The principal known localities of Moa-bones in swamps, river alluvions, in caves and on the sea-beach on both Islands are:

Upon *North Island:*

a) The limestone caves on the Upper Waipa and Mokau; among them the caves *Te ana ote moa* and *Te ana ote atua,* in which Dr. Thomson was collecting in 1852.

b) The Tuhua district West of Lake Taupo, and Mt. Hikurangi in the same district. The Rev. Mr. Taylor has gathered many interesting bones in that neighbourhood.

c) The plateaus of the Taupo country in the centre of North Island.

d) Opito between Mercury Bay and Wangapoua. Mr. *Cormack* found here, in 1859, Moa-bones around of the cooking places and between the very cooking stones of the Maoris.

e) The eastern coast districts between East Cape and Hawkes Bay, especially in the alluvium of smaller rivers and creeks (Wairoa, Waiapu etc.); Rev. Williams and Collenso collected here.

f) The environs of the Lake Tarawera. Here an extensive area was found literally strewn with Moa-bones, after the trees upon it had been burnt down.

g) The *Ngatiruanui District* near Rangatapu on Waimate Bay, South East of Cape Egmont, especially on the Waingongoro River, where Mr. *W. Mantell* made up a large portion of his famous collection, and where he found a mound, in which Moa-bones were interred promiscuously with human remains and dog-bones, the offals of extensive feastings in days gone by.

h) The *plains* of the Wanganui River.

The Moas, consequently, seem to have been distributed all over the southern part of North Island; but are totally wanting upon the narrow northwestern peninsula, North of Auckland, where to my knowledge no trace of Moa-bones has as yet been found. This circumstance would moreover serve to explain, why there is no mention made of Moas in the traditions of the Ngapuhis, who inhabited this northern peninsula.

Upon *South Island:*

a) The limestone caves of the Aorere Valley in the Province Nelson, especially the Moa Cave and Hochstetter's Cave. Whole skeletons of *Din. elephantopus, didiformis* and *Palapteryx ingens* have been discovered in these caves. The lower strata with *Din. elephantopus* are probably diluvial.

b) The plains of Canterbury; where it scarcely ever happens that a ditch of any length is dug without bones coming to light there which chiefly belong to the species *Din. dromioides, struthioides* and *robustus;* and about 35 miles north of Christchurch there is a large swamp near the Glenmark home station which is literally interlarded with Moa-bones; Dr. Haast in 1866 found here no less than twenty-five skeletons of the Din. elephantopus and Din. crassus.

c) The vicinity of Timaru South-west of Banks peninsula; the caves and swamps adjacent to this coast-point are said to be full of Moa-bones.

d) Near Ruamoa, 3 miles South of Oamaru Pt. ("First Rocky Head"), Mr. *W. Mantell* found a skeleton of *Din. elephantopus,* buried in the sand near the coast and close tho the same spot circular pits with charcoal, half burnt Moa-bones and round stones, such as the natives use in cooking, in fact regular Moa cookingstoves (Hangi Maori); the same gentleman also found some old stone-knives of obsidian.

e) At the outlet of the Waikouaiti, 17 miles North of the Otago Peninsula, there is a swamp, which in time of highwater is flooded by the sea; this is the famous spot from which Mr. *Percy Earl,* Dr. *Mackellar* and Mr. *W. Mantell* have enriched their collections.

f) At the mouth of the Clutha River, South of the Otago Peninsula, and Moa Hill, 15 miles farther in the interior.

7. Accumulations by the hand of Man.

There are found in different parts of the North and South Islands:

a) heaps of muscle-shells, — Cardium, Ostrea, Mytilus, Patella, Venus, Haliotis, Mesodesma, Turbo, Monodonta etc. — particularly in places formerly occupied by pahs and villages, analogous to the Kjökkenmöddings of Danemark;

b) cooking stones, char-coal, and wood-ashes in the cooking places of the Maoris;

c) stone implements of the Maoris; prepared of Aphanite, Nephrite, Chert;

d) human bones; bones of dogs, sea-mammalia, fishes, and various birds close by the cooking places.

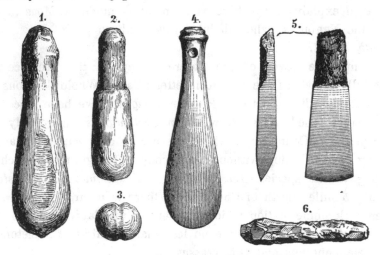

Stone-implements of the Maoris.

VI. Volcanic Formations.

1. Older Volcanic Formations

of the tertiary and older quartary period. Compact or fissured mountain cones without distinct craters and lava streams; thick and far extending beds of breccia, conglomerates and tuffs.

Upon *North Island:*

a) North of Manukau Harbour (Prov. Auckland) along the West coast extensively developed andesitic and doleritic breccias; decomposed farther in the interior into conglomeratic clays, together with dikes of anamesite and basalt.

b) South of Manukau Harbour on both sides of the Waikato and thence as far as Aotea Harbour basaltic conglomerates and basalt; without any distinct cone and crater formation.

c) The volcanic table-land between the Upper and Middle Waikato Basin; deposits of trachytic and pumicestone tuffs, together with extinct volcanic cones, consisting of trachytic, andesitic, and doleritic rocks. Examples: Karioi, Pirongia, Kakepuku, Maunga Tautari, Aroha etc.

Upon *South Island:*

a) eruptions of Quartz-trachyte at the foot of the Southern Alps (Prov. Canterbury), compact domes and conical mountains such as Mount

Sommers (5240 feet), Mount Misery, Survey Peaks, Mount Grey and others, together with extensive tuff strata. With these it may perhaps be proper to class also the Inland Kaikoras.

b) The group of the extinct trachyte and andesite volcanoes of the Banks Peninsula.

c) The volcanic "Traps" of Dunedin (Province Otago); according to Dr. *Lindsay:* basalts with columnar, spheroidal, and tabular segregation on Stoney Hill, Mount Cargill, Saddle Hill, Signal Hill Range, Flagstaff and Kaikorai Hill etc. Trachytes and volcanic tuffs, the latter used in building (quarries on Anderson's Bay).

2. Newer Volcanic Formations

of the recent period with acid (siliciferous) and basic products of eruptions. Cones with opened and unopened tops, partly still active; distinct lava streams.
Upon *North Island:*

a) *Taupo Zone.* Rhyolitic and trachytic lava formation, obsidian and pumice stone developed on a grand scale. Two active volcanoes, Tongariro, 6500 feet high, and Whakari or White Island (863 feet), in the state of solfataras; numerous extinct volcanoes, among them the highest peak of North-Island, Ruapahu, capped with perpetual snow, about 10,000 feet high.

b) The *Taranaki District*, with Mount Egmont (8270 feet), an extinct trachyte volcano belonging perhaps to the older volcanic period.

c) *Auckland Zone.* Basaltic lava formation upon the Isthmus of Auckland. 63 points of eruption. Tuff cones, lava cones, and scoriæ or cinder cones with distinctly preserved craters and lava streams, all extinct.

d) *Bay of Islands Zone* between Hokianga Harbour and Bay of Islands; basaltic lava formation as on the Isthmus of Auckland; a number of small extinct cinder-cones from which basaltic lava streams have issued.

Upon *South Island:*

a) Basaltic and doleritic cones with lava streams at the eastern foot of the Southern Alps, among the Malvern Hills (Prov. Canterbury). Palagonite tuff at the foot of Mount Sommers.

b) Portions of the volcanic system of Bank's Peninsula, for example the basalt eruptions of Quail Island.

3. Hot springs.

Upon *North Island:*

a) The hot (intermittent and non-intermittent) springs, boiling mudpools, solfataras and fumaroles of the *Taupo Zone*, or the Ngawhas and

Puias of the natives, with deposits of silicious incrustations, alum, gypsum and sulphur. Formation of small mud cones.

b) The hot springs of the *Bay of Islands Zone.*

Upon *South Island:*

The hot springs of the Inland Kaikoras.

From the preceding synopsis it appears, that the history of the geological development of New Zealand can be traced to the very remotest periods of the earth's history.

At the time, when the neighbouring Australia, one of the oldest continents of the earth — at least, as regards the eastern and western portions of it, the latter consisting principally of palæozoic strata, — arose from the depths of the ocean, there were also portions of New Zealand already projecting above the mighty main as rugged, barren masses of land; their shape, of course, was very different from the present appearance of the archipelago; they perhaps stood in connection with larger continental bodies, that long ago have been submerged again in the depths of the watery abyss. But while the eastern and western portions of Australia since the close of the palæozoic period have been quiet, and the soil rarely disturbed, upon which plants and animals found ample chance to grow and propagate themselves in an uninterrupted succession up to this present time; New Zealand, on the other hand, was till within the latest period a scene of the grandest revolutions and convulsive struggles of the earth which, continually changing the original form of the land, gave it by degrees its present shape.

Numerous observations made on the North and South Islands lead to the conclusion, that not until within the most recent period of the earth, after the tertiary period (probably with the commencement and during the time of the volcanic action on both islands) large portions of the land were raised by quite 2000 feet; some parts even by 5000 feet above the level of the sea; not all at one time, but by slow and gradual secular elevations, perhaps with longer and shorter intervals of perfect stagnation. To this height ascend the tertiary strata upon the North and South Islands with numerous imbedded shells; and the same height is reached by the deposits of the drift formation and the peculiar terrace formations in all the larger river valleys of both islands, as well as by the shingle and gravel deposits upon the broad plains on the East side of South Island.

Yet, while the land was greatly enlarged by elevation, by alluvial deposits and by the eruptions of volcanoes; other portions, on the other hand, were simultaneously submerged in the deep. It is to such an event that the formation of Cook and Foveaux Straits probably owes its origin.

A mere glance at the map of New Zealand shows us the very peculiar shape of the northern part of North Island: on the East coast steep promontories, peninsulas, numerous cliffs and islands, inlet upon inlet, bay after bay; and on the West coast, where the predominant West wind piles up long rows of sand-hills, instead of bays or inlets the dammed-up estuaries of the rivers. All this conspires to make the impression of a land once of a far greater extent, of which only the higher parts, the mountain ridges and peaks, are still towering above the sea, while its low lands "bottoms" and valleys are over flowed, coming forth in time of low water in the shape of shallow mud flats. The peculiar features of the northern peninsula of North Island are only to be accounted for by adopting the theory of a gradual sinking of the land.[1] This sinking process, however, seems to have extended more or less over

[1] Zoological facts speak likewise in favour of this opinion. Little Barrier Island in Hauraki Gulf North of Auckland is the haunt of numerous Kiwis *(apteryx)*. How should these wingless birds have got into this little island, unless it was formerly contiguous to and part of North Island? In the same manner, the former distribution of the now extinct species of Moa *(Dinornis* and *Palapteryx)*, — birds, like the Kiwis wholly without any organs of flight, — over the North and South Islands, is an evident proof of the former contiguity of the two islands. Doubtless, Norfolk Island also in a north-westerly direction was once attached to North Island. According to Capt. King's statements, the *bottom* is found everywhere between Cape Maria van Diemen and Norfolk Island; and various peculiarities of the flora and fauna of Norfolk Island (for example the occurence of the New Zealand flax plant, the occurrence of the Nestor species etc.) speak in favour of a former cohesion of the parts mentioned.

the whole western coast of the North and South Islands, whilst the East coast was raised at the same time, so that an imaginary line drawn parallel with the West coast of South Island, — not at too great a distance from it, prolonged through North Island and terminating on its East coast in Tauranga Harbour, would represent the axis of elevation on one side, and of depression on the other.

Striking proofs of the elevation of the East coast are presented especially in the environs of Bank's Peninsula. This peninsula seems to have been an island till within a very recent age. It is connected with the main land only by a very low neck, the numerous lagoons of which, and especially the great *Lake Ellesmere* (Waihora) on the S. W. side of the peninsula, may be considered remnants of the former sheet of water covering the intervening tract of land. Two miles inland, within a semicircle extending from the mouth of the Waimakariri to Lake Waihora, lies a chain of old sand hills, that once constituted the coast. [1]

Whether New Zealand formed part of other larger bodies of land previous to these late catastrophes, which gave the Archipelago its present form, is a question, that can hardly be answered in the affirmative, interesting as such an answer, if based on geological facts, would be for a full understanding of a great many particularities presented in the flora and fauna of the islands. Supposing such a contiguity to have really existed, say with Australia or America or some continent now buried in the South sea, the separation would have taken place already at a very remote period.

If from the identity of fossil plants found upon Iceland, Madeira, the Azores, the Canaries, and the Cape Verd Islands, the inference was drawn of a former contiguity of all the Atlantic Isles, of one vast continent *Atlantis*, that once connected Europa, Africa and America; all and every evidence is at yet wanting for a similar inference of a former contiguity between New Zealand and

---

[1] Ch. Forbes, Quart. Journal Vol. XI. p. 526. Under these sand-hills the remnants of former woods are found (W. Mantell, in Quart. Journal Vol. VI. p. 321), an indication of repeated fluctuations.

the neighbouring continents. Neither the fossil flora and fauna, as far as they are now known, nor the geological structure of New Zealand intimate such a contiguity; on the contrary, various geological facts speak in favour of the opinion that New Zealand, situated in the middle of a vast and very deep sea, has been an island from time immemorial, remote from all larger continents and existing in an isolated position, although perhaps of a form somewhat different from its present one.

Another interesting question is, whether New Zealand, after the many volcanic and neptunic struggles of which the soil bears unmistakable traces, does now rest fully at peace with nature? Most certain it is that since Cook's times the forms of the double island have suffered no material change. But even since the establishment of the first European settlements there, the ever active powers within the bowels of the earth have shocked and knocked and shaken the land so strongly in various corners of the islands, especially on both sides of Cook Strait, on the Wanganui (1843), in the vicinity of Wellington, and right opposite to it in the Wairau plains, and near Cape Campell (1848 and 1855), that the awestruck settlers naturally asked, whether the ground beneath their feet was really safe, and whether they had not ventured by a few centuries too soon to entrust themselves to a new-born baby of our mother earth?[1] With regard to this, the fearful may well feel at ease, bearing in mind the consoling fact that long before the first European set foot on New Zealand's shores, the land was the abode of numerous and populous tribes, that could boast of a long succession of time-honoured ancestry, and that it has been clearly proven by observations and facts that the volcanic forces below, which are now no longer capable of discharging fiery liquid lava, are visibly decreasing and dying out.

By this, however, we do not mean to imply that the ease and quiet of the inhabitants may not be disturbed hereafter by many a violent earth-quake, by many a "subterranean storm", and

[1] An English Author in the "Rambles at the Antipodes" describes New Zealand as "a geological baby, an infant fire-born, still pulling in his mother's arms."

even perhaps by occasional faint volcanic eruptions (Tongariro). There is especially one sore, suspicious spot in Cook Strait. The violence and frequent occurrence of earth-quakes on both sides of this strait is an undeniable fact. From the observations of the last twenty-two years, an earth-quake may be anticipated every sixth or seventh year. The first great earth-quake since the founding of the town of Wellington occurred in 1848, seriously damaging the buildings of the place. Besides various slight shocks, there were three violent shocks felt, the first on the 16. October at 1. 40 a. m., the second on the 17. October at 3 p. m., the third on the 19. October at 5 a. m. The second and last earth-quake on the 23. January 1855, attended with the most extraordinary phenomena, was felt throughout New Zealand. A powerful surge rolled from Cook Strait into Wellington Harbour. Manuka Point near Wellington was suddenly raised nine feet, whereas the elevation in the town itself was only two feet, and on the opposite side of Cook Strait, at the mouth of the Wairau River, depressions took place. In the Awatere Valley the soil received large fissures and crevices. One such fissure was traced full forty miles, and as late as 1859 my friend Dr. Haast found some of those fissures three feet wide and several feet deep. Near Cape Campbell parts of the mountains fell exposing white rocks, so that the sailors spread the report of having seen fresh fallen snow, and two days after the earth-quake the surface of the sea was seen covered with dead fishes. All the phenomena hitherto observed point to a central point in Cook Strait, and it is a generally prevailing opinion that a submarine volcano lies there, with the eruptions of which the earth-quakes are intimately connected. And, in fact, it has been proven by soundings made by English Naval Officers,[1] that in front of the entrance to the harbour of Wellington, 41° 25' South lat. and 147° 37' long. East of Greenwich, there is a crater-shaped hole at the bottom of the sea, over which the sea has never been seen quite calm.

[1] Comp. No. 2054 English Admiralty Maps. A similar deep hole at the bottom of the sea is at the West coast of North Island a little North of Kaipara Harbour in lat. 36° 20' and long. 173° 40'.

Earth-quakes are very frequent also in other parts of New Zealand, and especially on the volcanic line between Tongariro and White Island, where, — on Lake Tarawera, — not a single month passes without at least one slight shock. Upon South Island Dusky Bay, according to older accounts, seems to have been visited by numerous earth-quakes in 1826 and 1827; and 80 miles North of it a little bay named "jail", formerly much frequented by seal-hunters on account of the safe shelter it affords, is said to have been laid perfectly dry by an earth-quake.

[1] I am indebted to Mr. Triphook in Port Napier for the following list of earth-quakes from 1856—1858:

Year.	Date.	Time.	Number of Shocks.	Place.	Remarks.
1856.	Dec. 11.	6 45 a. m.	2	Wellington.	Wind S. W.
"	13.	8 45 p. m.	1	Hutt Valley near Wellington.	
1857.	Feb. 21.	3 20 p. m.	1	Wellington.	Undulatory motion.
	Mar. 18.	5 15 p. m.	1	Pourua Harbour.	
	Apr. 9.	10 0 a. m.	2	Wellington.	Undul. motion.
	July 30.	12 55 a. m.	2	"	Violent Vertical Shock.
	Aug. 23.	9 50 a. m.	1	"	Fine calm day.
	Sept. 6.	1 20 a. m.	1	"	Fine calm day.
	" 27.	12 25 a. m.	1	"	Wet day.
	Oct. 8.	8 10. a. m.	2	"	Wind S. W. rainy and cold.
	Dec. 27.	1 20 a. m.	1	Port Napier.	Changeable weather.
1858.	Jan. 6.	4 15 a. m.	1	" "	
	Apr. 16.	2 0 a. m.	1	" "	Showers of rain.
	" 18.	1 0 a. m.	1	" "	Showers of rain.
	July 20.	1 0 a. m.	1	" "	Fine weather.
	Aug. 2.	2 0 p. m.	1	" "	Fine weather. Swelling sea.
	Oct. 23.	6 10 p. m.	1	" "	Wind S. E. Swelling sea. Vibrative motion.
1859.	July 1.	7 0 a. m.	1	" "	Showers.

1863. Feb. 23. an earth-quake was felt in New Zealand (Hawkes Bay) which destroyed several houses and left various crevices in the soil.

CHAPTER IV.

The Mineral Riches of New Zealand.

Mineral Coal.

Government Geologists in the English Colonies. — Demand for coal-supplies upon the Southern Hemisphere. — The Australian Newcastle. — Brown coal upon North Island. — The coalfield in the Drury and Hunua Districts near Auckland. — Ambrite, a new fossil resin. — Fossil plants. — The coalfield on the Lower Waikato. — On the Middle Waikato and on the Waipa. — Recent lignite. — Coal upon South-Island. — Jenkin's Colliery near Nelson. — Motupipi brown coal. — Pakawau coal. — Coal on the Buller and Grey Rivers. — Coal in the provinces of Canterbury and Otago. — Conclusive remarks. — Appendix. Synopsis of the results from analytical investigations of New Zealand coals.

———————

It is not only their manners and customs, but also their excellent institutions that the sons of Albion carry with them across the ocean. The farfamed Geological Institution in Jermyn street has its members all over the world. Under the excellent direction of Sir *Roderick Impey Murchison*, the liberal patron of geographical and geological science, — to whom I also am greatly indebted for various proofs of kindness, — that Institution has become a nursery of worthy young geologists, who carry on their researches and explorations in all parts of the world. The English Colonial Governments, convinced of the great benefit, that can be conferred on the industrial arts by an accurate study of the crust of the earth, have spared no expense to secure zealous labourers in the field of science and by their help to carry through the geological and mineralogical explorations of the colonies. Thus we have seen for years past the Government geologists busy at their work in India, in Australia, in Canada, at the Cape of Good Hope and in Tasmania; and science hand in hand with practice for the purpose of exploring and bringing to light the mineral treasures of those countries.

New Zealand alone, the youngest of the colonies, was made an exception. While its remarkable animal and vegetable products were made known long ago through the elaborate treaties of celebrated naturalists, its soil, in a geological point of view, remained a *terra incognita*. It is only within the last few years that a geological and mineralogical survey, conducted under the auspices of the Government, has been established. The first impulse to this proceeded from a practical want, — *the want of coal.*

Although this question has been for the present cast into the shade by the dazzling discoveries of gold in the southern provinces; yet it will arise anew and gain in importance, in proportion to the rising swell of immigration, to the rapid increase of the population, and to the speedy development of navigation. The coal question in its turn will supplant the gold question, and the coal fields of New Zealand will at any rate long survive its goldfields. The demand for coal upon the Southern Hemisphere in the Pacific Ocean, although almost exclusively for navigation purposes, is already a colossal one. How much greater will this become, when once in the various countries of that hemisphere manufacturing is fully developed! The whole immense coast extent bordering the Pacific Ocean from the Cape of Good Hope to the Behrings Strait, and from Cape Horn to the Aleutian Islands is likewise in want of coal. The supplies are furnished from England and North America, — millions of tons, and millions in value! As yet, there is but one point in the whole immense territory of the Pacific, which has begun a very modest competition with the American and English coal trade; I am speaking of the Australian *Newcastle.* [1]

As in Newcastle, England, so also here the vessels can

[1] Newcastle on Hunter River, North of Sydney, in the Colony of New South Wales.

receive the coal immediately from the coal mines at the mouth of Hunter River, which by structures erected on a grand scale has been turned into an accessible and safe harbour. The coal-fields lie close by the sea shore, some beds cropping out even on the steep coast bluffs, so that they can be distinctly seen from the sea on a voyage from Sydney to Newcastle. There are in the vicinity of Newcastle already eleven seams known extending over an area of about 6 miles along the coast and 20 miles into the interior, having a thickness of from 3 to 30 feet. The coal is an excellent pitch coal, on the geological age of which the learned have as yet not been able to agree. Professor *M'Coy*, on the University of Melbourne, for palæontological reasons, places these coal deposits in the oolite period; others regard them as representatives of the palæozoic coal formation. [1] About 900 miners are engaged in the coal mines and the average price of a ton at the mine is 15 shillings. The Australian coal is much used in steam navigation upon the Australian waters and is already shipped as far as China, Batavia, India, California and South America. The demand far exceeds the quantity produced, which in 1860 did not amount to above 350,000 tons, but since has increased to one million a year.

There is reason to hope that ere long the Australian Newcastle will find its mates upon Tasmania and New Zealand, so that in course of time the navigation upon those seas will become independent of the English and American fuel. How far the prospects upon New Zealand justify such anticipations, the following pages will show.

For a series of years past carboniferous deposits have been known to exist in various portions of the North and South Island; and several attempts have been made to work them. In the vicinity of Auckland, of Nelson, on Golden Bay and in the Malvern Hills near Christchurch small mines have been opened for several years past, which repeatedly gave rise to very sanguine hopes;

---

[1] This controversy is being decided more and more in favour of the palæozoic age of the Australian eoal. The fossil plants consist principally of species of *Glossopteris, Phyllotheca, Pecopteris, Taeniopteris* and *Zamites*.

but which as often were given up as worthless and hopeless. The coal obtained was frequently tried upon steamers, and at one time found to be excellent material at another utterly unfit for use; the enterprise, however, stopped at these few isolated experiments, and a regulated persevering working of mines was nowhere undertaken despite the general urging demand for fuel, [1] and despite the extravagant prices, which serviceable coal would have commanded (£ 2 to 3 per ton). The reasons for such neglect are to be sought partly in the considerable working expences, partly in the scarcity of labourers and of the means for conveying the coal; but chiefly in the nature of the coal obtained — *brown coal* —, the inferior quality of which, being totally different from the English coal, did not suit the consumers. The closer investigations of the last years, however, have proved that upon North and South Islands coals greatly varying in quality, and of very different geological ages, are to be found in beds extending over large areas, and well worth a regular systeme of mining. It is not only *lignite,* of a comparatively inferior value that is known, but also thick beds of excellent *brown coal* of a tertiary (cainozoic) age, which comes up to the best German brown coal; and moreover *coal* of a probably secondary (mesozoic) age, — resembling the Australian coal, — which in quality is scarcely second to the best English coal. [2]

---

[1] Despite the immense forests covering the interior of the country serious complaints have been made in Auckland for years past about the high price of firewood, likewise in Christchurch. In the latter place there has been even a "Coal and Fire-wood Society" established, a company got up for the purpose of providing the labouring classes with fuel at the lowest possible prices. All these grievances will be thoroughly disposed of by a regulated working of the rich coal-fields in the various districts of the country. — Auckland is said to consume at present annually coal to the amount of £ 30,000. The quantity of coal imported into New Zealand in 1865 was 86,172 tons, the value £ 159,160.

[2] In New Zealand no coal seams have yet been discovered that can be referred to Palæozoic coal measures, such as occur in the Northern Hemisphere, as all the formations, which contain workable seams, belong to the mesozoic or cainozoic epochs. Dr. *J. Hector* in his able Report on the Coal Deposits of New Zealand (1866) has adopted a provisional classification of the New Zealand coals into two groups, under the terms *Hydrous* and *Anhydrous Coal,* or, those which still contain a large percentage of water chemically combined with them, and those which we may assume have been deprived of that water by a chemical change. The hydrous

The first brown coal field examined by me lies upon North Island, 20 miles to the South of the city of Auckland in the *Drury* and *Hunua Districts*. The vicinity of the Capital and of the Waitemata and Manukau Harbours, with which a communication may be very easily established by means of a rail road, render this coal field very important upon North Island. The merit of its discovery in 1858 belongs to my worthy friend the Rev. Mr. Purchas.

The coal bearing strata are situated on the western declivity of the wooded ranges bordering the plains of Papakura and Drury in the East and South East, at a height of 200 to 300 feet above the level of the sea. In the more elevated portions of these ranges, in deep water courses, a formation of old primary slates, of dioritic aphanites, and of gray sandstones comes to light, which have their continuation southward in the Taupiri range; towards the North near Maraetai (East of the mouth of the Wairoa), on the coast of the Hauraki Gulf, they splinter into more or less vertical strata, and are afterwards again continued on the Island Waiheke. Upon these rocks the coal bearing strata are deposited, while the tabular limestones, which are obtained from the quarries in the Wairoa District not far from Drury, and likewise the basaltic conglomerates covering the chains of hills South of Drury between the Manukau Harbour and the Waikato, belong to a higher horizon above the coal deposits.

The coal, at the time of my visit, had been laid bare partly by natural openings on the sides of the ravines, partly by diggings of the settlers. Not being able to enter here into the whole detail of my observations, I have nevertheless to remark, that I could not convince myself of the existence of several coal seams one above the other, although I am willing to admit the possibility of such being the case. On the contrary, it appeared to me on the various points of observation always as *one and the same bed* with an average thickness of 6 feet, which, however, owing to faults and

coals are corresponding with our brown coal, the anhydrous with the coals of a probably mesozoic age.

dislocations, appears on the slope of the hills at a different level, usually with a dip of 10 to 20 degrees towards South West or West.

The coal, the quality of which in the various portions of the bed in one and the same locality, and also in the various localities observed, suffers but little change, partakes, according to the varying gloss upon the surfaces of fracture, sometimes rather of the character of "glance coal", at others more of "pitch coal". It has a conchoidal fracture and black colour; but by the brown colour of its powder it at once betrays the character of *brown coal*. It is but slightly contaminated by iron pyrite, or by intermediate strata of bituminous shale; when recently broken it has a considerable consistency; yet by desiccation on exposure to the air and especially to the sun, it crumbles into small fragments.

A fossil resin is very frequently found imbedded in the coal, sometimes in pieces from the size of a fist to that of a man's head, but usually only in smaller groups. It is transparent, very brittle, and has a conchoidal, and quite glossy fracture. The colour changes from a bright yellow to a dark brown. It is easily ignited, much more so than the Kauri gum, burns with a steady, fast-sooting flame, and develops a bituminous rather than aromatic smell. Although originating probably from a coniferous tree related to the Kauri pine, it nevertheless has been erroneously taken for Kauri gum. [1] The fossil leaves found in the shales and sandstones flanking the coal belong to dicotyledon plants, which leads me to infer a *tertiary age* of this brown coal formation. I much regret to say, that my collection of these fossils remains but very

---

[1] Mr. Richard Maly found as a mean result of three chemical analyses of this fossil resin:

Carbon	76,53 computed	 76,65
Hydrogen	10,58 „	 10,38
Oxygen	— „	 12,78
Ash	0,19 „	 0,19
		100,00.

yielding the formula $C^{32} H^{26} O^4$. It shows great indifference to solvents. By friction it becomes electric. Hardness 2, specif. grav. 1,034 at 12^0 R. It is sufficiently characterized to deserve a special name; but it comes so near to real *amber* that, it deserves the name *Ambrite*.

small, as the shales and sandstones containing them were but little opened. [1]

Owing to the dense woods covering the country, the extent of the Drury coal field could not be fully determined. There is, however, no doubt, that the extent of the coal field is such as to justify continued and decisive mining enterprises. Already during my stay in Auckland, a company was formed under the name of "Waihoihoi Coal Company", which undertook to open the mines, and by constructing a tramway from the diggings to Slippery Creek to connect the coal field with Manukau Harbour. [2]

My observations furthermore led me to believe, that these brown coals have been deposited on the old shores of a tertiary basin, in the middle of which purely marine strata with sea shells, only here and there containing sporadic carbonified pieces of drift wood, are found, as in the cliffs near Auckland, on the peninsula Wangaparoa North of Auckland, and in other places. This opinion I found corroborated, when on my travels into the interior of North Island, on the N. W. slope of the Taupiri and Hakarimata ranges, where I met with the same brown coal formations

---

[1] Not a single one of the fossil plants was to be compared with European tertiary plants, nor has the New Zealand Flora of to-day any distinctly cognoscible representatives in them. The following species were determined by Prof. *Unger* of Vienna: *Fagus Ninnisiana Ung.* (named after Capt. Ninnis of Onehunga so highly meritorious for the opening of the coal-mines of Drury). Among the living species is the nearest *Fagus procera* Pöppig from Southern Chili. *Loranthophyllum Griseliana Ung.*, resembling *Loranthus Forsterianus Schult.*, but principally resembling the *Griseliana lucida Forst.* belonging to the family of the Carneæ and diffused throughout New Zealand. *Loranthophyllum dubium Ung.* The leaf but imperfectly preserved shows some identity with *Loranthus longifolius Deso.* *Myrtifolium lingua Ung.* has no analogon either among the fossil leaves or the plants of to-day. Of the Myrtaceæ, among which it might be classed if anywhere, only *Angophora cordifolia Cav.* of Australia has some resemblance.

Phyllites Purchasi Ung.   ⎫ fossil leaves in a very imperfect state of preser-
   „   *Ficoides Ung.*   ⎪ servation, the genus of which cannot be deter-
   „   *Novae Zealandiæ Ung,* ⎬ mined, and which therefore can only be very
   „   *Laurinium Ung.*   ⎭   generally designated.

[2] This tramway, a horse-road of $3^{1}/_{4}$ miles in length, was opened in May 1862, and the company indulges in the hope of being now able to so fix the price of the Drury coal, that the ton will cost at the mine 15 s., in Onehunga 25 s., and in Auckland 30 to 32 s.

likewise stratified upon old clay-slates and gray sandstones; at Kupakupa on the Waikato the coal bed showed a thickness of at least 15 feet. [1] I designate this coal field which probably surpasses the Drury coal field in area, as the coal field of the *lower Waikato Basin.* It will not begin to be of importance until the beautiful country on the banks of the Waikato which, as yet, is in the possession of the natives, shall have passed into the hands of the enterprising European settlers. A third, probably very extensive, but hitherto entirely untouched field of brown coal is found on the western and south-western borders of the *Middle Waikato Basin.* [2] From the existence of seams of brown coal on the Mokau River and on most of the tributaries of the Wanganui River, it is probable that another extensive brown coal field exists in the south-western part of the island, while there are also indications of a third extension of this same formation on the East coast.

Of older coal deposits nothing was known besides the thin coal strata connected with fossil ferns on the West coast, South of the mouth of the Waikato, which are most probably of a secondary age. [3]

On the other hand the youngest clay and sand deposits of a post-tertiary age on the shores of Manukau Harbour, in Drury and Papakura flats, and likewise the alluvial plains of the Waikato, the

[1] See Chap. XIV.

[2] The places in the Middle Waikato Basin, where coal was found, are the following: in the Hohinipanga-range West of Karakariki on the Waipa; near Mohoanui and Waitaiheke in the Houturu range on the upper Waipa; and in the Whawharua and Parepare Hills on the northern slope of the Rangitoto-range.

[3] A coal formation of a probably secondary age has only recently been detected in the northern districts of the Province of Auckland in a North and South line fifty miles North of Auckland to Wangaroa Harbour. A large portion of the isolated mass of hills at the North Cape is composed of this formation; and on the North side of the Harbour of Parenga-renga natural sections show these strata to be more than a thousand feet in thickness, although the coal is here only represented by very thin and worthless seams. Attempts to open mines have been made in the vicinity of the Harbour of *Wangarei,* seventy miles North from Auckland, and on the *Kawa-Kawa River,* which enters the Bay of Islands. The bituminous coal from Kawa-Kawa is of superior quality to any other of the Auckland coals. Its color black, lustre resinous, fracture and cleavage very irregular and granular, powder and streak dark brown and glistening. (According to Dr. Hector.)

New Zealand Thames, and the larger rivers in the North of the island, contain half carbonized and turflike strata of *lignite*, which must not be mistaken for deposits of brown coal.

Of much greater variety and extent seem to be the indications of coal beds upon *South Island*, and if anywhere it is reasonable to hope that upon this island a Newcastle will spring up some day, which will furnish the desired quantity and quality of coal for the Pacific steam navigation. I myself was able to examine only a few of the coal seams in the Province of Nelson on the shores of Blind Bay and Golden Bay.

The first point I examined is Ennerglyn, four miles South of the City of Nelson and about 200 feet above the level of the sea, close by the foot of the steep rising ranges of clayslate. Mr. *Jenkins* has the merit of having proved to the inhabitants of Nelson by several trial-shafts, that here they have coal in the immediate vicinity of the city. Through a tunnel driven about 250 feet in an easterly

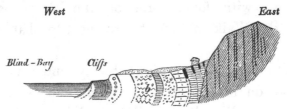

West   *East*

Blind – Bay   *Cliffs*

Jenkin's Colliery near Nelson.
a) Clayslate. b) Brown Coal Formation.

direction into the slope of the mountain through sandstones, conglomerates and shales, several brown coal seams from 3 to 6 feet thick were cut, which with a very steep inclination of 50 to 60 degrees dip towards the East, apparently under the older slate formation. The structure of the coal, however, indicates violent disturbances in the stratification of the layers in consequence of a pressure from the East, which has wholly bent and inverted the strata. In consequence of this pressure the coal has lost all consistency; it is specular-cleft and crumbles in small, very glossy scales or lamina. Between this crushed coal there are isolated nests of a remarkable jet-black coal with perfectly conchoidal fracture and a bright gloss, presenting the appearance of obsidian In the ferruginous sandstones flanking the coal indistinct petrifactions-are found, fossil leaves of dicotyledon plants,[1] similar to those at Drury. The disturbed state of stratification and the crumbled appearance of the coal were not favourable to the mining enterprise, which was soon given up again. But various indications lead one to suppose that further South on the outskirts of the Waimea plain, in the direction of Richmond, there are also beds of coal. Experiments by way of boring would probably be most apt to decide the question. Still further South near Mount Arthur, and on the Wangapeka and Batten Rivers, coal seams have likewise been discovered.

Far more favourable for the working of a mine are the localities on Massacre Bay,[2] West of Nelson. Near *Motupipi*, close by the sea shore, there are some coal mines which were opened under the able direction of Mr. *James Burnett;* but by a juncture of untoward circumstances were again left to decay. There several strata of coal are known from one to five feet thick. The bed which has been worked lies just at the level of the highwater mark and almost horizontal. The lower beds are below the highwater mark. The Motupipi brown coal stands next to the Drury coal; it contains the same fossil resin, Ambrite; its fracture, however, is less glossy. It crumbles easily when exposed to the atmosphere, and burns with a yellowish-red flame. The strong bituminous smell has frequently deterred consumers from using this coal for domestic purposes. But, as Mr. Burnett informed me, it was used with good effect for a whole year (in 1854 to 1855) upon the steamer "Nelson" plying on the coasts of the Province of Nelson, intermixed with Australian coal. At the time of my visit a lime-kiln in the neighbourhood was carried on with it. The coal would, no doubt, have met with a better sale, if the shipping of it had been less difficult and the production of it less expensive. But on the shallow

---

[1] Professor Dr. Unger has named the very imperfect specimens of those fossil plants, contained in my collection, as *Phyllites Nelsonianus, Ph. Brosinoides, Ph. quercoides, Ph. eucalyptroides* and *Ph. leguminosites.*

[2] The bay has repeatedly changed its name. "Massacre Bay" was changed to "Tasman's Bay", when coal was discovered to "Coal Bay" (Arrowsmith's map), and upon the discovery of the Aorere gold-fields to "Golden Bay".

shore only small vessels, at most of ten tons freight, are able to land, and in consequence of the extravagant rate of wages the price came so near to the price of the English coal that a sale of it on a large scale was entirely out of the question. Mr. Burnett was of opinion, that by establishing a coal depôt upon the Tata Islands, — situated along the coast not far from Matupipi, — where larger vessels can anchor with safety, and by the putting up of a steam engine the whole matter may be set right at once, and the price of the coal be reduced to from 15 to 20 shillings per ton.

The extent of the coal field near Motupipi is considerable. The coal has been found in various places up the Takaka river as far as Mr. Skeet's farm. At low water the coal seams can be traced upon the shallow muddy bottom far out into the sea. Furthermore on the *Rangiheta Point*, several miles West of Motupipi, I observed the same brown coal formation again. The vertical bluffs along the shore present here moreover a very interesting geological section. The coal seams, likewise containing Ambrite, are again partly on, partly below, the high water mark; they are covered with bituminous shales, sandstones, conglomerates and solid quartzites, such as are frequently met with in Germany and Bohemia within the range of the brown coal deposits. The top-most layer consists of tabular limestones, the banks of which protrude at the top of the cliffs. Here, therefore, the curious observer has ample opportunity to convince himself that the limestone formation so extensively met with upon the North and South Islands overlies the brown coal.

Totally different from the brown coal hitherto described is the coal of *Pakawau* in the western corner of Golden bay. The carboniferous strata rest on the metamorphic slates of the Whakamarama range, and are laid bare on both sides of the Pakawau Creek by natural sections and by small trial-shafts. They consist of micaceous sandstones, conglomerates and shales, with several coal seams. The strata dip at an angle of 20^0 towards South West in the direction of the Harbour of West Wanganui, — a distance of six miles, — where the outcrop of the carboniferous strata is again visible.

It is to be ascribed chiefly to the inconsiderable thickness of the coal seams, — the principal seams being only four feet thick, — and to their being contaminated by the frequent intermixture of bituminous shale, that the mines opened on the bank of the creek, at a distance of one mile from the sea shore, met with little or no success; for the coal itself surpasses in quality all the other New Zealand coals hitherto described. Its color is black, powder and streak black brown. It has a resinous lustre, a slaty fracture along its cleavage, but across it conchoidal. It is moreover characterized by its extraordinary consistency. Large pieces, which for years had been exposed to rain and sunshine, continued to present the same consistency as newly broken pieces. It burns freely with a bright, long flame; it cakes to a very coherent mass and would certainly prove an excellent gas coal. The fossil plants imbedded in a micaceous sandstone flanking the coal seams are wholly different from those near Drury and Nelson; but the pieces I found were too indistinct for a specific determination. [1] On the other hand, the character of the strata leads me to suppose, that to the Pakawau coal a secondary age is to be ascribed. How far the Pakawau coal field extends, and especially how far South it extends, is not as yet known, and I am inclined to believe that future examinations will establish the existence of beds well worth working, which will realize all the hopes that have been attached to the Pakawau coal on account of its superior quality.

Thus I have briefly described the various traces of coal which I had an opportunity to examine in person. But since my return to Europe, my friend Dr. *J. Haast* in 1860 and 1861 has carried through highly important researches in the western portions of the Province of Nelson and in the Province of Canterbury, the main results of which I cannot refrain from communicating such as I have gathered them from his reports and letters.

On an expedition through the inhospitable regions of the New

---

[1] It is only with difficulty, says Professor Unger, that in the coarse-grained sandstone *Neuropteris*, *Equisetites* and a palm (*Phoenicites?*) are to be recognised.

Zealand Alps to the West coast of the Province of Nelson, carried through with a great deal of courage and perseverance, extensive coal fields were discovered on the *Buller* (or *Kawatiri*) and *Grey* (or Mawhera) Rivers. On July 28th 1860 my friend wrote to me from the mouth of the Ngakuwaho River, 18 miles North of the Buller: "Since the date of my last letter to you, I have made an exceedingly interesting and instructive tour into the Papahaua range North of the Buller. — Once before I had travelled about ten miles up the Buller River, and found on both sides granitic and syenitic rocks; I was therefore not little surprised on finding this range to consist of coarse sandstone, conglomerate, and bituminous shale, and I was really lucky enough to discover a magnificent bed of coal 8 feet thick, at a height of about 1500 feet above the level of the sea, the strata resting on granite, and forming in some places perpendicular walls 1000 feet high. The coal extends over an area estimated at 8 miles in width and 15 miles in length. This coal field is sure to grow into importance during the future development of New Zealand. Is is quite probable that there exist numerous other beds; for the boulders of rivers and creeks, which empty into the sea North of the Buller, consist nearly all of coal and bituminous shale."[1]

Of still greater importance are the discoveries on the *Grey River* emptying South of the Buller on the West coast. The untiring Mr. *J. Mackey* already, on his excursions through the Southern Alps and along the West coast, had found coal there. But it was my friend *Haast* who succeeded in proving the existence of four workable seams, the main seam being fifteen feet three inches in thickness. The coal lies between micaceous sandstones, coarse, hard sandstones and between shale. The latter is said to be abounding in fossil plants. Dr. Haast speaks of leaves of dicotyledons, and also of Cycadeæ, *Zamites, Pecopteris* and *Equisetum*. The coal, as to quality and appearance, is said to be so much like the Australian coal of Newcastle, that it can be scarcely distinguished

[1] Mr. Burnett in 1862 found in the neighbourhood of the Buller River four seams with a thickness of 5 feet to 12 feet.

from the latter, and even to be quite comparable with the best English coal. [1]

The discovery of coal deposits of such thickness close to the sea coast has greatly roused the attention of the population in the two Provinces of Nelson and Canterbury, which are the first concerned in it. [2] In 1865 only one mine had been opened in this field on the North side of the Grey, at seven miles from its mouth. From this point the coal is supposed to extend on the rise for about four miles, in an easterly direction. Arrangements are also made for working the coal on the South side of the river, which lies in the Province of Canterbury, [3] and for the construction of a railway to convey it to a point on the river where it can be shipped in vessels of six to seven feet draught.

At Preservation Inlet, in the Province of Otago, a bituminous black coal occurs of the same description as the Grey coal, but no workable seams have been discovered as yet.

The *Pakawau*, *Buller* and *Grey coal*, — as far as I am able to judge, — I take to be of one and the same, probably *mesozoic, age*. There are reasons for comparing that coal formation with coal deposits such as are found to exist in Europe in the Oolite of England (Yorkshire and Sutherland), or in the Wealden of northwestern Germany. These geological questions, however, will be fully

---

[1] The Grey coal is described as very compact; its color black, lustre dull. The fresh fracture has a glistening appearance. It possesses a slatey cleavage. Powder black; ash light brown. The coal puffs up slightly when heated, and gives 68.37 per cent of metallic coke.

[2] The Grey or Mawhera River forms the boundary of the two provinces.

[3] At the present time, says Dr. Hector (Report on Coal Deposits 1866), the great obstacle to the development of these coal mines is the high price of labour, owing to the attractions offered by the gold-fields, but when this excitement has subsided, a large portion of the present population of that district, which is estimated at 15,000 persons, will doubtless become permanent, and, notwithstanding the present temporary check, the importance of these coal-fields will be much sooner established than if the auriferous deposits had not been discovered, and the Westland District had still remained as inaccessible as it was twelve months ago. At present the output from the coal mine is about 250 to 300 tons per week and is taken down the river in canoes and in 16, ton barges, and delivered alongside the vessels at 40 s. per ton.

and clearly understood only when the successive order of the Australian and New Zealand strata with their organic remains shall have been as clearly ascertained and determined as the European series of formations.

In addition to the coalfields on the West coast of South Island extensive tracts of carboniferous deposits have been known to exist for several years past also on the East side of the mountains. Already since 1857, a coal mine has been worked near the *Malvern Hills*, in Canterbury, about 30 miles from Christchurch, which supplies the neighbouring sheep stations with fuel. My friend Dr. Haast examined this locality more closely in the summer of 1861, and found in the bed of the Kowai River seams of a hard black glistening coal three to six feet thick and very much like anthracite. This Kowai coalfield is easy of access from Canterbury; it is situated just between hill and dale, and the construction of a railroad would meet here with no difficulties whatever. Deposits of brown coal are found to underlie the tertiary rocks of the higher portions of the great Canterbury plains, and occur in the valleys of the Selvyn, Upper Waimakariri, Rakaia, Rangitata, Ashburton, Northern Hinds, Potts, Tenawai and other rivers, in the Malvern Hills, Mount Somers, Big Ben Range, Thirteen Mile Bush and many other places.

In the *Province of Otago*, as Dr. *Hector* writes, the best known and most important deposit of the brown coal formation is that on the Southeast coast, northward from the Molyneux River, where it extends continuously over at least forty-five square miles, forming hills 500 to 1000 feet in height. In this formation there are several seams of good coal varying from six to twenty feet in thickness, their aggregate thickness in a section three miles in length, exposed on the sea shore, amounting to 56 feet. The total quantity of coal in this district has been estimated at 100,000,000 tons. Two large mines have been opened in this coalfield. The *Clutha Mine* is on the sea coast, about three miles from the mouth of the Clutha River. The principal consumption of this coal has been for local use and for the steam navigation of the river; but as it is proposed to construct a railway from Dunedin to Clutha Valley, a distance of sixty miles, this valuable fuel will be brought into more general use than at present. The only other mine of importance opened in this coalfield is in the vicinity of *Tokomairiro*. The seam is 9 feet thick and is worked by a level drive in the side of a hill.

The *Green Island* and *Saddle Hill Basin*, which is within six miles of Dunedin and about 5 to 6 square miles in extent, is another important development of this same coal formation. Two of the principal seams, 7 and 9 feet in thickness, are worked at four different collieries, and though the quality is slightly inferior to that of the Clutha coal, yet, owing to the proximity to Dunedin, its consumption is very considerable. [1]

A third great extent of brown coal formation on the eastern seabord of Otago underlies the level country to the eastward of the Kakanui Mountains. Coal of the same description is also very extensively distributed throughout the interior of the Province of Otago, as local deposits in hollows of the surface of the older formations. Such deposits, though generally of inferior quality, have great local value, and are essential to the working of the goldfields, on which there is a great deficiency of timber.

From the number of localities in the Province *of Southland* where brown coal is excavated for local use, its extensive development over the greater part of that province is probable.

From the preceding descriptions and from the facts mentioned we infer that New Zealand, while it presents so many other points of comparison with Great Britain, is equal to the latter also in its richness of coal. The discoveries and openings made heretofore are to be considered merely as a beginning of what future years will bring to light far more abundantly. As we have already seen, New Zealand has coal of very different geological ages, and of very different qualities. All stages from the turf-like *lignite* to the anthracitic *black coal* are represented, and while the pitch coal (anhydrous

---

[1] In 1864 the quantity produced from the mines in this district was about 4000 tons.

coal) of North and South Island will be of the greatest value for marine purposes, where the best heating fuel occupying the least space is required, the brown coal, on the other hand, is well adapted for manufacturing and domestic purposes. That those brown coals, — being, as they are, of a nature and quality far different from English coal, — should in many instances be deemed in New Zealand far inferior to what they are, is easily accounted for; and years will pass away before prejudices will be overcome with a people that hitherto have only known and used the excellent coal of their mother country. Time and experience, however, will show that the brown coal in New Zealand can be used for the same purposes for which just the same coal, and sometimes of a far inferior quality, is used on a most extensive scale in various parts of Germany and especially in Austria, in whole provinces of which (Styria, Krain and Northern Bohemia) it constitutes the almost exclusive fuel for manufacturing and railroad, as well as for domestic purposes.

Appendix.

Comparative Tables of Analyses of New Zealand and Australian Coals.

Table I.

Locality.	Elementary Analysis (in percentages).					Value as fuel determined according to Berthier.				Analysist.	Remarks.
	C Carbon.	H Hydrogen.	O Oxygen.	Ash.	Water.	Per cent of Coke.	Reduced parts of Lead.	Units of Caloric.	Equival. of a 30 inch. cord of soft wood in cwts.[1]		
Tertiary. Drury near Auckland (North Island)	58.0			2.9	8.0[2]		19.57	4423	11.8	C. v. *Hauer* (Lab. of the Imp. geolog. Survey Office Vienna.	specif. grav. 1.38, "glance-coal" powder brown.
Motupipi, Prov. Nelson (South Island)	55.57	4.13	15.67	9.04	14.12[3]	(1.15 Nitrogen;		0.36 Sulphur)		*Ch. Tokey* (Labor. Mus. of Practical Geology, London). C. v. *Hauer.*	sp. gr. 1.48, comp. Percy, Metallurgy p. 90.
	55.0			5.3	23.1		18.80	4248	12.3		sp. gr. 1.37, "pitch coal," powder brown.
Secondary. Pakawau (Prov. Nelson South Island)	66.72			8.4	1.[7]	56.6	22.63	5119	10.2	C. v. *Hauer.*	sp. gr. 1.31, "caking coal" and probably a good "gas-coal," powder dark-brown.
West-coast of Pr. Nelson at the Mouth of Grey River	79.00	5.35	7.71	3.50	1.05	64.32 (0.89 Nitrogen;		2.50 Sulphur)		Dr. Percy (Govt. School of Mines, London.)	comp. Percy Metall. p. 100 "caking coal" and probably a good gas-coal.
Australian Coal from New Castle (New South Wales)	78.0			5.0	1.6	63.0	26.72	6038	8.6	C. v. *Hauer.* Talton.	sp. gr. 1.29, "caking coal," powder dark-brown. comp. *J. Haast,* "Report etc." p. 106.
	74.13	25.87									

[1] The cords of soft, dry wood computed at 18 (Vienna) cwt., with 52,497 Units of Caloric.
[2] Expelled at 100° C.
[3] Expelled at 120° C.

Table II.

Extracted from the first general report on the Coal Deposits of N. Z. by James Hector M. D. 1866.

Locality.	Percentage of Water.	Specific Gravity.	Percentage of — Fixed Carbon.	Volatile matter.	Water.	Ash.	Color of Ash.	Percentage after deducting Water and Ash — Fixed Carbon.	Volatile matter.	Coke.	Sulphur.	Color of Powder.	Nature of Coke.	Percentage of Water when first examined.
Auckland:														
Kawa-kawa, Bay of Islands	4.94	1.309	57.20	36.00	4.60	2.20	light red	61.37	38.63	59.40	4.90	black	dull, sinters	4.60
Ditto 2nd. sample	4.68	1.284	55.40	38.50	4.40	1.70	light red	59.00	41.00	57.10	5.10	black	dull, sinters	4.40
Nelson:														
Pakawau; 1 yard seam	4.03	1.330	50.10	38.08	3.56	8.26	light red	56.82	43.18	58.36	1.04	brownish black	semi-metallic	3.56
Ditto 1 foot seam	1.89	1.409	38.00	35.20	1.60	25.20	light gray	51.91	48.09	63.20	—	dark brown	dull, cakes	—
Buller R., Coalbrook Dale	2.69	1.244	65.45	31.55	2.60	.40	dark buff	67.48	32.52	68.55	1.20	dark brown	dull, cakes	—
Ditto	1.85	—	57.20	40.20	1.80	.80	buff	57.20	40.20	58.00	—	dark brown	puffs up	—
Ditto	1.10	1.250	62.10	31.55	1.05	4.70	white	66.52	33.48	67.40	1.85	brownish black	semi-metallic, cakes	—
Ditto	.83	1.260	58.74	35.97	.70	4.55	white	62.02	37.98	63.29	—	brown	semi-metallic, cakes	—
Grey River	1.99	1.280	61.00	33.90	1.90	3.20	gray	64.28	35.72	64.20	—	dark brown	dull, puffs up	—
Ditto	1.72	—	55.40	37.20	1.60	5.80	white	59.83	40.17	61.20	—	dark brown	puffs up	—
Ditto	1.60	1.300	54.11	33.19	1.40	11.30	white	61.98	38.02	65.41	1.85	brown	puffs up	—
Canterbury:														
Grey River	2.18	1.333	62.37	29.44	1.99	6.20	light brown	67.94	32.06	68.57	—	black	cakes	—
Malvern Hills	1.66	—	61.10	33.40	1.60	1.90	white	63.32	36.68	63.00	1.80	dark brown	dull	—
Otago:														
Shag Point	13.68	{1.250 / 1.260}	42.16	37.60	10.90	9.33	chocolate	52.86	47.14	51.49	4.78	dull brown-black	dull	12.30
Waikawa	8.53	—	44.34	32.95	6.60	16.11	white	57.37	42.63	60.45	not much	brown	dull	—
Preservation Inlet	5.50	—	54.58	25.62	4.20	11.20	light buff	66.05	31.95	65.78	very much	tinge of brown	unchanged	8.60
Ditto 2nd sample	4.94	—	66.43	22.53	4.40	2.24	light buff	69.01	30.99	68.67	very much	tinge of brown	unchanged	8.86
Ditto 3rd sample	4.84	1.290	61.83	28.90	4.40	5.14	light buff	69.03	31.00	66.97	very much	brownish black	dull	8.80

Anhydrous Coals of N. Z., probably of mesozoic age.

Table III.

Constructed chiefly from materials contained in a Report by Dr. Hector of date 13. April 1864.

Tertiary hydrous Coals of Otago. 1

Locality.	Physical Characters of Coal-Specimen Analysed.	Character of			Specific Gravity.	Percentage of					
		Powder.	Ash.	Coke. 2		Coke.	Fixed Carbon.	Volatile Hydro-carbons.	Ash.	Sulphur.	Water.
I. Clutha-Coal Point.	1. Common Brown Coal. — *Fracture* moderately bright — conchoidal; becomes dull on exposure; *streak* deep brown; *burns* freely with a very slight *odour*; contains lamina of *Jet*.	Deep brown.	White or light grey.	Dull.	1.280	43.28	38.99	37.28	5.46	–	16.10
Do. *Top of Main Seam.*	2. Brown Coal, apparently of very superior quality. — Distinctly laminated; traversed irregularly by thin plates of *Jet*; contains abundance of *Mineral Resin*, scattered through its substance in small specks, or occurring in large nodulated masses, also *Iron Pyrites* in its cracks.	Dull brown.	White.	Do.	1.258	42.79	43.23	33.87	3.36	3.63	12.94
Do. *Middle of Main Seam.*	3. Quality and characters as in No. 2.	Do.	Do.	Do.	1.267	41.10	39.12	35.67	4.43	–	16.35
Do. *Bottom of Main Seam.*	4. Blacker; more compact and massive; *Fracture* conchoidal in all directions; *streak* chocolate brown; possesses higher illuminating power than Nos. 2 and 3; and contains fewer impurities, with exception of nodules of *Iron Pyrites*.	Light chocolate brown.	Do.	Do.	1.275	40.46	39.71	32.24	7.28	3.85	16.96
Do. *Associated Sandstones.*	5. *Jet*; bright; lustrous; compact; *streak* black; *burns* with very great difficulty, caking slightly.	Black.	Light red.	Iridescent	–	50.35	59.04	27.61	11.31	–	16.43
II. Greenisland.	6. Common Brown coal. — Dull; earthy; very moist when first extracted from the mine; absorbs water readily, with a creaking noise; laminated; *fracture* subconchoidal; no trace of woody structure; *streak* dull, blackish-brown; contains thin plates of *Jet*; ignites slowly, and rather smoulders than burns; forms a useful fuel, when *mixed with a better class of coals*.	Dark brown.	Very light; almost like that of *wood* - Colour light buff.	Slightly metallic or iridescent.	1.256	39.72	36.83	33.83	4.32	2.24	19.33
Do. from another Pit.	7. Quality and characters as in No. 6.	Dull brown.	Light buff.	Dull.	–	–	43.12	34.37	2.25	–	20.26
III. Saddlehill.	8. Earthy Brown Coal. — Soft; dull; very moist; structure homogeneous; no trace of vegetable tissue; *streak* dark brown; in burning gives off a fetid smell.	Do.	Very light — Colour pure white.	Do.	1.294	38.97	39.07	35.93	2.78	2.22	15.95
IV. Shag Point.	9. Pitch Coal. — Brownish black; very compact; *fracture* conchoidal and splintery; *lustre* fatty or resinous; *streak* lustrous; does not absorb water; does not soil the fingers; *burns* with a rich oily flame and a slight odour; on distillation at high temperatures yields a rich *gas*, and at lower ones a large amount of *oils*. Probably one of the oldest tertiary coals of Otago. The freest burning quality of Brown Coal yet found (1864) in that province. Its characters correspond with Dr. Percy's definition of "*Pitch Coal*," and it holds the same relative position among the Brown Coals of Otago that Cannel does among the household coals of Britain.	Dull brownish black.	Bulky; lighter-gilliaceous - Chocolate brown.	Light; dull or-semi-metallic.	1.258	50.68	41.83	37.31	9.25	4.78	10.91
V. Tokomairiro.	10. Dull Brown Coal. — Physical characters resemble those of Clutha and Greenisland Coal. *Streak* brown; contains occasional *quartz* pebbles.	Dull brown.	Light grey.	Dull or semi-metallic. — Does not cake.	1.299	41.38	39.93	37.55	5.91	–	11.40
VI. Waitahuna. — Flat or Junction.	11. Lignite, or Woolly Brown Coal. — Dull; friable; slightly laminated; contains a considerable amount of *ligneous tissue*; also *mineral resin* in small masses; *burns* slowly, like turf; with a heavy, fetid odour. Perhaps of more recent geological age than the Brown Coals proper.	Light brown.	Light buff.	Imperfect; lustrous; slightly iridescent.	–	43.74	35.30	37.63	11.15	6.21	11.06
	Mean of the foregoing 11 Classes, which include varieties of *Jet, Lignite, Brown Coal,* and *Pitch Coal* . . . . . }	–	–	–	1.273	43.24	39.64	34.84	6.13	3.82	15.24

1 Dr. Hector regards the Otago Brown Coals as of three distinct ages, the age being in each case indicated by the relative percentage of water of constitution.

2 None of the tertiary coals of Otago coke.

The material originally positioned here is too large for reproduction in this reissue. A PDF can be downloaded from the web address given on page iv of this book, by clicking on 'Resources Available'.

CHAPTER V.

Gold.

The gold-riches of Australia. — Incentive to prospecting parties in New Zealand. — First discovery of gold near Coromandel Harbour in the Province of Auckland, in 1852. — Scanty result. — The Nelson goldfields. — The Motueka diggings in 1856. — The Aorere gold-field in 1857. — Satisfactory results. — The Western ranges of the Province of Nelson. — The auriferous formations. — Aorere diggings. — Parapara diggings. — Takaka diggings. — Discoveries of gold in the Southwestern part of the Province of Nelson. — Discovery of the Eldorado in the Province of Otago on the Tuapeka, 1861. — Gold fever. — Diggers flocking in from Australia. — Great extent of the gold-deposits. — Gold in the Province of Marlborough 1864. — The Westcoast of the Province of Canterbury. — Hokitika, the metropolis of the Westland goldfields. Appendix. Produce of goldfields.

The Australian Colony Victoria sent to the International Exhibition in London, in 1862, a gilded obelisk of 10 feet square at the basis and 45 feet high. This obelisk, if it were a solid mass of gold, would have a weight of 800 tons; it represents the volume of gold produced by the colony from October 1, 1851 to October 1, 1861, valued at 104 million pounds. The population of Victoria, numbering in 1851 only 70,000 inhabitants, has increased since to 550,000, and Melbourne which but a few decades ago consisted of some wretched shanties on the sea coast, is now a large, splendid city, a capital with 100,000 inhabitants. Such a rise is without precedent in the history of colonies, and the Eldorado of the South Sea, of which the Spaniards of the fifteenth century were dreaming, has here become full reality in the nineteenth century.

The discovery of the golden riches in Australia reacted very sensibly on adjacent New Zealand, which had scarcely overcome the difficulties attending the commencement of its colonization; the latter losing a great number of its labouring population, who all went flocking towards the new gold-land. But, still, there were hopes also for New Zealand. People began to prospect for gold here, and already in October 1852, a "Reward Committee" was formed, which promised a reward of £500 to the discoverer of a valuable goldfield in the northern district of New Zealand. Within less than a week, the reward was claimed by Mr. Charles Ring, a settler, recently returned from California, who asserted that he had discovered gold upon Cape Colville peninsula, 40 miles East of Auckland, in the vicinity of Coromandel Harbour. The specimens produced by Mr. Ring were pieces of auriferous quartz, and some minute particles of gold dust, which he had found on the Kapanga, a creek flowing into the harbour. The commissioners sent out to investigate the matter also confirmed the existence of gold, leaving it however doubtful whether there was a goldfield extensive and rich enough to pay for the working.

This was the first discovery of gold upon New Zealand. There was a general rejoicing in Auckland over the lucky event; the people indulged in the most sanguine hopes, and at once arrangements were made for working the goldfield. As the land upon which the gold was found belonged to the natives, an agreement with the latter on the part of the Government had first to be brought about. The Maoris agreed for a certain payment to cede the acquisition of gold upon their own land to Europeans, and already in November 1852 a treaty was made with the Coromandel chiefs for the term of three years, in which the Government pledged itself to pay the natives for each square mile of land upon which gold was being dug one pound sterling annually, and for each gold digger two shillings per month. In consequence of this the Government was, of course, obliged to lay a tax upon the gold diggers. Granting an exemption for the first two months, it afterwards exacted from each digger 30s. per month for a digging license.

About 3000 diggers set to work. On the Kapanga River towards North the *"Coolahan Diggings"* promised favourable results, and likewise the *"Waiau Diggings"*, a short distance from the former, on the Matawai Creek, a tributary of the Waiau River which flows southwards into the Coromandel Harbour. The ore produced was sold in Auckland by public auction. But when the taxes were to be paid there were only about 50 diggers who took licenses. These also, however, were not able to subsist under the heavy taxes demanded; and as moreover nothing at all was heard of any encouraging results on a grand scale, and more and more difficulties arose on the part of the natives, the whole enterprise died out after about 6 months. The simple verdict was, that the gold mines were too poor, and the promised reward was withheld from the discoverer. The whole produce upon the first New Zealand goldfield up to the time when the enterprise was given up was computed at £ 1200 in gold value, and the largest nugget found was a spheroidal piece of quartz of the size of an egg which contained gold equivalent to about £10.

Despite various trials and movements in later years, and although the natives brought from time to time small quantities of gold to Auckland for sale, no serious trial was ever afterwards made upon the Coromandel goldfield; and the natives at last denied the Europeans even the right to make experiments.

Such was the state of affairs when in June, 1859, I visited the goldfield in company of Mr. Ch. Heaphy, the late gold commissioner. — What the traveller observes on entering Coromandel Harbour and examining its shores does in no way correspond with what a geologist expects of a gold region. The Coromandel Peninsula consists mainly of a mountain ridge, running nearly North and South, the mountains having a bold serrated outline and varying in height from 1000 to 1600 feet. The most noteworthy point is *Castle Hill* (1610 feet high) a rocky peak resembling the ruin of a castle. The valleys between the spurs given off laterally by this main or dividing range are of the character of ravines or gorges, occupied by mere mountain streams; the flats or alluvial tracts at

View of Castle Hill and Coromandel Harbour.

their mouths and on the coast are inconsiderable. The coast consists of nothing but trachytic breccia and tuff, in the most varying colours and in the most different state of decomposition, from the hardest rock to a soft clayish mass, and in various places broken through by doleritic and basaltic dykes. Silicious secretions in the shape of chalcedony, carnelion, agate, jasper, and the like, are a very frequent occurrence in these tuffs and conglomerates, likewise large blocks of wood silicified and changed to wood-opal. By local geologists those trachytic rocks were erroneously taken for granite and porphyry, and by a gross mistake the most sanguine hopes were based upon the notion that these silicious secretions might be auriferous quartz veins.

The Coromandel gold originates from quartz reefs of crystalline structure, belonging to a palæozoic clayslate formation,[1] of which under the cover of trachytic tuff and conglomerate the mountain range of Cape Colville peninsula consists. The mountains are so densely wooded that it is only here and there in the gorges

---

[1] The same clayslate formation constitutes the Hunua range in the brown coal district South of Auckland near Drury and Papakura, and continues towards South and North to a great distance. It is but very recently (May 1862) that traces of gold are also said to have been discovered in the Hunua district.

of the streams that sections of these slates may be examined. In these sections the clayslates are frequently found to resemble Lydian stone; they are arranged more or less vertically, their irregular upturned edges affording the most convenient and abundant "pockets" for the detention and storage of the alluvial gold washed from the higher grounds. The most gold was found in the narrow valleys, where, after digging to a depth of four to five feet through boulders and shingle, the "bed-rock" is struck. Where the valleys extended into broader alluvial plains, there was always but little and very light gold found.

At a small branch of the Kapanga in the vicinity of the "Coolahans Diggings", not far from Mr. Ring's mill, at a place pointed out to me by Mr. Heaphy as especially rich, I went to work myself to make an experiment in washing. We dug, partly from the bed of the small creek, partly from the banks, several shovelfulls of quartz gravel intermixed with earth and clay, which, after removing the larger pieces, we washed in round tin dishes. The result of the very first trial was a considerable number of extremely fine scales of a light yellowish green gold, [1] which glistened among the black magnetic iron sand that had remained after the washing process, and some small pieces of ochrey quartz, in which fine scales were seen imbedded. Each successive trial yielded the same result, nor was there a single dish full of "dirt" that did not show the "colour", so that I had to acknowledge to myself that, if those deposits of detritus should extend over a larger area, and could be worked on a large scale with the necessary machinery, the result must doubtless be a very remunerative one. But in regard to the former point I had no opportunity to convince myself, and as to the latter the natives would not have consented at that time. The pieces of quartz, among which there were many violet coloured or amethystine, all being angular fragments, could not have been brought from any considerable distance, although in the creek itself we found nowhere a quartz vein *in situ*. On

[1] On the Waiau Creek the gold found is said to have been heavier and more rounded.

the slope of the hills I saw large blocks of quartz lying, which from all appearances originated from "reefs" or veins, that — according to the statement of Mr. Heaphy, — protrude on the top of the dividing ridge in various places like walls, eight to ten feet high and ten to twenty feet thick. I much regretted the inclemency of the weather at the time, which frustrated our intention to examine these quartz reefs more closely It is worth mentioning that gold was also found on the creeks flowing from the east side of the Cape Colville range on Mercury Bay into the sea: on the Arataonga, Waitekuri, Cook's river and others. The traces of gold, therefore, seem to extend over a larger district, and the Coromandel gold-fields — such was my opinion in 1859 — bid fair to grow into importance in future years, when the country as yet covered with dense woods, shall have become more accessible, when the auriferous quartz reefs themselves shall have been discovered, and the difficulties, which the natives have hitherto opposed to every undertaking on a more extensive scale have ceased.[1]

Besides those mentioned, there had been at the time of my travels no other discovery of gold made upon North Island; although it is not improbable that the hitherto wholly unexplored mountain range on the S. E. side of the island, forming the continuation of the Alpine chains of South Island, still harbours many hidden

[1] From Auckland newspapers I learn, that the discovery of the rich gold fields in the Province of Otago, in 1861, has given a new impulse to entreprise. The Coromandel gold-field was again worked, and in April 1862, 248 gold diggers — among them about 100 Australians who had come from Dunedin, — are said to have been assembled in the Coromandel Harbour, to try their chances on Cape Colville peninsula. The latest results seem fitted to inspire brighter hopes. On the Matawai and Tiki creeks pieces of gold-quartz weighing 30 to 40 ounces, and even of 11 pounds weight, are said to have been found containing 50 to 60 per cent of gold; Murphy and Co., who began to work a quartz reef on the Kapanga, are said to obtain from one ton of gold-quartz by crushing and washing an average result of 2 1/2 ounces of gold. But it is known as a matter of fact, that quartz crushing in Australia with good machinery yields profit even when the ton of quartz contains not more than one ounce of gold. The Coromandel gold occurs in the form of dusty scales or nuggets — frequently as scaly nuggets or "pepites", but still more generally dendritically disseminated in quartz, which is usually ochrey or brownish in colour. While there is a very limited and insignificant field for alluvial digging, there is ample scope for quartz mining.

treasures.[1] But, at any rate, nature has lavished her favours more bountifully upon South Island.

Let us direct our attention first to *Golden Bay* (formerly Massacre Bay). Already in 1842, on the occasion of an exploring expedition undertaken by Captain Wakefield from Nelson to Massacre Bay, Mr. M'Donald is said to have found small scales, which were supposed to be gold. But no further notice was taken of them. Ideed, who would at that time have thought of, or believed in gold-fields, before the discoveries in California and Australia had rendered the thing familiar. At length in 1856, the stirring news of the discovery of gold 18 miles from Nelson, in *Bigg's Gully*, Motueka district, aroused in the minds of the colonists a general excitement. About 300 diggers rushed to the place; but they soon left their diggings again, the produce being too inconsiderable; and it was not until 1859, that these first "Nelson Diggings" were taken up again by a few diggers, and with pretty good success.

The principal event of the year 1857 was the discovery of gold in the *Aorere District* on Massacre Bay. In the beginning of the year Mr. W. Hough, a Nelson storekeeper, having some land at the Aorere, went over there in company with Mr. W. Lightband, a young man who had some experience as a gold digger in Australia. They commenced prospecting in some of the gullies. Mr. Hough shortly returned to Nelson, leaving Lightband to prosecute his labours, with the assistance of some Maoris. They continued steadily at work until they had obtained about three ounces of gold, which was forwarded to Nelson. This was the beginning of the development of the rich mineral resources of the province; and the greatest amount of credit is due to William Lightband, whose steady perseverance at the diggings induced others to go over and try for themselves. The first reports received from Lightband were to the effect that his average earnings were about ten shillings per day. Although this could not be called a splendid profit,

---

[1] In 1862 there were to be seen in the London Exhibition Terawiti Gold and Wairiki Gold from the Province of Wellington.

it allured a good many others to follow his example. Fresh gullies and creeks were discovered, yielding better returns; and when the encouraging fact became known that three men, working on the Slate river, a tributary of the Aorere, had obtained one hundred ounces of gold in seven weeks, the diggings might be said to promise hopefully.[1] The number of diggers increased from day to day, and it was estimated that on the 1st of May 1857 there were no less than one thousand men at work upon the Aorere diggings. On the mouth of the river arose the fast thriving little town of Collingwood. As the winter approached, the expense of transporting provisions from the port of Collingwood (owing to the want of roads) became greater; and as the floods in the rivers at this period destroyed the dams and other works of the diggers, many of these became disheartened and left the place, and although many remained throughout the winter, earning good wages, and others returned in the spring, the diggings have never regained their former population.

When I visited the gold-field, in August 1859, there were in all only about 250 diggers at work. Although the work is frequently interrupted by the overflowing of the rivers, and although much time is lost in the difficult transportation of provisions,[2] the average gain of a digger at that time was nevertheless computed at 12 shillings per day. But such pay, although sure and permanent, seems after all too small to allure a larger number of men to engage in the laborious work of gold digging. What caused the headlong rush of thousands of persons to other gold-fields was less the certainty of a reward for their labour to all than the enormous, lottery-like gains of some lucky individuals. Such prominent cases of good luck, however, never occurred upon the Nelson goldfields; they consequently continued to be only scantily worked, and yielded a comparatively small, although permanent produce, which they will continue to yield for a long series of years to come. The largest piece of gold (which was found in the Rocky

[1] D. L. Bailey, Nelson Directory. 1859. p. 19.
[2] Pack-oxen were used for transportation.

river) weighed not quite ten ounces; a second eight ounces,[1] and by August 1859 the total amount of the produce was estimated at about £150,000.

The mode of occurrence of gold in the Province of Nelson is quite different from that in Australia, in the Colony of Victoria. The Australian gold is originally derived from quartz reefs passing through fossiliferous strata of Silurian[2] age, which are but very little metamorphosed; and the gold is obtained partly as alluvial gold from deposits of gold drift[3] ("wash-dirt" of the miner), partly from the quartz veins themselves, by crushing the quartz, and by subsequent washing and amalgating processes. As the gold alluvias are already nearly all washed over, an extensive system of quartz mining has been begun within the last few years, and the vital question still awaiting its final and decisive settlement is, whether the quartz veins, — which close to the surface were sometimes found to be unusually rich (the auriferous quality of them has, however, hitherto been tested only to a depth of 300 to 400 feet) — will continue at a still greater depth to be so rich in gold as to pay for the mining.[4]

---

[1] In Australia nuggets have been found of more than 1 cwt. in weight. The "Welcome Nugget" found on the 11th January, 1858, on Bakery Hill near Ballarat in Victoria, the largest of all nuggets hitherto found, weighed 184 pounds 9 ounces 16 dwts., and was valued at £10,000.

[2] A large portion of the gold-fields of Victoria falls within the range of the so-called Bala beds (lower Silurian), containing numerous fossils, especially remarkable graptolites (Diplograpsus, Didymograpsus, etc.) and crustaceæ (Hymenocaris Salteri). The slates of Castlemaine and Bendigo are full of them.

[3] The gold drift deposits are divided by the Australian geologists into old-pliocene, new-pliocene, and post-pliocene deposits.

[4] Experience seems more and more to confirm the views of the Australian geologists, Messrs. A. Selwyn and G. Ulrich, that the reefs of gold-quartz in Victoria are real mineral veins, which render a permanent system of mining in a downward direction possible, as on the veins of silver, lead, tin and copper ore in Great Britain and Germany, while the prevailing opinion had been, that the gold decreased in proportion to the increasing depth of the mine. G. Ulrich has proved upon the quartz reefs of Victoria the most different ores, such as iron pyrites, arsenical pyrites, copper pyrites, galena, grey antimony ore, copper glance, bismuth glance, native copper, and native silver. The greatest depth hitherto reached in the gold mines in Victoria is 460 feet, and at this depth quartz has been obtained containing over 5 ounces gold per ton.

Upon the Nelson gold-fields the gold has been originally derived from quartz veins, which occur in non-fossiliferous crystalline (or metamorphic) schists. A section from East to West through the mountain ranges between Blind Bay and the West coast of the Province Nelson, presents to us the succession of the crystalline schists. The western shores of Blind Bay from Separation Point to the mouth of the Motueka River consist of granite, which towards West is flanked by gneiss. This granite and gneiss-zone can be traced towards the South along the Motueka valley to the junction of the Wangapeka river. It is intersected farther South by the Buller river at its entrance into the gorge of "Devil's Grip", and continues on the eastern escarpement of the range as far as Lake Rotorua (L. Howik).

Proceeding from the granite and gneiss towards the West, we find upon the top of the Pikikerunga range a broad zone of hornblende schist, which alternates frequently and regularly with quartzite and crystalline limestone in vertical strata, with a strike almost due North and South. These ranges continue to the Westward as far as beyond the Takaka valley, where they are intersected, on Stony Creek and on the Waikaro, by dioritic porphyry and serpentine. A characteristic feature of the limestone of this zone is numerous funnel-shaped pits and caves, reminding us of

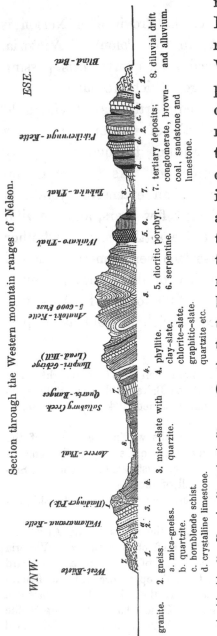

Section through the Western mountain ranges of Nelson.

ESE. Blind-Bай
Pikikerunga-Relle
Takaka-Thal
Waikaro-Thal
Anatoki-Relle 5-6000 Fuss
Wangiri-Gebirge (Lead-Hill)
Salisbury Creek Quartz-Ranges
Aorere-Thal
(Haidinger Pk.)
Wakamarama-Relle
WNW. West-Küste

8. diluvial drift and alluvium.
7. tertiary deposits; conglomerate, brown-coal, sandstone and limestone.
5. dioritic porphyr.
6. serpentine.
4. phyllite. clay-slate, chlorite-slate. graphitic-slate. quarzite etc.
3. mica-slate with quarzite.
2. gneiss. a. mica-gneiss. b. quarzite. c. hornblende schist. d. crystalline limestone.
1. granite.

the cavesof the Karst Mountains in Austria. The interesting phenomenon of the Waikaromumu Springs in the Takaka valley, which
send forth a powerful gush of water, is explained by the supposition
that the water after a long subterraneous course breaks suddenly
forth. This zone also can be traced in a southerly direction as
far as Lake Rotorua. It is followed by the mica-schist and clay-
slate zone. Garnet-bearing mica-schist, alternating with quartzite,
constitutes the highest, sharply serrated crests of the Western
ranges in the Anatoki mountains with peaks reaching a height
of 6000 feet above the level of the sea, while farther to the West
the mica-slate passes imperceptibly into clay-slate (phyllite). The

View of the Haupiri range and of the Aorere gold-field.
Slate-River Hill.   Lead Hill.   Mt. Olympus.

Aorere valley, and the mountains 4000 to 5000 feet high on its
east side, such as the Slate River Peak, Lead Hill, Mt. Olympus,
and the whole of the Haupiri range belong to the clay-slate zone.
The strata are very much inclined throughout the whole mica-slate
and clay-slate zone, and variously bent. On Mt. Olympus the strata
diverge from below towards the serrated edge of the mountain top
like the folds of a fan.

In the as yet but little explored Wakamarama coast range the
succession of the crystalline slates seems to repeat itself in an inverted order, but with less thickness, while on the West coast

granite again appears. The mica-schist and clay-slate zone, — which, in a breadth of 15 to 20 miles, includes principally the Anatoki and Haupiri ranges, probably continuing in a southern direction through the whole chain of the New Zealand Alps, — contains in its quartz veins and beds the matrix of the gold. The gradual denudation of the mountains, continued through countless ages, has produced masses of detritus, which were deposited on the declivities of the mountains in the shape of conglomerates, and in the river valleys in the shape of alluvial gravel and sand. In this process of deposition, carried on under the influence of running waters, nature herself has effected a washing operation, during which the heavier particles of gold contained in the mountain detritus collected themselves at the bottom of the deposits and close to their source, so that they can now be obtained by digging and washing. The conglomerates accumulated on the slopes of the mountains are the proper field for the *"dry diggings"*, while from the gravel and sand of the beds of rivers and smaller streams the gold is obtained by *"wet diggings"*. The latter were those first worked. Nearly all the rivers and creeks running from the Anatoki and Haupiri Ranges, either East to the Takaka valley or West to the Aorere valley, or like the Parapara, towards North into Golden Bay, have been found to be more or less auriferous.

The *Aorere Diggings* are situated partly in the main valley itself, partly in the numerous side valleys intersecting deeply the slate rock,[1] at a distance of 5 to 12 miles from Collingwood. The gold is washed from the alluvium of the rivers by sluice boxes and cradles. It is a scaly gold, with rounded particles, which prove that it has been exposed to the action of running water, and brought thither from a greater or less distance. Yet, nearly every

---

[1] The principal ones of those gold bearing rivers and creeks are: Apoos River with Apoos Flat, Lightband's Gully, Cole's Gully, Golden Gully, Brandy Gully, Doctor's Creek, Bedstead Gully, Slate River with Wakefield Creek and Rocky River, Little and Great Boulder River, Salisbury Creek, Maori Gully; all of them tributaries from the right and their side branches rising in the Haupiri Range and its spurs. But recently gold was discovered also on Kaituna Creek, coming from the Wakamara Range as a tributary from the left.

valley and every creek contains gold of somewhat different appearance. [1] While most of the gold is very pure, that of the Slate River, for example, has always a brown ferruginous coat. On the Apoos River the gold is accompanied by crystals of iron pyrites, which remain behind in the process of washing; in other places magnetic iron or titaniferous iron is found in company with gold. The fact that the heaviest gold is found in the upper parts of the streams points clearly to the mountains as the original source of the metal.

But it would be improper to speak of an Aorere gold-field if the gold were confined to the deep and narrow gorges of the streams, cut down into the clayslate rocks. The whole region of the eastern side of the Aorere valley, rising from the river beds towards the steep sides of the mountains at an inclination of about eight degrees, and occupying from the Clarke river to the South, to the Parapara on the North, a superficial extent of about 40 English miles, is a gold-field. Throughout this whole district, on the foot of the range, we find a conglomerate deposited on the top of the slate rocks, attaining in some places a thickness of twenty feet. Pieces of driftwood changed into brown coal, as well as the partial covering of the conglomerate with tertiary limestones and sandstones at Washbourne's Flat indicate a probably tertiary age of this conglomerate formation. Where a ferruginous cement binds the boulders and the gravel together, this conglomerate is compact; in other places only fine sand lies between the larger stones. Quartz aud clayslate boulders are the most commonly met with. This conglomerate formation is not only cut through by the deep gullies of the larger streams, but in some places washed by the more superficial action of water, and is thus

Section through the Quartz Ranges.
a. clayslate. b. auriferous conglomerate. c. alluvial sand.

divided into parallel and rounded ridges, of which that portion of the district called the Quartz Ranges is a characteristic example.

[1] According to a test made at the mint office in Vienna, the Nelson gold averages 89 per cent fine gold, and 0,145 per cent fine silver.

This conglomerate formation must be regarded as the real gold-field, prepared in a gigantic manner by the hand of nature from the detritus of the mountains for the more detailed and minute operations of man.

While the less extensive but generally richer river diggings afford better prospect of gain to the individual digger, the dry diggings in the conglomerate will afford remunerative returns to associations of individuals who will work with a combination of labour and capital. The intelligent and energetic gold digger Mr. Washbourn, was the first person who has proved the value of the dry diggings in the Quartz Ranges, and he has demonstrated the fact that gold exists in remunerative quantities in the conglomerate. I am indebted to Mr. Washbourn for the following interesting details. He writes to me: — "In the drifts into the conglomerate of the quartz ranges, the average thickness of dirt washed is about two feet from the base rock, and the gold produced from one cubic yard of such earth would be, as nearly as I can calculate, worth from twenty-five to thirty shillings. This includes large boulders; so that a cubic yard of earth, as it goes through the sluice, is of course worth more, as the boulders form a large proportion of the whole. When the whole of the earth from the surface to the rock is washed, the value per cubic yard is much less; not more, perhaps, than from three shillings to six shillings per yard, but it would generally pay very well at that." These are the words of one of the most expert Nelson diggers, who payed his men for working in the Quartz Ranges wages of from ten to twelve shillings a day, and still made a considerable profit for himself. With these data, while at Nelson I ventured to make the following calculation in order to encourage the public to a more extensive enterprise in the working of the gold-field. If we reckon the superficial extent of the Aorere gold-field at 30 square miles, the average thickness of the gold bearing conglomerate at one yard, and the value of gold in each cubic yard of conglomerate at five shillings, the total value of the Aorere gold-field amounts to £22,500,000; or in other words, each square mile of the gold-fields contains gold to the amount of £750,000. Of this the above mentioned sum of £150,000, already obtained, is of course only a very small part.

The *Parapara diggings* are the northern continuation of the Aorere gold-field at the mouth of the Parapara river, four miles East of Collingwood on the shores of Golden Bay. A striking phenomenon at the Parapara Harbour is the large masses of sandy brown iron-ore protruding from the white quartz boulders in the form of rugged rocks of a dark-brown appearance, and giving rise, from their striking resemblance to volcanic scoriæ, to the erroneous supposition that volcanic forces had been active on the Parapara.

On the eastern slope of the Haupiri and Anatoki Ranges it is principally the Anatoki, the Waikaro (or Waingaro) and the Waitui, branches of the Takaka river, as well as the upper Takaka valley itself, that are found to be goldbearing; those together constitute the area of the *Takaka Diggings*. Of professional diggers I met but few in those parts; but farmers and wood-cutters settled in the Takaka valley, finding markets bad, occasionally exchanged their usual avocations for gold digging, thus finding in "hard times" among the wildernesses of their mountain heights a sure source of gain. Mr. S., one of these farmers, settled upon the fertile, wooded, alluvial plains of the Takaka valley, in whose house I found hospitable quarters, was in the habit, whenever he projected a trip to town for the purpose of making purchases, to send his sons for a few days previous into the mountains to wash for gold; and they would always return with their pockets filled. The heaviest nuggets were found in the Waitui River, which takes its rise from the Mount Arthur range. A characteristic feature of the Takaka diggings is the occurrence of *Osmiridium* and *Platiniridium*, which is washed out together with the gold in small grains of a whitish tin colour; likewise grains of titaniferous iron, and magnetic iron, and very numerous garnets, — not rubies, as the diggers generally thought-are found there. On the southern slope of Mt. Arthur

---

[1] To Mr. Hacket of Nelson I am indebted for a piece of Platiniridium, weighing 4,57 gr., which is likewise said to have been found on the Takaka River. The spec. gr. of the piece is 17,5.

range, on the sources of the *Todmore*, *Wangapeka* and *Batten*, tributaries of the Motueka river, very promising traces of gold were found.

These are the facts, as far as they were known up to August 1859, — at the time of my stay in the Province of Nelson, and which sufficed to convince me that the Nelson Gold-fields, although not like those of Australia or California, were nevertheless well worth a more extensive system of digging. On the other hand, there could be no doubt whatever that the auriferous formations continue to strike in a southerly direction probably through the whole of the South Island, and I was perfectly right in positively asserting in my Nelson Report, [1] "that, what is at present known is but the beginning of a series of discoveries which future yearswill bring to light."

I have been therefore always greatly delighted by the news furnished me, since my return to Europe, in letters from friends, and in New Zealand papers as to the favourable progress of all the enterprises on the Nelson gold-fields and the new gold discoveries. Numerous associations were formed which commenced their labours in 1860 and saw them crowned with the best success. [2] There were parties working with 20 hands, paying them wages of 10 shillings each per day, and still realizing £80 a week for themselves, while individual diggers averaged only £1 per day. The Takaka diggings especially met with a most satisfactory success, and in January 1861 the news coming in from the Wangapeka and its tributaries, where individual diggers realized as much as 10s. per day, roused the whole of Nelson into a general excitement. [3] Farther explorations

[1] New Zealand Govt. Gazette of the 6th December 1859.
[2] The Nelson Company, Collingwood Comp., Devil's Hill Comp., Tunnel Party, Metallurgic Comp. etc. The Nelson Examiner of Nov. 10. 1860 says: "Our gold diggings are going on steadily and well, the companies still realizing a regular profit, which gives them a good return for the capital invested; and the Takaka valley in particular bearing additional testimony to the truth of Dr. Hochstetter's assertion, that the whole range of mountains is auriferous, and the gold generally diffused all over their lower slopes and the valleys at their base. On Bell's diggings, situated between the river and the hills on the West, there are now about 70 men at work, all doing well, and averaging, it is said, a pound a day per man."
[3] The gold on the Batten River near Wangapeka is said to originate from decomposed hornblende-granite, and to be found in small grains of the size of gunpowder, thus differing from the leaf-gold coming from the slate mountains.

towards the South confirmed the supposition that the auriferous formations continue in that direction. Dr. Haast in his expedition to the West coast found traces of gold in the rivers forming the outlets of the Lakes Rotoiti and Rotorua, and also in the Owen and Lyell river. On the West coast, the precious metal was discovered in the Wakapoai (or Heaphy river), in the Karamea (or Makay river), in the Waimangaroha, seven miles North of the mouth of the Buller, [1] in the Buller and Grey river district. But the Nelson gold-fields were totally eclipsed by the surprising discoveries and splendid results obtained in the Province of Otago in 1861.

To Mr. Ligar, the former Surveyor General of New Zealand, the merit is ascribed of having first proved the existence of gold in the *Province of Otago;* already in 1857 and 1858 reports were current of various gold discoveries on the Mataura river, upon the Waiopai plains and on the mouth of the Tokomairiro, moreover on the Tuapeka, Pomahaka and Lindis, tributaries of the Clutha river; also near Moeraki and various other points. Even in the immediate vicinity of Dunedin, the capital of the province, in the Northeast Valley, reports of gold discoveries were noised abroad; and claims were made by two parties, in 1859, to the reward of £500 fixed for the discovery of a paying gold-field. Yet all these discoveries were not sufficient to rouse the general attention. The existence of gold in remunerative quantities was not established as a fact, and there were many among the colonists who would not even consider the discovery of a rich gold-field as a peculiar blessing to the young colony, but deemed the quiet and steady development of agricultural pursuits and the breeding of cattle more beneficial to the common weal than the richest gold-diggings.

It was not until 1861 that the gold-fever broke out. Thousands

---

[1] According to the testimony of Dr. Haast the granular or gunpowder-like gold in the beds of the rivers Rotoiti and Rotoroa is derived from the decomposition of rocks of a granitic and syenitic character. In the Waimangaroha diggings (in the Papahaua range North of the Buller) the gold is nuggety and angular. A few Maoris washed with a tin-dish, in two weeks, 80 ounces! In August 1861 there were about 60 diggers, mostly Maoris. — In the Grey River district in 1864 there were about 700 diggers.

of persons, who despite storm and rain, in the middle of winter, along almost impassable roads, flocked towards the Eldorado situated about 80 miles West of Dunedin on the Tuapeka river, [1] established by the produce of their labour in a few months, the

[1] The following letters received from diggers at Tuapeka are taken from the Nelson Examiner:

Tuapeka Goldfields, Aug. 14. 1861.

Dear Friend — I now write you a few lines about Tuapeka gold-fields, so that you will be able to inform your friends there from Dunedin. The roads are in an awful state. It takes 20 bullocks to pull one dray with a 12 cwt. load. The diggings lie in a gully between two ranges of hills, and a small creek runs through, where the diggers have their tents pitched close alongside. The gully is very narrow in some places, and not over 200 yards in the widest; all the best claims are taken up. Any man, however, who can stand up to his knees all day in water can get 25 s. a day and his grub. The average findings on these diggings are about £2 a man per day; but there are plenty not making more than their grub, and there are some working for 10s. a day and their grub. There are a few lucky ones making £40 per day, but that is only one party consisting of six. My advice is for people to stay where they are for two months; this gully will then be worked out, and Mr. Reid says when it is he will show them another. I would also advise parties of six to form themselves at home, and provide themselves with a California pump, sluice, picks, shovels, tent, blankets, etc., and whatever they need in the way of provisions can be obtained in Dunedin. I have had a cold ever since I came; at present I would not go and dig for any man under £3 per day. It is killing work, and every man deserves to do well.   F. B. T.

Dunedin, Aug. 1861.

Surprise, I expect, will be stamped on your countenance when you see my letter headed "Dunedin", but such is the case, and here I am. The day we wrote to you we all started for the diggings, and have been out ever since on the snowy ranges, lost our way, got short of food, and had to find our way back as best we could, nearly famished. I never went through so much fatigue and hardship in my life. I thought we were done for. Three days and as many nights we were forced to clear among the snow, which was three to four feet deep, and lie down with our blankets thrown over us. There was not a whole biscuit among us for the last two days we were out. Every man in the party (and there were 35 altogether) was completely knocked up, and all are either in town here or at the accommodation house, bringing themselves round again. My shoulders are cut to the bone carrying my swag, and all my teeth are loose with the intense frost; but my spirits are not a bit broken. I feel quite sanguine, and intend going to work here at once for some time till the roads become practicable. The weather is very severe at the diggings at present, and, as I have no pump or tent, it will be just as well to wait two or three weeks till the weather gets fine . . . . The news from the gold-fields is very good, but the weather severe. I will write to you every mail if you write to me; when you do so, direct your letters to the post-office, to be left till called for. By the next mail I will write you more fully, but I am now in a great hurry, as I am going out to see after some work.   N. J.

Gabriels Gully,
on the Tuapeka Goldfield in the Province of Otago.

fact that New Zealand is one of the richest gold countries of the known earth. The first news was dated in June. Any one who can stand the weather, the report says, can gather from one to two ounces of gold per day (£3 to £6). Such profits, of course, were enticing. On the 28th June 1861 the Tuapeka district was announced to the provincial assembly as a gold-field by the superintendent, and Mr. *Gabriel Reid* received a considerable public remuneration as its discoverer. Towards the close of July there were about 2000 diggers congregated in Gabriel's Gully on the upper Tuapeka, turning up the ground in all directions. A tented town of not less than 600 tents had sprung up in a region hitherto wholly desolate. The excitement in the Province of Otago spread rapidly also over the other provinces, and from Canterbury and Nelson, [1] from Wellington and Hawke's Bay, and even from Auckland, they came flocking by hundreds and thousands to the gold-promising South. The *"News from the Waikato"*, and of the *"Maori War"*, that hitherto has been the daily theme of all the New Zealand papers, was supplanted by *"News from Dunedin"* and *"Latest news from the Otago goldfields"*, and nurses would lull the children to sleep with:

„Gold, gold, gold, bright, fine gold!
Wangapeka, Tuapeka, — gold, gold, gold!"

Despite the immense rush of people to the gold-fields, reports continued to be favourable. Along side of Gabriel's Gully, Munroe's Gully was discovered, in which the Wilson Party obtained in one day 38 ounces, and Weatherstone's Gully, where four Cornish miners gathered in four weeks £1000 each, and other parties £90 each man per week. Gold was found everywhere, in the valleys and on the slopes of the mountains, so that on the 1st August the whole area of 51,000 acres, which is bordered on the North, East and West by the mountain ranges encircling the Tuapeka basin,

[1] The Nelson gold-fields were deserted in consequence of it; every body rushed to Otago. In September 1861 there were scarcely over 100 diggers left on the Aorere, who, however, found themselves amply rewarded by abundant profits for having withstood the temptation to emigrate for the South.

was declared by the Government to be a gold-field, to which the laws of the Goldfield Act were applicable. Mining statutes and digger licenses were issued, gold commissioners were installed, and escorts established, who conveyed the gold produce every fortnight under guard to Dunedin. At the close of August and in the beginning of September the gold-fields numbered already 4000 diggers, who together with their wives and children represented a population of 12,000 to 16,000 inhabitants. Already in the middle of August the weekly produce of gold was estimated at 10,000 ounces. The "dirt" was obtained from superficial boulder and shingle deposits in pits four to five feet deep, and the gold was mostly in thick leafs; larger nuggets, however, were of rare occurrence. The success of some individual diggers, and of parties of 4 to 6 men, exceeded even the most sanguine expectations. [1]

No wonder that the golden tidings from the Tuapeka resounded also beyond the sea. The Victoria diggers upon the gradually exhausted gold-fields of Australia [2] replied to the call, and two months after the arrival of the first news, which had spread abroad with incredible rapidity, the rush to Otago from Australia was general. Diggers bound for New Zealand thronged in the streets and on the quays of Melbourne; sailors deserted from their ships, and speculators of every kind saw a new field open in New Zealand. Victoria papers from the middle of September 1861 reported no less than 23 vessels, all bound for Otago, among them the best

---

[1] I quote here from Otago papers of July and August 1861 the following facts: "Stuart's party of four men averaged from 8 to 16 ounces per day. John Crammond brought in 32 ounces as the result of five days labour of 5 men. A party of 7 men gathered in three weeks 270 ounces; three other diggers, 93 ounces in two weeks. One man and his son made about £500 per month, and of Peter Lindsay's party each member had gained £1000 in two months, from the beginning of the diggings to the close of August. Five pounds (Sterl.) were considered as a middling result for one man per day.

[2] Since the "surface diggings" in the Australian gold-fields are being more and more exhausted, the individual digger no longer finds his expectations realized as formerly. But numerous mining companies have established themselves, who invest considerable capital for the construction of larger works, to be carried on with machinery, and are realizing splendid results, especially by mining the quartz reefs so that the golden days of Australia are yet far from being passed.

Australian steamers, and the most magnificent Liverpool and London clippers. It was calculated that this fleet would bring about 12,000 persons, which number would exactly double the former population of the Province of Otago. Not only gold diggers embarked, but also other enterprising men of all kinds, who hoped to secure their share of gold indirectly, were of the party. At the close of September the number of immigrants daily arriving from Melbourne was estimated in Dunedin at 1000. The busy hum and bustle, the noise and confusion caused by this sudden rush of people, were something unheard-of on the hitherto quiet shores of New Zealand. But while thousands came with golden hopes, there were many who soon quitted the country, sadly disappointed; and after having sacrificed to the thirst for gold, the little which they possessed, were glad to earn their passage back to their former homes by working on board some returning vessel.

The arrival of experienced professional diggers, however, was of great importance for the working of the newly-discovered goldfields. The Australian diggers soon found out, that hitherto only the superficial deposits had been worked, and that they had not yet come to the "bottom", where according to their experience, gathered in Victoria, the richest treasure was to be expected. They began "deep sinkings", and after having dug to a depth of about 130 feet through a tertiary deposit of clay and marl, struck a second gold-bed, which proved rich beyond all expectation. In this way Gabriel's Gully was worked a second time, at a greater depth, and the result was even more brilliant than before.

Additional researches at greater or less distance from Tuapeka continued to lead to new discoveries. Similar treasures were found seven miles Southeast of Tuapeka-Camp in Mansbridge Gully, and in all the side valleys of the Waitahuna river, likewise in the Waitahuna flat itself. There was room for 1000 to 1500 diggers, where they could average from 30 shillings to 3 pounds a day per man. Northeast of the Tuapeka rich gold-beds were discovered on the Waipori river and its various branches; and the upper range of sources of the Tuapeka, Waitahuna and Waipori now consti-

Hochstetter, New Zealand. 8

tutes together one gold-field of about 400 square-miles, an area affording room enough for 50,000 diggers. With the extension of the gold-fields and with the always increasing number of diggers, which at the close of 1861 was already estimated at 12,000 to 15,000, the amount of gold obtained increased, so that in December 1861 weekly escorts were established, which, exclusive of considerable quantities of gold, that remained in the hands of private individuals, conveyed each time from 10,000 to 12,000 ounces to Dunedin. By the middle of January, 1862, the total amount of gold obtained upon the Otago gold-fields was already about 250,000 ounces, or in round numbers about one million pounds sterling in value.

The gold from the Otago gold-fields is partly granular or gunpowder-like, partly in scales or nuggets, or crystallized; it exhibits every degree of intermixture and variety of each of these forms or kinds in different localities. It is associated with titaniferous iron-sand, iron pyrites, tin-ore, topaz, garnets and other minerals. The original matrix of the gold is quartz, and the latter occurs interbedded in, or associated with metamorphic slates, especially of the gneiss, mica-, talc-, chlorite- and clay-slate families. These auriferous schistose formations, which form the geological base of the greater part of Otago, are the source of the gold drift so abundantly distributed over the lower parts of the province. Quartz-reefs are confined to the upper arenaceous schists, but there are very few instances of true fissure-reefs having been discovered. Only one reef as yet is being worked in the same manner as those in Victoria, and the yield is about one ounce to a ton. Dr. Hector states, that he has nowhere seen in the Province of Otago the exact mineralogical equivalents of the auriferous slates of Victoria or California, but they resemble much more those of British Columbia. [1]

In the beginning of 1864 gold was discovered on the northern end of the island, in the Province of Marlborough, on the banks of the *Wakamarina* and its tributaries, a river running into the

---

[1] See Dr. Hector on the Geology of Otago (Quart. Journ. of the Geol. Society of London, 1865), and W. L. Lindsay, on the Geology of the gold-fields of Otago (Proceed. of the Geol. Section of the British Association at Cambridge, 1862).

Pelorus at a point now known as Canvas Town; while the Pelorus itself runs into the Pelorus Sound near the township of Havelock. It was at the juncture of the Wakamarina and Pelorus, that Mr. Wilson and his party first made the discovery of the existence of gold. In the month of May, 1864, nearly three thousand men were on the ground up the river, all actually engaged in digging. Many of the claims were wonderfully rich, others turning out satisfactory, but the great bulk yielded little more than fair wages.

The last, though not the least important discovery of gold has been made in the Province of Canterbury. In the North on the Teramakau river, on the borders of the Provinces of Nelson and Canterbury, the bold and enterprising James Mackay found traces of gold as early as in 1859; and Lindis Pass, where in January 1861 there were already 500 diggers collected, is situated quite close to the southern boundary of the Province of Canterbury. The Canterbury Government therefore in 1861 set a reward of £1000 for the discovery of a productive gold-field within its territory. Such a gold-field has at last in 1864 and 1865 been discovered on the West Coast of the province. All the numerous streams which run into the ocean from the Southern Alps are charged with deposits of the precious metal, and the whole beach from Grey-mouth in the North to the Wanganui river in the South, and beyond, is prospected in hundreds of places with varying success. The Wahi punamu of the Maoris (the greenstone coast) may now truly be called the New Zealand Gold-coast.

1 Of the finds recorded — Henry Wilson and party divided two hundred ounces for their week's work. Rutland and his party disposed of eighty ounces of gold for the same period. Three men, in felling a tree at the edge of the river, in order to work their claim, came suddenly in a pocket, from which in a few hours they washed thirty ounces of gold. Matthews, an old man, with his two sons, obtained eighteen ounces of gold near the Fork in four days. Many others have been more or less successful. A correspondent of the "Havelock Mail" writes on the 2d June: "However lucky the digger is on the Wakamarina, and the tributary creeks, he is certainly not to be envied in the labour he has to perform in getting at the gold. The fields of Victoria are gardens in comparison, and even the bleak inhospitable ranges of Otago are preferable. Here the sun seldom penetrates the forest, and is never seen by the men at work in the creeks. There is not a man who has visited those parts but will endorse these remarks."

The country at the mouth of the Hokitika (Okitiki) river about twenty-one miles South of the Grey river, and fifteen miles South of the Teramakau now became the centre of immigration. A township was set out at the mouth of the river; a gold-field was proclaimed, magistrates and wardens appointed, and already in the month of March, 1865, 3000 men were at work over some twenty miles of the country around Hokitika. The extraordinary impulse given to trade by the discovery of gold has never been more fully illustrated than in the case of Hokitika. In spite of the depressing influence of a climate as tempestuous certainly as that of any habitable part of the globe; in spite of a harbour perhaps the most dangerous to enter in New Zealand,[1] a populous township has arisen within a few months, which now is the important metropolis of the Westland gold territory.[2] A spit or mere bank of sand a mile and a half in length, and, for the greater part of its length, scarcely ten chains wide, formed by the conflicting actions of the river Hokitika and the ocean, is the deposit of capital, the amount of which, could it be stated, would be considered fabulous even in the annals of gold-digging communities. This discovery of gold on the West coast will be of the greatest importance to the colonisation

[1] The tongue of sand on which stands Hokitika is bounded by a very narrow channel of deep water on the outside of its lower extremity, which is entered at a sharp angle by vessels running the bar. Unless there is a good breeze, a sailing vessel cannot preserve sufficient steerage-way through the rollers to round the corner; she drops bodily off before the sea, and two or three tremendous broken seas lift her broadside on to Hokitika spit, there to be pounded to pieces by the following tide. The entrance to this river is exactly represented by the letter S of which two-thirds form a channel not 100 feet wide, and exposed to the whole drift of the Pacific Ocean. The wreck of a sailing vessel attempting to enter has really come to be considered a certainty here, safety the exception. This state of things is producing its natural results. Running the blockade, as it is termed here, will not be attempted twice; no offer, however lucrative, is sufficient to induce a fortunate skipper to dare the danger he has once escaped. The beach is strewn with wrecks; eighteen sailing and seven steam ships were hurried to destruction here within five months in 1865. This great destruction of shipping makes it clear that the trade of the gold-fields cannot be carried on with security until a road fitted for heavy traffic has been completed, connecting the eastern with the western side of the Province.

[2] A second township, Kanieri, is situated on the right bank of the Hokitika, two and a half miles from its mouth.

and future development of South Island, and no doubt the West-land gold-fields of Canterbury are quite equal, if not richer than those of Otago. [1] Concerning the geology of the West coast I may refer to the instructive reports of Dr. J. Haast. [2]

May New Zealand in its golden age, thrive and flourish in a manner never before dreamed of, and by the opening of its coal treasures and the yet hidden mineral veins may the bronze and iron ages of art and manufacture follow in rapid succession. This is my wish in concluding this chapter.

[1] In Sept. 1865 the following estimate of the population now on the West Canterbury gold-fields has been made: —

Grey } Arnold }	2,000
Greymouth	600
Saltwater Creek	300
Greenstone	400
Teremakau	100
Waimea	1,000
Hokitika	2,500
Kanieri	3,000
Totara	4,000
Further south	2,000
	15,900.

The population was increasing at the rate of three thousand per month.

From a return in the New Zealand Gazette, published on June 20. 1866, we find the quantity and value of gold exported from the colony from the 1st April, 1857, to the 31st March, 1866, to be as follows: — From Auckland, 15,794 ounces, value £48,512; from Marlborough, 32,898 ounces, value £116,465; from Nelson, 188,669 ounces, value £731,188; from Canterbury, 351,913 ounces, value £1,369,256; from Otago, 1,938,837 ounces, value £7,512,995; from Southland, 22 ounces, value £85; making the total export from the colony 2,528,133 ounces, value £9,788,501.

[2] Report on the Geological Exploration of the West Coast by Dr. J. Haast, Christchurch 1865; Lecture on the West Coast of Canterbury by Dr. J. Haast, Christchurch 1865. I may be allowed to mention here the words of the Super-intendent, Mr. J. Bealy, in his address on opening the Provincial Council on the 21st Nov. 1865: — "It is with much satisfaction that I recognise the importance of the scientific researches undertaken by the Provincial Geologist; and the fact of his having two years ago — in papers laid before this Council — accurately defined the gold districts on the West Coast is a strong proof of the practical value of his labours, and leads me to expect that the Province will derive great benefit from his knowledge as to the auriferous nature of the lands which it may be advisable to survey for sale."

Appendix.

A. Dr. J. Haast gives the following list of ores and minerals occurring in the Province of Canterbury.

Mineral Coal:

a) Carboniferous Anthracitic and bituminous coal, River Kowai, Mount Harper, Clent Hills, etc.

b) Secondary coal, bituminous, River Grey, West Coast.

c) Tertiary brown coal and lignite in tertiary formations, all over the Province, Malvern Hills, Mount Somers, Rakaia, Coal Creek, Rangitata, Northern Hinds, River Potts, Ashburton, Tenawai, etc.

Selenite, in crystals on the surface of tertiary shale, Tenawai, etc.

Calcite (calcareous spar) in cavities of volcanic and in veins of sedimentary and metamorphic rocks, abundant all over the Province.

Travertine, deposited from water having carbonate of lime in solution, Weka Pass.

Marble, Malvern Hills.

Stalactite, ⎫
Stalagmite, ⎭ caves of Mount Somers, etc.

Arragonite, lining fissures and cavities of volcanic rocks, Banks' Peninsula.

Dolomite (Magnesian limestone), Malvern Hills, interstratified with augitic greenstone.

Quartz, in veins, in metamorphic and palæozoic rocks, all over the Province. This mineral occurs also in the following varieties: —

Rock Crystal, ⎫ in amygdaloidal trap lining, geodes, and cavities,
Amethist, ⎭ Malvern Hills, Mount Somers, etc.

Milky Quartz, in granites, West Coast.

Prase, small deposits in quartzose porphyritic trachyte, Gawler's Downs.

Chalcedony, in mammillary and botryoidal forms in amygdaloidal trap and quartzose trachytes, Malvern Hills, Clent Hills, Mount Somers, etc.

Chrysoprase, filling cavities, ditto, ditto.

Carnelion, in small geodes and filling cavities, ditto, ditto.

Agate, in geodes, often of very large size, ditto, ditto.

Flint, filling cavities in the rocks, ditto, ditto,

Aventurine, ditto, ditto.

Onyx. Some horizontally arranged chalcedonies in different colours show a tendency to become onyx and sardonyx, ditto, ditto.

Plasma, filling fissures in tertiary quartzose trachytes, and occurring principally in Gawler's Downs.

Heliotrope, in tertiary quartzose trachytes in small pieces, Snowy Peak, Malvern Hills.

Jasper,
Basanite,
Chert, } in different varieties, Malvern Hills and elsewhere.
Lydian Stone,

Silicified Wood (petrified), in creeks in many localities where siliceous rocks are decomposing.

Ferruginous Quartz, Gawler's Downs.

Semi-opal, filling small cavities in quartzose porphyritic trachyte, Malvern Hills, and Mount Somers.

Opal, ditto, ditto.

Quartz, in pseudomorphs, imitative crystals of calcite, Snowy Peak, Malvern Hills, Gorge of Rakaia, Clent Hills, etc.

Hyalite, in small masses lining cavities, Snowy Peak, Malvern Hills.

Apophylite, in amygdaloids, Rangitata.

Ichthyophthalmite (zeolite), in felsite porphyry, Rangitata, Turn-again-Point.

Serpentine, in veins, Mount Cook Range, and some other localities in the Alps.

Diallage, in Gabbro, Mount Torless Range and Upper Rakaia.

Delessite, in amygdaloides, Rangitata and Malvern Hills, etc.

Chlorite, in laminæ, metamorphic schists, West Coast.

Nephrite (punamu of the Maoris), in rolled pieces of the beach of the West Coast.

Augite, in trachydolerites and in fine twin crystals imbedded in agglomeratic tufa, Banks' Peninsula.

Hornblende, in basaltic and doleritic rocks, Banks' Peninsula, Malvern Hills, Timaru, etc.

Hypersthene, in hypersthenite, Malvern Hills.

Actinolite, in metamorphic schist.

Chrysolite, in grains of basaltic rocks, Banks' Peninsula.

Bole, filling cavities in lava streams, Banks' Peninsula.

Pimelite, filling cavities in amygdoloidal rocks, Malvern Hills, Clent Hills, etc.

Palagonite, in angular fragments in palagonite tufas. Harper's Hills, near Selwyn, and Two Brothers, Ashburton. Another variety changing insensibly into a

Pitchopal, inclosing leaves and stalks, silicified, occurs in the same localities.

Heulandite (Zeolite), in amygdaloidal traps associated with felsite porphyries, Turn-again Point, Rangitata.

Stilbite (Zeolite), in amygdaloidal traps associated with felsite porphyries, Turn-again Point, Rangitata.

Natrolite, filling cavities in volcanic rocks, Banks' Peninsula.

Mesotype, in needles, in fissures of volcanic rocks, Banks' Peninsula.

Chabasite, in trachytes,   do.   do.

Orthoclase (potash felspar), in granites and other crystalline rocks at the West Coast, etc.

Sanidine, or glassy felspar in trachytes and trachy dolerites, Banks' Peninsula, and quartzose porphyritic trachytes, Malvern Hills.

Obsidian, on the sides of trachytic dikes, Banks' Peninsula.

Pitchstone, associated with quartzose porphyritic trachytes, Snowy Peak, Mount Somers.

Albite, in dioritic porphyries, River Wilkin and Makarora Ranges.

Oligaclase (soda felspar), in quartzose porphyritic trachytes, Mount Misery, Malvern Hills.

Labrador, felspar in lava streams, Banks' Peninsula.

Saussurite, in Gabbro, Mount Torlesse.

Garnet (Almandine), in quartzose porphyritic trachytes and pitchstones, Malvern Hills and Mount Somers.

Pistacite, in diorite, Mount Torlesse Range.

Potash Mica (muscovite), in granites and schists, West Coast.

Magnesia Mica (Rubellan), in volcanic rocks, Banks' Peninsula.

Pearl Mica (margarite), in gneiss and metamorphic schists, West Coast.

Tourmaline, in granite, Mosquito Hill, West Coast.

Marcasite (white iron pyrites), in clays and tertiary rocks, in many localities.

Pyrites, partly mundic, in older palæozoic rocks as well as in brown coal and shale, ditto.

Mispikel, in diorites, Malvern Hills.

Clay Iron Ore, } in tertiary strata assotiated with brown coal and
Sand Iron Ore, } lignite.

Ilmenite, titaniferous magnetic iron ore, in grains in melaphyres, Clent Hills.

Magnetic Iron Ore, in grains in dolerite, Malvern Hills.

Green Earth, in amygdoloidal trap, Malvern Hills, Ashburton, Rangitata, etc.

Spathic Iron (carbonate of iron), found in large boulders coated with black psilomelane, near the sources of the River Kowai, Mount Torlesse.

Sphærosiderite, in small crystals, or lining cavities of volcanic rocks, Banks' Peninsula, Malvern Hills.

Vivianite, coating cavities in melaphyres, Clent Hills.

Hausmannite (red oxyde of manganese), coating joints in rocks and in rolled pieces in River Selwyn.

Psilomelane, in veins, Upper Waimakariri.

Glauconite (green sand), as small grains in the pepperstones, middle tertiary series, Malvern Hills, Coal Creek, Rangitata, Weka Pass, Ash hurton, etc.

Copper Pyrites, in grains imbedded in quartzose schists, Moorhouse Range, etc.

Green Carbonate of Copper, in a rolled piece, from Mount Somers Range, River Stour.

Gold, south-eastern part of the Province near Waitaki, and south-western near Lake Wanaka, and western side of the main range generally.

Retinite, in brown coal (fossil gum).

———

B. At the international exhibition at Paris, 1867, a golden obelisk was again exhibited by the Commissioners of Victoria. The Pyramid, representing the aggregate bulk of gold obtained in the colony of Victoria (Australia) from October 1851 to October 1866, a period of 15 years, is ten feet square at the base and 62 feet $5\frac{1}{2}$ inches high; its bulk is equal to $2081\frac{1}{3}$ cubic feet. The gross quantity of gold extracted is 36,514,361 ounces, value £146,057,444 Sterling.

C. Summary of the Quantity and Value of Gold Exported from New Zealand, from 1st April, 1857, to 31st December, 1865, extracted from the Statistics of New Zealand for 1865.

Port of Export.	Produce of Gold Fields in the Province of	During the Year 1865.							From 1st April, 1857 to 31st December, 1864.		Total Exported from New Zealand to 31st December 1865.	
		To Great Britain. Ozs.	To New S. Wales. Ozs.	To Victoria. Ozs.	To Tasmania. Ozs.	To other Places. Ozs.	Totals Quantity. Ozs.	Totals Value. £	Ozs.	£	Ozs.	£
Auckland	Auckland	—	5,449	—	—	—	5,449	21,115	9,524	29,875	14,973	50,990
Nelson	Nelson	300	11,043	794	—	—	68,860	266,883	76,238	295,424	145,098	562,257
Hokitika		—	37,459	247	—	—						
Greymouth		—	19,017	8	—	—						
Havelock	Marlborough	12	2,443	330	—	1	7,952	30,814	24,838	96,241	32,790	127,055
Picton		—	3,668	135	—	—						
Nelson		—	1,355	10	—	—						
Nelson	Canterbury	1	9,985	—	1	—	233,174	903,519	1,463	5,670	234,637	909,219
Hokitika		7,707	173,847	25,737	—	27						
Greymouth		—	15,847	12	—	—						
Hokitika	Otago	—	—	17	—	—	259,117	1,004,078	1,637,296	6,344,520	1,896,413	7,348,598
Dunedin		14,557	60,409	178,731	—	—						
Invercargill		—	2,614	2,714	—	—						
Bluff Harbour		—	—	75	—	—						
Bluff Harbour	Southland	—	22	—	—	—	22	85	—	—	22	85
Total	Total	22,577	343,158	208,810	1	28	574,574	2,326,474	1,749,350	6,771,730	2,323,933	8,998,204

1 The value was calculated in the years up to 1865 at the uniform (estimated) rate of £3 17s. 6d. per ounce, in all the Provinces, with the exception of Auckland, for which the ascertained value was given. But for 1865, the whole, including Auckland, has been calculated at the same rate (of £3 17s. 6d.)

CHAPTER VI.

The Flora.

Investigation of the flora since Cook's time. — Dr. Hooker's classical work on the flora of New Zealand. — Number of the species of plants known. — News from the Alps. — New tropical ferns in the vicinity of hot springs. — Peculiarities of the flora. — Abundance of cryptogamic plants. — New Zealand a botanical province of its own. — Affinity to the flora of Australia, South America, Europe, and the Antarctic Islands. — The mother-flora. — Hypothesis of a former continent. — Physiognomical character of the vegetation. — Scarcity of flowers. — Fern-heaths. — Grass-plains. — The forest scenery. — Appendix. List of alpine plants. — The "vegetable sheep."

The peculiar flora of New Zealand is far better known than its geological structure and mineral riches, and far more completely than its fauna. For the earliest account of the plants of New Zealand we are indebted to two of the most illustrious botanists of their age, and to the voyages of the greatest of modern navigators; for the first and to this day the finest and best illustrated herbarium that has ever been made in the islands by individual exertions is that of Sir *Joseph Banks* and Dr. *Solander*, during Captain Cook's voyage in 1769. Upwards of 360 species of plants were collected during the five months that were devoted to the exploration of the East Coast, at various points between the Bay of Islands and Otago, including the shores of Cook's Straits. Captain Cook was, on his second voyage, accompanied by three scientific men, all more or less conversant with botany, namely the two *Forsters* (father and son) and Dr. *Sparrmann*. Queen

Charlotte's Sound in Cook's Straits and Dusky Bay on the Southwest side of South Island were the chief points botanized. Mr. *Anderson*, surgeon to Cook's third expedition, undertook the botanical department on that voyage; but though Dusky Bay was visited a second time, nothing of importance was added to its botany. It remained for Mr. *Menzies*, the surgeon and naturalist of Captain Vancouver's voyage, to discover the cryptogamic riches of New Zealand. That naturalist devoted himself to the collection of Mosses and Hepaticæ, and this at a time when these objects were scarcely thought worthy of attention. In 1824 and 1827 followed the french expeditions under Captain *Duperry* (Corvette "Coquille") and Admiral *D'Urville* (Corvette "Astrolabe"); the combined collections of the two voyages, amounting to 200 species of flowering plants and ferns, were published by Professor *A. Richard* as numbering about 200 species.

On the establishment of colonial gardens and botanists at Sydney, New Zealand became an object of especial interest to the latter; the Bay of Islands was visited by Mr. *Charles Frazer* in 1825, and by the two brothers *Allan* and *Richard Cunningham* in 1826, 1833 and 1838, while at the same time in New Zealand resident missionaries and colonists, such as Dr. *Logan*, the Rev. Mr. *William Colenso* and others were making collections. Mr. *Bidwill* (in 1839) was the first to ascend the Tongariro Volcano, 6500 feet high, and Dr. *Dieffenbach* was in 1839 the first European that ascended Mt. Egmont, 8270 feet high. These gentlemen furnished the first contributions to the most interesting sub-alpine and alpine flora of New Zealand. Then followed the Frenchman *M. Raoul*, who accompanied the French Frigate "L'Aube" in 1840 and 1841, and again the Frigate "L'Allies" in 1842 and 1843, during which voyages he made a very complete botanical exploration of Banks' Peninsula and the Bay of Islands.

With the Antarctic Expedition (1839—1843) under Captain James Ross, Dr. *J. D. Hooker* came to New Zealand. It is to this eminent botanist that the palm is due; for to him science is indebted for the classical work on the Flora of New Zealand, in which all the botanical material known up to the year 1853 has been collected and worked up, and the botany of New Zealand — we may almost say — brought to a close.[1]

The total number of species brought together in Dr. Hooker's Flora amounts to nearly 1900; among them the Phanerogamic Plants are to the Cryptogamic as 1 to 1.6, or about two to three. These 1000 species are divided among the principal families as follows:

Phanerogamic Plants.	Species.	*Cryptogamic Plants.*	Species.
Compositæ	90	Ferns including the Ly-	
Cyperaceæ	66	copodiaceæ and num-	
Gramineæ	53	erous varieties, formerly	
Scrophularineæ . . .	40	treated as species .	117
Orchideæ	39	Hepaticæ	118
Rubiaceæ	26	Musci	250
Epacrideæ	23	Fungi ⎫	
Umbelliferæ	23	Lichenes ⎬	388
Coniferæ	12	Algæ	300
The remaining families .	358	Number of species	1173.
Number of species	730.		

[1] To the detailed knowledge of the flora of New Zealand, as it is laid down in Dr. Hooker's celebrated work, numerous private collectors have also contributed a considerable share. I mention Dr. *Lyall*, who in 1847 accompanied Captain *Stokes* (H. M. St. Acheron), and like Menzies devoted himself most zealously to collecting cryptogamic plants; furthermore, Capt. *Drury*, Mr. *Jollife*, Lieut. Colonel *Bolton*, Rev. *W. Taylor*, Mr. *Th. H. Hulke*, Dr. *Andrew Sinclair*, and Mr. *Knight* in Auckland; Dr. *Monroe* and Capt. *Rough* in Nelson. The principal works on the Flora of New Zealand are the following: —

1776. *J. Reinh. Georg Forster*, Characteres generum plantarum quas in itinere ad insulas maris australis collegerunt, descripserunt, delinearunt annis 1772—1775.
1786. *Forster*, Florulæ insularum australium prodromus. Göttingæ.
1832. *A. Richard*, Essai d'une Flore de la Nouvelle-Zélande. Voyage de découvertes de l'Astrolabe. Botanique, I. Paris.
1838. *Allan Cunningham*, Floræ Novæ Zelandiæ Præcursor; Companion to the Botanical Magazine, Vol. 2, and Annals of Nat. Hist. Vol. 1. 2. 3.
1846. *M. E. Raoul*, Choix des plantes de la Nouvelle-Zélande. Paris.
1853—1855. Dr. *J. D. Hooker*, Flora Novæ Zelandiæ, Botany of the Antartic Voyage. 2 Vols. London.
1864. Dr. *J. D. Hooker*, New Handbook of the New Zealand Flora.

We must not, however, consider the above list as a limit to the Flora of New Zealand. There still remained upon the three islands large districts to be explored. It is only on the North Island that the botanical collectors have penetrated farther into the interior, while upon the South Island the lofty heights of the New Zealand Alps, extending through the whole length of the island, had formerly been left wholly unexplored. It was not until within the last years that from Nelson, Christchurch and Dunedin scientific expeditions were undertaken into those alpine regions; and to the untiring zeal of my botanical friends in Nelson, Dr. *Monro*, *W. F. L. Travers*, and Captain *Rough*, and last but not least, to my friend Dr. *J. Haast*, science is indebted for numerous and highly interesting contributions to the knowledge of the Alpine Flora of New Zealand. [1] It is therefore to be expected that the Flora of New Zealand will be increased by a considerable number of species. Already Dr. Hooker was inclined to regard 4000 as the probable number of New Zealand plants, of which 1000 may be flowering plants. [2]

During my travels upon North Island in the Province of Auckland, and upon South Island in the Province of Nelson, I also devoted myself to botanical collections as far as the main object I had in view admitted. I limited myself, however, almost exclusively to cryptogamic plants and grasses. By the assistance of my travelling companions Dr. Haast and Captain Hay, and by considerable contributions of resident collectors, my collection increased to about 3000 specimens. [3]

What surprised me most was that it was among the ferns,

---

[1] See Appendix to Ch. VI.

[2] The proportion of the number of species for the several families of cryptogames is probably as follows: 130 ferns and lycopodiaceæ, 370 hepaticæ, 500 musci, 1000 fungi, 500 lichens, 500 algæ. In Australia the number of phanerogamic plants is estimated at 8000, but upon the whole earth 80,000 to 150,000.

[3] It becomes my duty here to return my sincere thanks especially to Colonel *Houltain* in Otahuhu, Rev. Mr. *Spencer* in Taravera, Rev. Mr. *Grace* on Lake Taupo, Mr. *William Young* and Rev. Mr. *Kinder* in Auckland; also to Dr. *Monro* and Mr. *W. T. L. Travers* in Nelson.

which are most eagerly gathered by numerous lovers of plants, and especially from the frequently scoured North Island, that I found some new species, — new, if not to science, at least upon New Zealand. They belong to genuine tropical forms, growing in New Zealand only on the margin of hot springs, upon a warm soil and in a constantly hot-moist atmosphere. Dr. Hooker [1] mentions the *Lycopodium cernuum* so frequently found everywhere about the hot springs in the interior of North Island, a species of general occurrence throughout all warm climates. Without the tropics, however, it is found only in the vicinity of hot springs heating the soil, for example on the Azores, upon St. Paul's Island in the South-Indian Ocean; a remarkable instance to prove how far the little spores of this lycopodium have been spread. To this I have the pleasure to add the *Nephrolepis tuberosa, Nephrodium unitum* and *Nephrodium molle*, tropical species, which are more or less general all over the torrid zone. These tropical species are found in the very heart of North Island near the hot waters of the Rotomahana and near the boiling springs of Waikite at the foot of the Pairoa Range between Lake Taupo and the Rotomahana. There they thrive with luxuriant growth in a soil equally warmed by the hot water, and in a constant, unchanging, hot and moist atmosphere. Their spores were most probably transported from the tropical countries of Australia or America to New Zealand. Other species, not tropical, such as *Pteris scaberula, Polypodium rugulosum, Gleichenia decarpa*, are transformed in those places into very peculiar varieties.

For the following remarks on the peculiarities of the New Zealand flora let us refer to the ingenious explanations of Dr. Hooker in his Introductory Essay to the Flora of New Zealand.

From the proportion of the number of *cryptogames* to that of *phanerogames* we first perceive that New Zealand is not only relatively but also absolutely abounding in cryptogames. [2] It would

---

[1] Introd. Essay to the Flora of New Zealand. London 1853. p. 27.

[2] Great Britain, almost equal in size, for example numbers 1400 phanerogames and only 50 ferns to 114 upon New Zealand.

be incorrect, however, to infer from this circumstance that the land remained behind in the history of the development of the earth, or to say that New Zealand is representing up to the present time the age of ferns, or, what is about the same, the age of the carboniferous period. This abundance of cryptogames is rather the mere consequence of the moist climate. Where there is a sufficient supply of moisture, the principal vital requisite for the lower orders of the vegetable kingdom, they grow to this very day as exuberantly as in former times.

The 117 species of ferns described by Dr. Hooker represent 37 genera. Only 42 species are peculiar to New Zealand; 30 it has in common with South America, 61 with Australia and Tasmania; 30 species have a cosmopolitan distribution, and 10 species are found also in Europe. Of the phanerogamic plants more than two-thirds are endemic, or absolutely peculiar to New Zealand, namely 26 genera and 507 species, principally of the families of Orchideæ, Coniferæ, Scrophularineæ, Epacrideæ, Compositæ, Araliaceæ, Umbelliferæ, Myrtaceæ, and Ranunculaceæ. The remaining third of the phanerogamic flora of New Zealand are species which are found also in other countries, illustrating the relations of the plants to those of other countries. New Zealand, therefore, appears now-a-days together with the neighbouring smaller groups of islands, the Chatham, Lord Auckland, and Campbell Islands, — the flora of which corresponds to that of New Zealand, — as a most peculiar botanical province in the temperate zone of the Southern Hemisphere, the independent and peculiar character of which originates from the isolation of the group of islands from all the larger continents. This botanical province has been named "*Forster's realm*". The non-endemic species belong to New Zealand in common partly with the extratropical portions of Australia and with Tasmania, partly with South America, partly with the antarctic flora upon Fuegia, the Falkland Islands, Tristan d'Acunha, Kerguelen Land, St. Paul and Amsterdam etc; and few cosmopolitan species also with Europe.

[1] Norfolk Island, despite its close proximity, shows in consequence of its being quite near the tropics a closer relation to the floras of the Pacific Islands and Australia.

The Australian affinity is illustrated by the large number of 193 absolutely identical species, and by the fact that upwards of 240 of the 282 New Zealand genera are Australian, and of these more than fifty are all but confined to these two countries. New Zealand, however, does not appear wholly as a satellite of Australia in all the genera common to both, for of several there are but few species in Australia, which hence shares the peculiarities of New Zealand, rather than New Zealand those of Australia: this is the case with *Pittosporum*, *Coprosma*, *Olearia*, *Celmisia*, *Forstera*, *Gaultheria*, *Dracophyllum*, *Veronica*, *Fagus*, *Dacrydium* and *Uncinia*, of which there are comparatively few species in Australia and America; on the other hand, *Stackhousieæ*, *Pomaderris*, *Leptospermum*, *Exocarpus*, *Personia*, *Epacris*, *Leucopogon*, *Goodenia*, and a few other large Australian genera are very scantily represented in New Zealand. If the number of plants common to Australia and New Zealand is great, and quite unaccounted for by transport, the absence of certain very extensive groups of the former country is still more incompatible with the theory of extensive migration by oceanic or aerial currents. This absence is most conspicuous in the case of *Eucalypti* and almost every other genus of *Myrtaceæ*, of the whole immense genus of *Acacia* and of its numerous Australian congeners, with the single exception of *Clianthus*, of which there are but two known species, one in Australia and the other in New Zealand and Norfolk Island. Considering that *Eucalypti* form the most prevalent forest feature over the greater part of South and East Australia, rivalled by *Leguminosæ* alone, and that the species are not particularly local or scarce, and grow well wherever sown, the fact of their absence in New Zealand cannot be too strongly pressed on the attention of the botanical geographer, for it is the main cause of the difference between the floras of these two great masses of land being much greater than that between any two equally large contiguous ones on the face of the globe. If no theory of transport will account for these facts, still less will any of a gradual variation of originally identical species.

With South America, New Zealand has 89 species and 76 genera in common. But 77 of those species and 59 of those genera appear also in Australia; they are consequently common to all three countries and full 50 of those species are found also in Europe, leaving but very few species such as *Edwardsia grandiflora*, *Veronica elliptica*, two species of *Coriaria*, *Myosurus aristatus*, *Haloragis alata*, *Hydrocotyle Americana* absolutely peculiar to the two countries. The peculiar development of the common genera *Edwardsia*, *Fuchsia* and *Calceolaria* makes it also here doubtful whether America or New Zealand was their primitive home.

The 60 species which New Zealand has in common with Europe consist chiefly of strand plants, salt-water and fresh-water plants, Grasses, *Cyperaceæ* and some *Compositæ*. Most of them are cosmopolitan, occurring also in Australia, South America, and upon the Antartic Islands, the general diffusion of which appears to have taken place in a way which no longer exists.

The 50 antarctic species inhabit the mountains and the southern extreme of New Zealand. They likewise are partly of a general distribution common also to Europe, partly they are found in Tasmania, and chiefly on its mountains, but not else where. Only, when the flora of the mountains of South Chili, New Zealand, southern Tasmania, the Australian Alps, and the Antarctic Islands shall have been properly explored, will the great problem of the affinity and distribution of the flora in the South Temperate and Antarctic zone be solved. Then will it be clearly proven, that between the floras of the extensive main lands in the Southern Hemisphere simular relations exist as between those of the North Temperate and the Arctic zones, affinities and peculiarities which cannot be accounted for by assuming an oceanic or atmospheric transportation, or by the variability of the species,[1] but which speak in favour of the existence of a once very widely-spread mother flora, broken and cut up by geological and climatic changes into separate botanical regions and provinces.

[1] New Zealand, it is true, has many very variable genera, such as *Coprosma*, *Celmisia*, *Epilobium*, *Ranunculus*, *Oxalis*, *Dracophyllum*, *Leptospermum*, *Podocarpus*, *Dacrydium*, etc.

Moreover, the extraordinary variety in New Zealand of Natural Orders in proportion to the genera, and of genera in proportion to the species[1] speaks in favour of its having formerly been contiguous to larger bodies of land; while, on the other hand, the peculiarity of the species is an evident proof that this contiguity must have ceased to exist for a very long time past. It would become the duty of geology to prove that former contiguity, and by it the original soil of the mother flora. But here the difficulties are crowding together, and it is only to daring hypotheses that we can take refuge. The hypothesis of only *one* continent, sunk in the vast gulf, would not suffice; it would be necessary to premise three or four such continents. First an extensive main land with a rough continental climate. At that time, the antarctic genera which now-a-days are found upon the higher ranges, covered the whole extent of the land. When the sea with its milder breezes took the place of the land, the frightened children of the South Pole fled into the cooler mountain regions, and a new flora took possession of the lower landscape. Simultaneously or in alternating succession with essential climatic changes, New Zealand was then united with Chili, with the tropical islands of the South Sea, and with Australia by means of a continental bridge over which the plants could conveniently immigrate, receiving representatives from

[1] The number of Natural Orders is large in proportion to the genera; being as 92 to 282, that is, about one to three, while the genera are to the species as 282 to 730, each genus having on the average only two and a half species; whence it follows that there are, on the average but 8 species to each Natural Order. In Great-Britain each order counts 14 species; but the probable proportion of the species to the known Orders throughout the globe is as 350 to 1. The extraordinary variety of forms upon a limited insular space would be a very singular occurrence, if it were not substantiated as a fact in Nature itself, that one and the same area is the more productive of life, the greater the variety of forms in that life, since homogeneous bodies mutually tend to destroy, but heterogeneous bodies to sustain each other. This variety of forms, moreover, is quite regularly distributed upon New Zealand, and the differences in the vegetation of different districts are fewer by far, than the extraordinary variety of the structure of the ground would lead one to infer. Considering these circumstances and the additional one, that very many of the Natural Orders cannot be recognised by the flower alone, by fruit alone or by habit or foliage, it may safely be said, that the New Zealand Flora, is, for its extent, much the most difficult on the globe to a beginner.

all those countries, until the connection ceased, and New Zealand stood isolated as it is found to-day. The plants now-a-days found upon New Zealand must consequently have lived through periods of extensive geological and climatic changes.

Many of the above-mentioned pecularities of the New Zealand flora are of importance also for the physiognomy of the vegetation. The traveller, from whatever country, on arriving in New Zealand, finds himself surrounded by a vegetation, that is almost wholly new to him, and two peculiarities will appear striking before all others, that is an abundance of bushes and ferns, and a want of meadows and flowers, which is to be accounted for by the scarcity of grasses and annual plants.

What from afar appears by the side of the immense tracts of forest covering hill and dale to be open land, or meadows, is on a closer observation found to consist of bushes of the size of a man; and where grasses, weeds and chequered flowerage were expected, only uniform ferns and bushes with some scanty white blossoms are to be seen. *Pteris esculenta* (Rarahue of the natives), the roots of which formerly yielded the chief aliment to the natives, covers nearly all the open land, upon the heights and in the low lands, wherever it is not replaced by swamp. Upon fertile soil it grows to the size of a man, and it is with great difficulty that the traveller works his way through the thicket where there is no beaten path, and even upon paths he finds himself from time to time most disagreeably hindered by the woody stems entangling his feet. Only Manuka and Rawiri bushes (species of *Leptospermum*), or Kumaharau *(Pomaderris)* and Kororiko *(Veronica)* are intermixed with *Pteris*. Here and there, in moist places, arises isolated the "grass tree" or "cabbage tree" (Ti of the natives; *Cordyline australis*), and, quite modest, like a recluse hidden among the bushes, blooms the tender blue Rimuroa *(Whalenbergia gracilis)*, the only bell-flower of New Zealand, and the Tupapa *(Lagenophora Forsteri)*, taking the place of our little daisies. The verdure of those ever-verdant bush-heaths is not a rich sap-green, but a dim brown-green; and when, after a longer journey in the interior of the country, the traveller at length arrives once more in the vicinity of European settlements, of fields and meadows, their rich verdure appears to him almost like a flash of glaring colours.

Upon the pumice-stone plateaus in the interior of North Island, and in the Alpine valleys of South Island, the bush vegetation is partly supplanted by a meagre grass vegetation which, however, is still sufficient to make those parts a natural pasture ground for sheep, horses and cattle, and which is easily and rapidly improved by the introduction of European grasses. Peculiar to these grass plains are certain species of *Acyphilla* and *Discaria*, rendering many tracts where they grow in larger quantities, wholly inaccessible. On account of their slender blades terminating in sharp spines the colonists have named them "spear grass", "wild Irishman", and "wild Spaniard".

On entering the "Bush" — as in New Zealand the forest is called — it is again ferns that principally meet the eye, magnificent Tree ferns,[1] their trunks as if coated with scales, and with neatly shaped crowns *(Dicksonia* and *Cyathea)*; Hymenophylla and Polypodia in the most different varieties, which cover with luxuriant growth the trunks of the forest trees; the singular form of the Kidney-fern *(Trichomanes reniforme)*, the round, kidney-shaped leaves on the edges of which are bordered with seed pods; ferns between the branches and twigs of the trees; ferns on the ground; bulbiferous Asplenia *(Asplenium bulbiferum)*, tender species of *Goniopteris* and *Leptoteris;* in short all sorts and varieties of ferns.

But in the woods also there are scarcely any gay flowers and blossoms; but few herbaceous plants, nothing but shrubs and trees; shrubs with obscure green flowers, and very often of obscure and little known Natural Orders.[2] Of the numerous Pines, very few

---

[1] Some of them reaching the size of 30 to 40 feet.

[2] More than 200 New Zealand species having, according to Dr. Hooker, either unisexual or polygamous flowers, or are otherwise incomplete in their reproductive organs, even when their floral envelopes are more or less developed. We are by this fact reminded of the imperfect organs of locomotion of numerous New Zealand animals, of the wingless birds and the wingless insects in the order of the Orthopteræ (see Chapters VIII. and IX). Of flowering trees there are upwards of 113, or nearly

recall by habit and appearance the idea attached either to trees of this family in the northern hemisphere, or to the Araucariæ of New Holland and Norfolk Island; while of the families that indicate the only close affinity between the New Zealand Flora and that of any other country (the *Myrtaceæ*, *Epacrideæ* and *Proteaceæ*), few resemble in general features their allies in Australia. From the above-mentioned peculiar proportion of the species to the genera and orders, on the other hand, it is evident why the New Zealand forest lacks every clearly and decidedly a prominent physiognomy. Only few trees grow gregarious, and are prominent in the landscape by their appearing either in closed forests or as separate clumps and groves. These are the Kauri *(Dammara australis)*, the Kahikatea *(Podocarpus dacrydioides)*, and the Tawai (black birch, *Fagus fusca)*. With the exception of the Kauri forests in the North, the Kahikatea forests on marshy and swampy river banks, and the black birch forest upon South Island, we find nothing that would suffer a comparison with the individual character of our pine, beech, and oak forests. The New Zealand trees mostly grow so intermixed, that more than a dozen varieties may be found on the same acre. The forests, therefore, have not any particular physiognomical character; and it is only by botanically analyzing the indefinite brown-green mass of the forest vegetation, that the beautiful and manifold objects constituting it are distinctly observed.

Among the chief ornaments of the mixed forest are the various species of pines. Totara *(Podocarpus totara)* and Matai *(Podocarpus spicata)* are large and beautiful trees found in every forest. Rimu *(Dacrydium cupressinum)* is distinguished by hanging leaves and branches; Tanekaha *(Phyllocladus trichomanoides)* by its parsley-shaped leaves. Alongside of them towers the poplar-shaped Rewarewa *(Knightia excelsa)*, belonging to the Proteaceæ; the Hinau *(Elaeocarpus hinau)*, the fruit of which is the favourite food of the parrots, and the bark of which is used by the natives for dyeing

one sixth of the phanerogamic Flora, besides 156 shrubs and plants with woody stems. In England there are not more than 35 native trees, out of a flora upwards of 1400 species.

New-Zealand: Province Auckland.

C.Fischer del.
A.Meermann sc.

Tuakura
Fern tree
(Dicksonia squarrosa)

Totara & Matai
(Podocarpus)

Nikau-Palm
(Areca sapida)

Konau
(Edaccarpus)

Ponga
Fern-tree
(Cyathea dealbata)

Rata
(Metrosideros robusta)

Karaakee
(Phormium tenax)

Forest in the Papakura District.

purposes. The Kowai *(Edwardsia microphylla)* also, with its magnificent yellow papilionaceous blossoms, grows in many districts to a considerable size. Among the largest forest trees there are in addition several representatives from the families of the Myrtaceæ and Laurineæ, and especially the Rata *(Metrosideros robusta)*, the trunk of which, frequently measuring 40 feet in circumference, is always covered with all sorts of parasitical plants, and the crown of which bears bunches of scarlet blossoms; also the Kahikatoa *(Leptospermum)*, Tawa *(Laurus)*, Pukatea *(Laurelia)*, Karaka *(Corinocarpus)* and a great many others. The under-wood is composed of bushes and shrubs of the most different kinds, especially species of *Panax* and *Aralia*, above which the slender Nikau palm *(Areca sapida)*, the sole representative of its genus upon New Zealand, rears its sap-green crown in picturesque majesty.

While this palm and the fern trees remind us by their forms of tropical forests, the New Zealand forest owes its tropical luxuriance to the countless parasitical weeds, ferns, to the Pandaneæ *(Freycinetia Banksii)*, and Orchideæ, covering trunks and branches, and to the creepers *(Ripogonum, Rubus, Metrosideros, Clematis, Passiflora, Sicyos,* etc.*)*, which cover the ground as with a natural netting, which coil round every stem, run up every limb, glide from head to head and entwine the topmost branches of a dozen trees in Gordian knots. Thus the forests become impenetrable thickets, which sun and air scarce can penetrate, and which have to be cut through with the knife or sword at every step the traveller makes into the untrodden wilderness. Through the narrow paths of the natives it is only with the outmost efforts that a way can be worked over the gnarled roots of trees and through the creepers which obstruct the passage at every moment. To the wanderer there are especially two kinds of creepers extremely molesting and troublesome, the so-called "supple-jack" of the colonists *(Ripogonum parviflorum)*, in the rope-like creeping vines of which the traveller finds himself every moment entangled; and *Rubus Australis*, the thorny strings of which scratch the hands and face, and which the colonists, therefore, very wittily call the "bush lawyer." In the interior of the Bush it is gloomy; everything as silent as the grave; neither gay blossoms nor gaudy butterflies and birds greet the eye or relieve the melancholy monotony of the scenery; all animal life seems extinct, and, however much the curious traveller may have yearned after sylvan beauty, it is with feelings of delight that after days of tedious plodding through the dreary solitude of those gloomy and desolate woods he hails once more the cheering daylight of the open landscape.

Appendix.

List of flowering alpine plants in the Mountain Ranges of the interior of the Province of Canterbury, growing from 4000 feet to the line of perpetual snow; according to Dr. J. Haast.

Ranunculus Lyallii (Hooker fil. 1863)
    „   Traversii   do.
    „   pinguis var. Monroi
    „   Buchanani (Hooker fil. 1864)
    „   Haastii   do. 1863
    „   sericophyllus do. 1864
    „   Sinclairii   do. 1864
    „   gracilipes (Hooker fil. 1864)
Caltha novæ-Zelandiæ
Notathlaspi rosulatum (Hook. fil. 1863)
Viola Cunnighamii
Stellaria Roughii (Hooker fil. 1863)
Gunnera prorepens
Epilobium macropus
    „   confertifolium
    „   crassum
Pozoa Haastii (Hooker fil. 1863)
    „   reniformis
Aciphylla Lyallii (Hooker fil. 1863)
    „   Monroi
    „   Dobsonii (Hooker fil. 1863)
Ligustisum Haastii (F. Muller, 1861)
    „   piliferum (Hooker fi. 1863)

Ligustisum carnosulum   do.
    „   aromaticum   do.
Coprosma depressa
Olearia Haastii (Hooker fil. 1863)
Celmisia densiflora (Hooker fil. 1863)
    „   discolor
    „   Haastii (Hooker fil. 1863)
    „   Lyallii (Hooker fil. 1863)
    „   viscosa (Hooker fil. 1864)
    „   petiolata (Hooker fll. 1863)
    „   laricifolia
    „   Hectori (Hooker fil. 1863)
    „   sessiliflora   do.
    „   bellidioides   do.
    „   glandulosa   do.
Brachicome Sinclairii (Hooker fil. 1864)
Cotula atrata (Hooker fil. 1864)
    „   pectinata   do.
    „   pyrethrifolia   do.
Ozothamnus microphyllus
Raoulia australis
    „   tenuicaulis
    „   Haastii (Hooker fil. 1864)

Raoulia Monroi (Hooker fil. 1864)

„ subulata (Hooker fil. 1863)

„ eximia do. [1]

„ subsericea

„ grandiflora

„ mammillaris (Hook. fil. 1863)

Gnaphalium bellidioides

„ Youngii (Hooker fil. 1864)

„ Traversii (Hooker fil. 1864)

„ grandiceps (Hook. fil. 1863)

Haastia recurva (Hooker fil. 1863)

„ Sinclairii do.

Senecio Lagopus

„ bellidioides

„ Haastii (Hooker fil. 1863)

„ Greyii

„ elæagnifolius

„ cassinioides (Hooker fil. 1863)

Forstera sedifolia

„ tenella

Helophyllum clavigerum

„ Colensoi (Hooker fil. 1864)

Stylidium subulatum do.

Lobelia Roughii (Hooker fil. 1864)

Pratia macrodon (Hooker fil. 1864)

Leucopogon Fraseri

Pentachondra pumila

Epacris Alpina

Dracophyllum uniflorum

Dracophyllum rosmarinifolium

Gentiana saxosa

Myosotis Traversii

Veronica Hectori (Hooker fil. 1863)

„ tetrasticha do. 1864

„ Haastii do. 1863

„ epacridea do. do.

„ macrantha do. do·

Pygmea pulvinaris (Hooker fil. 1864)

Ourisia macrophylla

„ sessiliflora (Hooker fil. 1863)

„ cæspitosa

Euphrasia Monroi (Hooker fil. 1863)

„ revoluta

„ antarctica

Drapetes Dieffenbachii

„ Lyallii

Fagus cliffortioides

Podocarpus ferruginea

„ nivalis

Dacrydium laxifolium

Caladenia minor

„ bifolia

Astelia nervosa

Luzula pumila (Hooker fil. 1863)

Oreobolus Pumilio

Agrostis canina

Poa breviglumis

Triticum Youngii (Hooker fil. 1863).

---

[1] Mr. John R. Jackson (Intellect. Observer 1867, p. 128) remarks on this very interesting plant: 'Peculiar looking patches are to be seen upon the sides and tops of the mountains, which in the distance look like so many sheep, and even upon nearing them their shaggy appearance helps rather to confirm the first impression, than to dispel such a notion. Upon a somewhat closer examination, however, we should possibly be ready to believe that these hemispherical masses are those of a gigantic moss. These tufts are in reality masses of plants belonging to the genus *Raoulia,* one species of which is known to the New Zealand settlers as the "vegetable sheep", *Raoulia eximia,* Hf. The shepherds themselves are often deceived by them when calling in their sheep from the mountains. It is found at Ribband Wood Range, Mount Arrowsmith, and Dobson in the Province of Canterbury at an elevation of 5500 to 6000 feet.'

List of Ferns, Lycopodiaceæ, and Marsileaceæ, from the Interior of the Southern Alps.

Cyathea Smithii
Alsophila Colensoi
Hymenophyllam unilaterale
　　　　　　　minimum
　　　　　　　multifidum
　　　　　　　rarum
　　　　　　　pulcherrimum
　　　　　　　dilatatum
　　　　　　　crispatum
　　　　　　　polyanthos
　　　　　　　demissum
　　　　　　　scabrum
　　　　　　　flabellatum
Trichomanes humile.
　　　　　　　Colensoi
Cystopteris fragilis
Hypolepis tenuifolia
　　　　millefolium
Cheilanthes tenuifolia
Pteris aquilina var. esculenta
　　　macilenta

Lomaria procera
　　　　　fluviatilis
　　　　　membranacea
　　　　　vulcanica
　　　　　alpina
　　　　　Banksii
　　　　　nigra
Asplenium obtusatum
　　　　　Trichomanes
　　　　　Hookerianum
　　　　　bulbiferum
　　　　　flaccidum
Aspidium aculeatum var. vestitum
　　　　　cystostegia
Polypodium australe
　　　　　sylvaticum
Leptopteris superba
Botrychium cicutarium var. australe
Lycopodium Selago
　　　　　clavatum
Azolla rubra.

CHAPTER VII.

Kauri and Harakeke.

The New Zealand Pine and the New Zealand flax plant.

Kauri, *Dammara australis*, the Queen of the New Zealand forest. — Limits of its range. — Devastation of the woods. — Physiognomy of Kauri forests. — Wood-cutter colonies. — Saw-mills. — Kauri gum. — Quantity and value of the annual export. — The New Zealand flax plant, *Phormium tenax*. — Its various uses. — Its varieties. — Production of flax. — Other fibrous plants. Appendix. List of esculent plants. — Table of timber trees.

———

"Flora and Pomona have dealt most niggardly with New Zealand. There is no indigenous flower equal to England's dog rose; no indigenous fruit equal to Scotland's cranberry." [1] On the other hand, the vegetable world of New Zealand is as harmless as its animal world; there are no poisonous plants with the solitary exception of the Toot-plant, [2] and a great variety of very useful plants. In this respect the two most important and at the same

———

[1] Hursthouse, New Zealand 1857. Vol. I. p. 136. A list of the esculent plants is given in the Appendix.

[2] The Toot-plant, Tutu or Tupakihi of the Maoris *(Coriaria sarmentosa* Forst. = *C. ruscifolia* L.), is a small bush, one of the most common and widely distributed shrubs of the islands. It produces a sort of "hoven" or narcotic effect on sheep or cattle, when too greedily eaten. It bears a fruit, which is produced in clusters, not unlike a bunch of currants, with the seed external, of a purple colour. The poisonous portion of the plant to man are the seeds and seed stalks, while their dark purple pulp is utterly innoxious and edible. The natives express from the berries an agreeable violet juice (carefully avoiding the seed), called native wine. — When boiled with Rimu, a seaweed, forms a jelly which is very palatable. The Toot poison belongs to the class of narcotic irritants.

time the two most characteristic plants of New Zealand are the Kauri and the Harakeke or the New Zealand Pine, *Dammara australis*, and the New Zealand flax plant, *Phormium tenax.*

The Kauri Pine is justly styled the Queen of the New Zealand forest. What the silver-fir is to the mountainous regions of Middle Germany; what the famous Libanon cedar used to be in those majestic forests of Asia Minor which in ancient times furnished the material for the vessels of the Phœnicians and the timber for Salomon's temple; or what to-day the Mammoth tree *(Sequoia Wellingtonia)* is, the giant among the wood-giants in California: the same is for the forests of the warmer northern regions of New Zealand the celebrated and beautiful Kauri *(Dammara australis,* Yellow Pine of the colonists).

From the very first beginning of the colonization of New Zealand, the Kauri forests of the North Island have proved a source of wealth to the settlers. They furnish the best ship spars and masts, excellent timber, and the gum of the Kauri Pine is a very important article of commerce. Even up to a most recent date, Kauri timber and Kauri gum have been estimated among the most important articles of export of New Zealand. [1]

The Kauri pine of all the coniferæ of New Zealand is the only one bearing a cone. All the other pines belonging to the family of the coniferæ, such as Totara, Kahikatea, Miro, Matai, Rimu, etc. (species of *Podocarpus* and *Dacrydium*) produce berries. The surname *australis*, might lead to the erroneous supposition that this tree is found also in the neighbouring continent of Australia. But, in fact, New Zealand is the only and exclusive home of the Kauri pine, and even here its dominion is a very limited one. Kauri forests are found only upon the long and slender northwestern peninsula of the North Island between $34\frac{1}{2}^0$ and $37\frac{1}{2}^0$ South latitude, and between 173^0 and 176^0 longitude East of Greenwich.

---

[1] In the year 1859 the amount of timber exportation from the Province of Auckland was £ 34,376; that of Kauri gum exported £ 20,776; together more than one half of all remaining exportations from the several ports of the province.

Rev. Mr. Taylor[1] believed that, from the presence of fossil gum in the brown coal of the North Island and South Island, — which he erroneously considered to be identical with the gum of the Kauri pine, — it was a natural consequence to infer a high geological age, and a formerly far-extending range of the tree. A closer investigation, however, proves plainly that this fossil resin is totally different from the Kauri gum. Since, moreover, the points, where Kauri wood is found half fossillized and imbedded in recent lignite layers, as at the harbours of Hokianga and Kaipara, are likewise comprised within the boundaries above designated, there is no reason to suppose that the range of the Kauri had originally been another than it is now. Three degrees of longitude and three of latitude, therefore, encompass the entire and the only range of this remarkable tree; and even within those narrow limits the Kauri has at all times been by no means a common tree; besides, extensive districts within that range which formerly had been covered with Kauri woods, are now totally destitute of such; and the extermination of that noble tree progresses from year to year at such a rate that its final extinction is as certain as that of the natives of New Zealand. The European colonization treatens the existence of both, and with the last of the Maoris the last of the Kauris will also disappear from the earth.

The vitality of this tree would appear to depend upon two conditions, namely the moist sea breeze and a stiff clay soil. Both are found united upon the narrow, northern peninsula. On the East Coast, the sea enters by deeply indented bays far into the land; and in a similar manner, on the West Coast the estuaries of the Hokianga and Kaipara rivers are formed by ramified and far-reaching arms of the sea. It was on the shores of these very bays and estuaries, that the first settlers found the most luxuriant Kauri forests. Tracts near the sea coast, exposed to the sea breeze, yet beyond the reach of salt-water itself, and places sheltered from violent blasts of wind, are most favourable to the growth of the tree; and it crows on such places most luxuriantly even

[1] "Te Ika a Maui" p. 438.

upon poor, stiff clay soil where, after the Kauri forests have disappeared, nothing else will thrive. Whether it is, the Kauri forests extract from the soil all the ingredients requisite for the growth of other plants, or that they really grow only upon a soil productive of nothing else: this much is a matter of fact, that those tracts in the vicinity of Auckland which formerly were covered with dense Kauri forests, and where large masses of Kauri gum are dug from the earth, present now nothing, but waste, dreary, sunburnt heaths of notorious sterility, upon the white or yellowish clay-soil of which nothing but dwarfish Manuka shrubs (*Leptospermum scoparium*), and scanty ferns (*Pteris esculenta*) can grow. The colonists therefore say that Kauri forests indicate a poor soil and a rugged non-agricultural country. This ought to prove a lesson for the future; individuals should not be suffered to ravage those precious woods, and to turn the country into a desert to the detriment of whole generations to come. For the sake of a few serviceable trunks, sometimes whole forests are burnt down and desolated, and what formerly had been employed in the wars of cannibal tribes as a stratagem to burn out the enemy, is done now for the sake of money. The woods are ransacked and ravaged with "fire and sword." During my stay at Auckland I was able to observe from my windows, during an entire fortnight, dense clouds of smoke whirling up, which arose from an enormous destructive conflagration of the woods nearest to the town. When the fire had subsided, a large, beautiful tract of forest lay there in ashes, the newspapers giving only this laconic notice: "No damage done to timberwood." That may be; but there will come a time, when the question will be not only about the timber, but also about the forest!

In order to overlook at a glance the principal complexes of the Kauri forests from North to South, it is necessary to begin at the extreme North of the island, at the "land's end", Mariwhenua of the Maoris. Some few scanty specimens at Cape Reinga, a short distance from Cape Maria Van Diemen, in that part of the country where the traditions of the natives fix the entrance to the lower regions, indicate that also the northernmost neck of New Zealand, — which, being now entirely destitute of sheltering woods, has been changed into a sandy desert by the quicksand, — was in ages past covered with woods. The first real Kauri forests, however, are found in the mountain-chain which, with the Manuga Taniwa, 2150 feet high, as its principal peak extends, a little South of the 35⁰ of South latitude, from the West Coast to the East Coast. Already here it becomes evident, how the Kauri loves to be near sea. While in the middle of the mountain-chain the woods are very heterogenous, the purer Kauri tracts are all situated more towards the coast in the vicinity of the Mongonui, and Wangaroa Harbours on the East Coast, and of the Hokianga river of the West Coast. Farther South, at 36⁰ latitude, the Kauri forests seem to attain their most luxuriant development. Most noted are the woods on the shores of the variously branching Kaipara Harbour on the West Coast, and along the Waiaroa river, which flows into the former. Here the richest Kauri woods are said to be found, and the trees attain the greatest size and height.

From the earliest periods of European colonization in New Zealand, sawing-mills were established here. Not only Europeans, but also natives have become rich by the lumber-trade, and to this very day the Kaipara Harbour has remained the staple-port for the exportation of Kauri timber. Thence the woods extend as far as the vicinity of Auckland to the Waitemata Harbour on one side, and to Manukau Harbour on the other. Upon the North shore, however, and upon the Isthmus of Auckland every trace of them has disappeared. The plains and hills once covered with forests, are now sterile fern-heaths, where the white clay-soil is everywhere exposed to sight. The gum which the natives dig from the ground, and fragments of upturned wood-giants half decayed, half changed into lignite, which are found here and there buried beneath mounds of earth, are the only remnants of the former forest. Along the West Coast, the North side of Manukau Harbour may be designated as the southernmost point, to which the Kauri woods extend. On the Waitakeri, near Henderson's mill, and in the Huia Bay, I found yet considerable tracts remaining. Farther South, though there are single trees and smaller groups met with as far as Kawhia Harbour, yet whole forests are no longer seen.

Along the East Coast it is the Katikati river, at the North end of the Tauranga Harbour in Plenty Bay, at 37° 30' South latitude, which has been designated as the southernmost point for the growth of the Kauri. Very rich in beautiful forests is still the Cape Colville Peninsula on the East side of the Hauraki Gulf, whence the timber is exported partly by Mercury Bay, partly by Coromandel Harbour. The woods disappear more and more where the island South of Auckland and South of Cape Colville Peninsula assumes breadth. Near Papakura and in the mountains on the Wairoa, there are yet some isolated Kauris, but on the Waikato they are already at an end. I believe that the total area comprising all the Kauri woods upon New Zealand can scarcely be estimated as exceeding 4000 Engl. square-miles.

I have not seen the large and magnificent forests on the Kaipara, but I have seen beautiful Kauri woods in the coast range West of Auckland, in the Titirangi chain, on the Waitakeri in Henderson's Bush, and in the Huia on the Manukau Harbour; likewise in the mountains of the Cape Colville chain on Coromandel Harbour; and will now briefly state a few observations which I have made in those places.

The soil in the parts above mentioned consits partly of a ferruginous clay which has originated in the decomposition of volcanic conglomerate and tuff; partly of a stiff white clay. The Kauri pine never grows single or isolated, nor does it form woods quite free from other large forest-trees, but it occurs in clumps upon places sheltered from the wind. These clumps impart to the forest its characteristic physiognomy. On looking over the whole mass of woods from a hill or a mountain, the Kauri groups are easily distinguished at a great distance by their dark-green foliage. The crowns of the Kauri pines rise far above the rest of forest-trees, and produce dark shades upon the slopes of the mountains and in the valleys, here and there intersected by the light-green stripes of fern-trees shooting up luxuriantly, wherever a small stream of springwater may be flowing through the wood.

Kauri-Woods on Manukau Harbour.

As the coarse-grained micaceous granite is interspersed by fine-grained light-coloured veins, so the fern-trees with their soft, dawsy, light-green bushes streak the dusky Kauri woods.

These Kauri groups vary greatly in extent. They often occupy several square-miles; sometimes there are only 30 or 40 trees clustered together which thus, mutually protecting each other, thrive splendidly. Upon cutting down the woods, however, and allowing but a few, single, trees to remain, the latter wither away. In vain have the colonists tried, to keep and preserve upon the extensive tracts which they wrest from the wild woods for agricultural purposes and for the raising of cattle, some few beautiful trees as ornaments of the landscape, and to grace their farms. The offspring of the shady, humid wild woods will always pine away, as soon as it is exposed to wind and sun; and in the same manner every trial hitherto made to plant and cultivate that son of the wilderness, has sadly miscarried.

Closely connected with this peculiarity that the tree thrives and grows only in groups, is the other, that the trees of one and the same group or grove are usually of nearly the same age. Hence there are "clumps" with trees of 100, 200, 400 and 500 years; and the grand impression made by the Kauri forest is chiefly based upon this circumstance, that it is a forest as of one cast; that tree by tree rises of equal thickness and of equal height, like pillars in the halls of a cathedral. In these clumps, the Kauri pine suffers no larger forest-trees by its side; only smaller trees and shrubs compose the undergrowth.

Young trees have a very different appearance from the older ones. In its youth, the Kauri pine resembles more our red pine; and in matured age, rather the full-grown white pine. Young specimens of from 60 to 100 years have sharp pointed conical crowns; the trunk runs perfectly straight from the root to the topmost end of the crown. In advanced age, the sidebranches grow stronger and form in continual duplication an irregularly ramified, umbrella-shaped crown. The trunk, on the other hand, perfectly cylindrical, and almost imperceptibly lessening in its ascent, presents, as far as the crown, a majestic pillar whose beautiful stature is disturbed neither by sidebranches, nor parasitical plants, such as usually cover other forest-trees. The eye follows unimpeded the beautiful line of the trunk from the root to the crown, where the powerful branches are twined into a dense, dark-green roof, through which, like golden stars on the roof of a vault, the light of day peers into the dusk of the woods. The bark of trees four feet thick is from one inch to one and a-half thick, scaling off as in our firs. The blooming season of the tree is in December; the cones are almost spherical and comparatively very small, their diameter not even amounting to the length of our pine-cones; when dry, they fall easily to pieces. When the cones towards the end of February are ripe, the Kauri woods are frequented by numerous birds which feed on the seeds.

The oldest and largest trunks attain a diameter of 15 feet[1], corresponding to a circumference of from 40 to 50 feet, and a height of 100 feet to the lowest branches; or from 150 feet to 180 feet to the top of the crown. Such trees are probably from 7 to 800 years old. Having examined several trunk-sections, I found, as the mean result, from 10 to 12 annual rings to one inch, although in some cases the rings attain a much greater thickness. In some few cases of rare occurrence, I have even observed single rings of a thickness of one inch. For the saw-mill, the wood-cutters generally pick out trees of four feet diameter with trunks measuring from 60 to 80 feet to the crown. Such trees are probably 250 to 300 years old. The trunks are sawed into logs of 10 feet to 20 feet in length upon the spot, where they are felled, — one tree generally yielding from 4 to 6 logs, — and these logs are then conveyed to the saw-mill. Since the plank-saws are often at a considerable distance, situated at points from which the timber can be immediately shipped, the transport of those logs is in fact the heaviest piece of work in the whole wood-business. First, a broad clearing leads from the interior of the forest, generally straight down the steep mountain-slopes, and forms a kind of road along which the logs are rolled down to the head of a regular tramway. Upon this road which, as in the Huia, is cut through the bush

---

[1] In the vicinity of Coromandel Harbour stands a specimen of 17 feet diameter; upon the Papakura flats another of 15 feet, and near Matakana a third of 14 feet diameter.

many miles long, the logs are conveyed to the floating-place, and thence to the saw-mill by canals with which, here and there, swell-ponds are connected.

The timber of the Kauri pine resembles the timber of our white pine or silver-fir. It supplies splendid ship spars, and first-rate wood for inside and outside house work; painted furniture, ship-planking, decks and fittings. The deals and boards are said to possess the peculiar quality of shrinking more in length than in breadth. In 1859 nearly the whole of Auckland with the ex-ception of a few stone buildings consisted of houses built of Kauri timber,[1] and it is especially to the Kauri pine, that the province is indebted for its first rise.[2]

In remote inlets of the sea, and branches of rivers which were formerly frequented only by the lonely canoe of the savage, there pre-vails now a brisk intercourse of vessels of all kinds. Extensive saw-works, constructed upon the best principles, are scattered along the banks of those rivers and bays. In the dark bush, over hill and dale and in ravines which were once hushed in deathlike silence, the ring-ing of the axe, the creaking of the saw, and the far-sounding "cooey" of the wood-cutters are to be heard. Men whose nerves and sinews had been hardened in the wild woods of California and Canada, — Scotchmen and Irishmen, and now and then also an ill-starred Ger-man, — they are the champions combatting those giants of the woods. Merrily the columns of smoke whirl up from their log-fires; and many a wonderful story is told, when in the hour of repose the pipes are alight and the gin-bowl is going the rounds.

But the Kauri pine yields also, as already mentioned, a second, very valuable product, the Kauri gum, Kapia of the natives. This resinous gum, as it oozes from the tree, is soft and of a milky turbidness,[3] not unlike opal; in course of time, however, it hardens, becomes more or less transparent, and assumes at the same time a

---

[1] Houses built of Kauri-timber are said to last 50 years.

[2] The forests of New Zealand furnish, however, besides Kauri several other kinds of excellent timber; see Appendix.

[3] In this form, the gum is often chewed by the natives.

bright yellow colour, [1] so that it quite resembles amber. As to the quantity of the gum produced, the Kauri pine equals probably the ancient coniferæ of the post-tertiary period, the *Abietineæ* and *Cupressineæ*, from which the amber originated. Twigs and branches are bristling with white drops of gum; but in larger lumps, the gum collects especially on the lower part of the trunk from whence the roots proceed. Hence it is always found in great quantities in the soil of those places, where Kauri-forests stood of old. Pieces of 20 to 30 pounds weight and even more, — sometimes of 100 pounds — are of no rare occurrence. The Novara collections are indebted to Mr. Petschler, a German merchant in Auckland, for a magnificent piece. Kauri gum is not soluble in water; it is easily ignited and' burns with a sooty flame. It froths up strongly at the same time, and produces an aromatic-balsamic odour. In passing over places, cleared in the Kauri-woods by burning, I was always reminded of the smell of frankincense and myrrh.

It is an article of commerce which is in great demand, and principally exported to England and North America; it is used in the preparations of lac and varnish, and said to be applicable to various other branches of industry. The value of a ton of gum fluctuates between 10 and 15 pounds Sterling; sometimes it commands even a higher price. [2]

---

[1] Dieffenbach is of the opinion, that Kauri gum assumes this beautiful golden-yellow colour only under the influence of sea-water.

[2] The following table will indicate the quantity and the value of this article of export of the past ten years.

Export of Kauri-gum from various ports (Auckland, Bay of Islands, Hokianga, Mongonui and Kaipara) of the North Island:

	Tons.	Value in Pounds Sterling.
1856	1440	18,591
1857	2521	35,250
1858	1810	20,036
1859	2010	20,776
1860	1046	9,851
1861	865	9,888
1862	1103	11,107
1863	1400	27,026
1864	2228	60,590
1865	1867	46.060.

As the gathering of Kauri-gum has been hitherto an occupation limited almost exclusively to the natives, it may be taken for granted that by far the largest share of the proceeds from the sale of Kauri-gum has accrued to them alone. Within the last few years they are said to have earned not less than £16,000 during each year by gathering this gum. Even from the distant South of the Province of Auckland, from the Taupo and Rotorua district, larger and smaller Maori parties come during the summer months to the North, especially to the vicinity of Auckland, for the purpose of gathering the gum which still continues to be dug upon the fern-heaths from the surface of the earth in large quantities. But I have never heard of the gum being obtained also in the bush from the trees. The question would arise whether a resinous tar might not be easily obtained from the powerful crowns of the Kauri pines and from the resinous bark of the trees by carbonisation in tar-kilns, instead of simply burning them up, as they now do.

The New Zealand flax plant, *Phormium tenax*, is quite peculiar to New Zealand, the adjoining Norfolk Island and the Chatham Islands; it is found nowhere else. The flax-like fibre, prepared from the leaves by the natives, the value of which was soon observed by Europeans, constituted the first article of barter in the trade carried on by the Maoris with the Europeans. What the bamboo is to the inhabitants of eastern and southern Asia, this plant is to the natives of New Zealand. The various uses it is put to are innumerable. Near every hut, every hamlet, on every way-side its bushes, whether wild or cultivated, are at hand for use.

Phormium tenax is a flag-like plant, the sword-shaped drooping leaves of which the natives call Harakeke; the flower-stalk, bearing pink blossoms and resembling agavas, is called Korari, and each constituent part of the plant can be used for some practical purpose or other. The blossoms contain a sweet honey-juice, much liked by the children, and which the natives are wont to collect in their calabashes. One plant will produce nearly half a pint.

At the root of the leaves is found a semi-liquid gum-like substance which serves the Maoris as a substitute for sealing-wax and glue, and is also eaten. The dried flower-stalks, the pith of which, when ignited, keeps glowing like tinder, are to the travelling Maori excellent slow-matches by means of which he is enabled continually to carry fire about him. The most different uses and benefits, however, are derived from the leaf. Green on the bush or cut, it serves the modern Maori that knows how to read and write,

Bushes of Phormium tenax.

the purpose of writing-paper; with a sharp-edged shell he engraves his thoughts upon it. Split and cut into broader or narrower strips, and bound together longer or shorter, it serves by virtue of the extraordinary tenacity of its fibre, instead of cords, ropes, straps, and all sorts of strings, lines and tows. As a universal means for binding and strapping, it is of invaluable service in New Zealand, and indispensable to the natives in the building of huts and canoes. The green strips of the leaves are plaited by the women into very neat baskets which at dinner serve as plates and dishes; the men manufacture lines, nets and sails of them. The natives knew also how to prepare and to dye the flax-like fibre, and thus to obtain the material for their mats and woven garments. The Weruweru, a kind of garment, was prepared out of the half-prepared leaf; the state-dress Kaitaka is interwoven with many coloured borders of the fine and carefully prepared fibre. For dyeing black, the bark of the Hinau tree *(Elæacarpus Hinau)* is employed; for red, the bark of the Tawaiwai (or Tanekaha) tree *(Phyllocladus trichomanoides).*

The Phormium plant is widely disseminated over New Zealand from North to South, and millions of acres of land are covered with it. It grows upon any kind of soil, whether moist or dry; in any locality, whether high or low. In the Alps of the South Island, phormium bushes are met with up to a height of 5500 feet above the level of the sea. The plant, of course, varies according to the locality, and the natives distinguish by different names ten or twelf varieties, which they use for various purposes according to the quality of the fibre.

The flax plant attains its most luxuriant growth in the vicinity of swamps and rivers upon moist alluvial soil. Here the leaves grow to a length of 10 to 12 feet, and the flower-stalks to a height of 16 to 20 feet with a thickness of 2 to 3 inches. Large phormium bushes, therefore, indicate always a very fertile soil, and the natives knew very well how to cultivate the flax upon such land in the vicinity of their kaingas. But from those places the flax plant spreads on one hand into the swamps, and grows in the water, and ascends on the other hand on the dry slopes of the mountains to a very considerable height, without however attaining the above stated size. We may distinguish about three principal varieties:

1) *Tuhara,* swamp flax, with a coarse, yellowish-white fibre; used especially for ropes, lines etc.

2) *Tihore,* a cultivated variety, the best kind with a fine, silken-glossy fibre of pure white colour; used for mats and garments.

3) *Wharariki,* mountain flax, with coarse fibres; little used.

Experiments made to test the strength or tenacity of the New Zealand flax-fibre have shown, that it is far superior in tenacity to the European flax and hemp-fibre. [1] Consequently, when the great value of the New Zealand flax-plant was fully known, trials were made to acclimatize it in England and France. The attempts made, however, appear to have miscarried. It is only in botanical gardens that the plant became domesticated. It seems exceedingly surprising that, considering the immense quantities of flax yearly consumed by England, and imported principally from Russia, [2] the New Zealand flax has not long since become one of the chief articles of export from New Zealand. The cause of this is that it is extremely difficult to prepare the fibre sufficiently pure for the market, and to produce large quantities of such flax at moderate prices. It is only quite lately, that this appears to have been accomplished with complete success.

The phormium leaf, — like the leaves and stems of other fibrous plants, such as hemp, flax, the American aloe (Agave), etc. — consists of cellular trusses, which run out over the whole length of the leaf and are wrapped in the green substance of the plant, the so-called parenchyma. The cellular trusses in their turn consist of two parts, the wood-part and the bast-part; the bast-part constitutes the serviceable fibre. In order to obtain the latter, it is necessary to sever the parenchyma and the wood-part of the cellular trusses from the bast-part. The cellular tissue of those parts being far more easily injured, than the spindle-shaped, thick-coated and elastic bast-cells, the separation can be brought about by destroying and removing that cellular tissue either by maceration, without injuring the bast-cells, or by mechanical force. A combined process is frequently applied as in the treatment of flax, which is first exposed to a kind of putrid fermentation on the dew- or water-steep; then dried, and finally braked, swingled and combed.

---

[1] Lindly states the tenacity of the New Zealand flax-fibre in a comparative synopsis as follows:

Silk	34
New Zealand flax	23
European flax	16
European hemp	11.

[2] In 1856 the value of flax and hemp imported is said to have amounted to nearly 6 mill. pounds sterling.

Similar operations have been tried with the phormium leaf, for the purpose of producing its excellent fibre in a pure state, and thus rendering it marketable for European commerce. The process of the natives, who use only the upper half of the leaf, — above the point, where the two constituent parts of the leaf are sheath-like grown together, — and only one side of it, simply consists in scraping off the parenchyma with a shell (generally *Mytilus*). This is a kind of work that formerly devolved on women and slaves; but which now-a-days, nobody likes to perform. In the beginning of this century there was still a chance to barter from the natives quite considerable quantities of flax prepared in said manner, and according to statistical statements the export amounted in 1828 to about 60 tons, with a value of £2600, and in 1830 already 841 tons, and in 1831 as much as 1062 tons. But since that time the export has grown less and less every year, in the last few years it scarcely amounted to more than 60 or 70 tons per annum.[1] There was no more flax to be got from the natives, and a proper method of manufacturing it at moderate prices was not yet known. Although the colonial Government, fully aware of the importance of this article of export, had set a reward of £4000 on the construction of a suitable machine for the production of the pure flax-fibre in quantities large enough for exportation, yet up to 1859 there was nobody to claim this reward.[2] It is true, there were some so-called flax factories, but their produce was inconsiderable as to quantity, and most deficient in quality.

The process employed in a small factory near Nelson, which I visited in September 1859, consisted in the following: the leaves were first boiled in lye-water; then, after having been dried and twisted together into a thick rope, they were made to pass between ribbed, wooden rollers, until the fibre was laid bare in a tolerable degree of purity. The dried and bleached produce the manufacturer sold at £25 per ton. This raw material is said to

[1] According to the statistical tables the export amounted in 1859: 77 tons, worth £1593; in 1860: 48 tons in 64 bundles, worth £1240.
[2] Thomson, Vol. II. p. 260.

be principally used for stuffing matresses. A finer product could not well be furnished at the high rate of wages paid for labour.

It was not until 1860, that my friend Rev. Mr. A. G. Purchas in Onehunga near Auckland, succeeded in devising a proper method for obtaining the flax-fibre in a state of perfect purity. [1] He found, that a sudden heavy stroke upon the leaf, when spread upon the cross-section of a block of hard wood, destroys all its parts except the bast-fibre, and that consequently by a series of such strokes the fibre may be obtained quite pure. After numerous trials he succeeded, with the cooperation of Messrs. J. Ninnis and J. Steward, in constructing a machine by means of which it i possible to obtain by one single operation the pure fibre from the leaf, so that a number of leaves cut fresh from the stalk and placed in the machine on one side of it, come out on the other side in less than a minute as a pure fibre, which requires only to be dried so as to be ready for the rope-maker's use.

This machine the principle of which is exceedingly simple, but the working of which nevertheless requires great care, consists, according to the communications of my friend Rev. Mr. Purchas, of two main-parts: first, of a large, solid cylinder or drum A of hard wood, revolving, and so put together that its surface all round presents the cross-section of the wood; and, secondly, of a row of long and thin iron-plates B, at the lower end of which a groove is cut. These iron-plates may be raised from nine inches to one foot, and fall by their own weight back upon the leaves, which are made to pass through between the revolving wooden cylinder and the iron pounders.

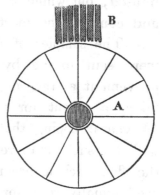

Machine for braking flax.

---

[1] Also Messrs. J. Ninnis, J. Probert, G. Webster, Neil Lloyd, T. Turnbull, G. Cole and Baron de Thierry, have distinguished themselves in this respect by forwarding beautiful samples of their manufacture to the exhibition in London, in 1862. Excellent samples of cable rope are manufactured out of New Zealand flax by Mr. Neil Lloyd; they are infinitely preferred to the Manilla ropes, it being both stronger and less liable to injury from exposure to water.

The pounders in each row, however, must not strike all at once, because the fibre would thus be torn, but the machine must be so arranged, that they rise and fall one after another, the leaf giving way in this manner to each successive stroke. The only additional requisite is moreover, that during the whole process plenty of pure water be kept running over the leaves, thus carrying off the particles smashed.

A steam-engine of 8 horse-power is sufficient to work such a machine, consuming daily about one ton of leaves, and yielding 3 cwts. flax, since 5 to 6 cwts. green leaves average 1 cwt. pure flax-fibre.

One acre of land grown with *phormium tenax* is said to bear 30 to 60 tons green leaves; it is, however, not yet established as a fact, how much of this quantity can be gathered every year. Purchas is of opinion, that about one half of the leaves may be taken every year without damage to the plants. Each bush is stripped of its outer leaves; the inner ones remain standing for the following year. Upon the machine the whole leaf may be used from the thick lower end, — which the natives formerly left unused, but which is the very part that is most easily to cleaned and contains the most fibres, — to the top.

The samples of flax produced with the new machine, which were sent to me by my friend Purchas, leave no room for improvement as regards the purity of the fibre. The inventors have taken a patent for the whole of New Zealand for the term of 14 years, hoping that before long the number of machines worked will be such as to render the flax a chief article of export. As the flax-machine according to experiments made with it, may be used equally well for obtaining the fibres of other fibrous plants, such as the American Aloe *(Agave)*, the so-called Manilla hemp *(Musa textilis)*, the *Ananassa sativa*, etc., its introduction into other countries might prove advantageous.

In comparison with the Phormium tenax the other fibrous plants of New Zealand are only of inferior importance. The only still noticeable plant of the kind is the Ti or Mauku of the natives, grass- or cabbage-tree of the colonists *(Cordyline australis)*, which is principally met with upon fern land and in swamps. The fibre of its leaf is very much like the phormium-fibre; it is only more yellow and has neither the gloss nor the tenacity of the flax-fibre; but it is prized very highly by the natives for its durability. It is said to resist the decomposing action of the atmosphere far better than Phormium. A little coarser than the Ti fibre is the fibre of a second species of Cordyline (perhaps *C. indivisa*) with larger, broader leaves, which the natives call Kapu or Ti Kapu; its fibre is said to be especially suited for making cables, because it does not contract in the water as much as Phormium. Similar is the fibre of a third species, Turuki of the natives *(Cordyline stricta)*, which grows in the bush. Mr. Probert has exhibited in London a liana Pikiarero (a species of *Clematis*) which is said to contain a fine silky fibre.

Appendix.

A. A list of some of the vegetable productions of New Zealand, available as food for man (extracted from A Leaf from the Natural History of New Zealand by Rich. Taylor, Wellington 1848)

Dicotyledones:

Hinau, *Elæocarpus hinau,* a large timber tree, producing a berry with a hard stone. The berry is edible, but unless prepared it has a very harsh taste.

Rimu *(Dacridium),* Matai, Miro, Kahikatea (Species of *Podocarpus*) belong to the Coniferæ and produce small fruits, which are much prised by the natives.

Karaka, Tawa, Kohekohe, Taraire, belonging to the *Laurineæ* also produce eatable fruits.

Rengarenga *(Tetragonia expansa)* New Zeal. Spinach; it was first brought into notice by Captain Cook, who found it useful as an antiscorbutic; the natives use it as foot.

Panapana, Hanea, Nau *(Cardamine),* New Zeal. cress.

Retireti, Tutaekahu *(Oxalis),* is a wholesome vegetable when boiled.

Kawa Kawa *(Piper excelsus);* the fruit is similar in shape and taste, before it is ripe, to the Jamaica long pepper; when fully ripe it has an agreeable flavour, the leaves are infused as tea, and when brewed, make a very refreshing beer.

Kahikatoa, Manuka *(Leptospermum scoparium)*; the leaves of this shrub are a very common substitute for tea.

Monocotyledones:

Ti, Whanake *(Cordyline)*; there are several varieties of this tree, all of which have long tap roots, which the natives cook; they have then a bitter sweet taste; the early Missionaries brewed beer from them; the tender shoots are also eaten.

Kiekie, Uriuri, Ori, Tiore, Patangatanga *(Freycinetia)*, this plant is found in forests, where it sometimes runs along the ground or climbs up the trees; it bears a male and female flower. In Autumn the pistils of the female flower, which are generally three, sometimes four in number, increase in size, until they attain a length of nearly a foot and a diameter of three inches; the outer skin is rough and bitter, but when scraped off, it exposes the pulp of the fruit, which when fully ripe, is very sweet and of an agreeable flavour; this may be considered by far the finest native fruit in New Zealand. It is called New Zealand's pine-apple.

Nikau, Miko *(Areca sapida)*, the tender shoot is eaten, either row or cooked; in the former state it has the taste of a nut.

Raupo *(Typha angustifolia)*, in swamps; the root, Korere, is white, tender and cellular, filled with a fine mealy substance which is eaten.

Acotyledones:

Mamaku, Pitau, Korau *(Cyathea medullaris)* an arborescent fern; the entire stem being peeled is eaten and when cooked is very good; it is a favorite dish of the natives.

Rarauhe *(Pteris esculenta)*, the common fern, the root of which (Aruhe) is eaten: when well beaten, roasted and deprived of its fibres, it is good eating; it is considered to be a preventive for sea sickness.

Many of the N. Z. Fungi, and most of the Algae are edible, and still occasionally used as food. The Rimu *(Chondrus crispus)* possesses all the properties of the Carrigeen moss.

B. Table of the chief Timber trees of New Zealand.

Maori name.	Trivial name among the colonists.	Botanical name.
Kauri . . .	N. Z. Yellow Pine . .	Dammara australis
Rimu . . .	N. Z. Red Pine . . .	Dacrydium cupressinum
Totara . . .	N. Z. Mahagony Pine .	Podocarpus totara
Kahikatea . .	N. Z. White Pine . .	„ dacrydioides
Matai or Mai .	N. Z. Black Pine . .	„ spicata
Miro	— — — — . .	„ ferruginea
Tanekaha or Tawaiwai .	Pitch Pine	Phyllocladus trichomanoides
Rata	N. Z. Oak-Elm . . .	Metrosideros robusta
Pohutukawa .	N. Z. Ash	„ tomentosa
Puriri . . .	N. Z. Oak or Teak . (Ironwood)	Vitex littoralis
Kohekohe . .	N. Z. Cedar	Hartighsea spectabilis
Rewarewa . .	Honey-suckle	Knightia excelsa
Hinau . . .	— — — —. . .	Elæocarpus hinau
Maire	N. Z. Sandalwood . .	Mira salicifolia.

Pines {

Magnificent pieces of New Zealand cabinet-ware were produced, in the year 1861, by Mr. Seyfert in Auckland, an immigrant joiner from Vienna for the London exhibition.

CHAPTER VIII.

The Fauna.

Remarkable scarcity of land-mammalia. — Introduced domestic animals. — Pigs. — Frogs. — Lizards. — A large salamander. — Sea-serpents. — Fishes. — Singing-birds. — The Nestor. — The night-parrot. — Swamp-fowl and sea-birds. — Mollusca. — Land-shells. — Insects. — The Wheta. — Mosquitoes and sand-flies. — Blatta. — The vegetating caterpillar. — Crustaccæ.

On looking over the Fauna of New Zealand, the almost total lack of land-mammalia appears, no doubt, to the observer as strikingly peculiar, as the singular substitution found in the shape of the wingless birds, some species of which, continuing in all probability into the present times, attained a gigantic size such as all the rest of the world has never produced.

Although from certain terms occurring in the Maori language, and from the most recent observations we may infer beyond a doubt, that New Zealand still harbours some few sporadic mammalia, which have thus far escaped the searching eye of science; yet, as regards the number of mammalia, this extensive insular country is surpassed by many far smaller islands of the South Sea. While upon islands of inconsiderable dimensions there are various gnawing animals, peculiar shrew-mice and bats living in trees; while upon the Marianas a deer even is found, — New Zealand possesses only two distinctly proven genera, the bat (*Pekapeka*, two species) and a small indigenous rat *(Kiore);* and even this little quadruped seems already to have disappeared before his alien congener from Norway. An other quadruped, the Kararehe, or native dog, which has likewise become almost extinct, is of doubtful origin. It was observed by the very first discoverers of the island. Some assert, that it is indigenous, others that it accompanied the natives in their first migration; others that it was introduced by some early Spanish ship. But even if it was not indigenous, at any rate it dates from the remotest ages of antiquity, since the Maori tradition knows for it a special creator, Irawaru. It is described as a small lurger-like animal, black, red or dirty yellow; and its look, gait and general deportment are decidedly hang-dog and vulpine. It is not wild like the ravaging dingo of Australia.

On continuing to trace the names used in the Maori language for animals, we find numerous names applied to domestic animals, imported since the natives have come in contact with Europeans: horse (hoio), ass (kaihe), heifer, sheep (hipi), goat (nanenane), pig (poaka from pork), dog (kuri, poipoi, peropero), cat (ngeru, poti, tori); besides these names we find the name Waitoreke, which has been only lately clearly defined, having been hitherto applied sometimes to an otter-like, and sometimes to a seal-like animal. According to the reports of Dr. J. Haast, the existence of this animal has been recently established beyond a doubt; it lives in the rivers and lakes in the mountain ranges of the South Island, is of the size of a large cony with a glossy brown fur, and is probably to be classed with the otters. [1]

The large maritime mammalia, whales and dolphins, likewise seals, were formerly very numerous on the coasts of the island.

---

[1] My friend Haast writes to me on this subject under date of June 6. 1861: "At a height of 3500 feet above the level of the sea I frequently saw its tracks on the upper Ashburton River (Prov. Canterbury, South Island), in a region never before trodden by man. They resemble the tracks of our European otter, — only a little smaller. The animal itself, however, was likewise seen by two gentlemen, who have a sheep-station at Lake Heron not far from the Ashburton, 2100 feet high. They describe the animal as dark-brown, of the size of a stout cony. On being struck at with the whip, it uttered a shrill, yelping sound, and quickly disappeared in the water amid the sea-grass."

There are eight kinds of whales, two of dolphins, and three of seals. The latter are growing scarcer from year to year. The sea-bear *(Kekeno)* has probably ceased to select the North Island for its home; it is only the rugged and uninhabited Southwest Coast of South Island, that still continues to afford it sufficient solitude for cubbing, and haunts sufficiently favourable for the gambols of the other seals.

Pigs and cattle are introduced into New Zealand and have rapidly propagated throughout the land. The pig was the most valuable gift made by the first discoverers to a people, whose chief food was fern-root *(Pteris esculenta)*, and who besides a few birds and fishes had no other animal food than the wild dog and the little rat. The pig lives with and by the side of man in his wildest, rudest natural state; it is a great addition to his means of subsistence, without interfering with his ordinary mode of living, while the possession of cattle depends on the existence of a more advanced stage of civilization. Cattle and swine run wild in various districts of the islands, and it is astonishing, to what numbers the wild pigs are multiplying. They find an excellent and everywhere plentiful food in the fern-roots, which formerly served the Maoris as a chief article of food. They retire shyly from the immediate vicinity of the settlements, because the settlers hunt them down energetically; but they congregate in the yet uninhabited valleys in a truly enormous number. The Wangapeka valley in the Province of Nelson I saw for miles up and down literally ploughed up by thousands of such wild pigs. They are nearly all black. Their extermination is sometimes contracted for by experienced hunters, and it is a fact that three men in 20 months upon an area of 250,000 acres killed not less than 25,000 of them; they moreover pledged themselves to kill 15,000 more. Where the wild pigs are very numerous, they do a great deal of damage to sheep-breeding.

Beside the domestic mammalia introduced by the Europeans, there are moreover some involuntarily imported vermin, that follow man most pertinaciously wherever he sets his foot. The mouse and the rat we found in the miserable huts of the fishermen upon the lone rock of St. Paul in the Indian Ocean as well as in the large sea ports on the coasts of all parts of the world. They will follow in the vessels all over the sea, through all climates, settle down with man wherever he settles, as unwelcome companions, and as a troublesome plague. Upon New Zealand, the European rat has totally exterminated the native rat Kiore, which used to be eaten by the Maoris.

Without following a sytematic order we pass over to the class of the *Amphibia*. The total absence of serpents, tortoises, and — with the exception of a frog but very recently discovered, [1] — also the absence of the batrachians is peculiarly striking. The lizards are represented by most harmless creatures despite the fables current among the natives about terrible dragon-like Ngararas. There are at present in all eleven species known; [2] five of them belong to

---

[1] The only place as yet known as the home of frogs are the environs of the Coromandel Harbour on the East side of the Hauraki Gulf (Province Auckland, North Island). There a very peculiar species is met with in the small creeks rising in the Cape Colville range; also in swamps, but always as a great rarity. The first specimens were discovered in 1852 (Edinburgh New Philos. Journal, 1853). I brought with me two specimens that had been collected by the natives; they have been described by Dr. L. J. Fitzinger in the Records of the Imperial Zoolog. Botan. Society in Vienna (series of 1861), as *Leiopelma Hochstetteri*. They come nearest to a Peruvian species, *Telmatobius peruvianus Wiegm.*, and belong to the water or common frogs. It is strange, that the natives formerly did not know this frog.

[2] The eleven species known are:

    Eulampus (Hinulia Gray) ornatus Fitz.
    Lampropholis (Mocoa Gray) Moco Fitz.
       „   „   „   *Smithii* Fitz.
       „   „   „   *grandis.*
    Hoplodactylus (Naultinus Gray) pacificus Fitz.
       „   „   „   *Grayi* Fitz.
       „   „   „   *elegans* Fitz.
       „   „   „   *punctatus* Fitz.
       „   „   —   „   *granulatus.*
    Dactylocnemis Wüllerstorfii Fitz., house-gecko, a new species named, after the chief-commander of the Novara Expedition.
    Hatteria punctata Gray, Ruatara or Tuatara of the natives, a *leguan*, the largest lizard in New Zealand known.
The discoveries, however, do not seem to be confined to the species hitherto known. Taylor (Te Ika a Maui p. 409) mentions a reptil of 4 feet length resembling

the neat genus *Naultinus*, peculiar to New Zealand, the sluggish character of which reminds somewhat of our salamander. With regard to the land-tortoise found on the Wanganui River in Cook's Strait, Mr. Ch. Heaphy, the author of the statement, informed me that he really had found the tortoise there, but that he is now fully convinced that it had got there by mere accident, having probably escaped from a whaler during its stay there. The natives had never been aware of the existence of such an animal. A venomous serpent is likewise said to have been left there by an English captain, intentionally, but luckily it does not seem to have found a living there, as nothing further has been hitherto dicovered of snakes.

The ringed sea-serpent *(Pelamys bicolor)*, which is found from the Indian Ocean to the eastern-most groups of Polynesia, has also been found about New Zealand; but whether New Zealand is its southern-most limit, or whether it was transported thither by accident, remains yet to be decided. We ourselves met on board the Novara with a very instructive case of transportation. One morning in one of the cabins such a snake was found, which, as unobserved as it had got into the ship, could have slipped out equally unobserved into the sea at some far distant place.

As to fishes, the bays and coasts of New Zealand are teeming with them; and there are about 100 species enumerated in the catalogues. But just as the forests are destitute of game, so

a salamander, which a man named Hawkins is said to have seen and even caught in the Greenstone Lake, and repeated reports are spread abroad about this large, black salamander. A very trust-worth sheep-keeper of the Province Canterbury related the following incident to my friend Dr. Haast. It was after the fall-overflows. — which usually carry large quantities of wood from the mountains into a lake close to his station, — that he was engaged in gathering such drift-logs to have a supply of fire-wood for the winter. He had pulled one of those logs, — which, as he observed afterwards, was hollow at the lower end — about half out of the water, when a black animal, 4 to 5 feet long, and resembling a crocodile, crept out, which immediately disappeared in the water. The narrator added as a special remark, that it was out of the question to suppose that what he saw was one of the large eels, such as are frequently found, sometimes 6 feet in length and weighing 20 pounds; he himself being a passionate eel-catcher, and consequently thoroughly acquainted with that animal.

are the fresh-water lakes and rivers destitute of fish. So far as I myself had an opportunity to become acquainted with the contents of the rivers and lakes, I only found eels and what the colonists call whitebait. According to Mr. Haast's testimony, eels are predominant also in the rivers and lakes of the Provinces of Nelson and Canterbury, where they grow to an astonishing size of more than 50 pounds weight. We may well take it for granted, that there are various species, as the numerous distinctions made by the natives in naming them also lead to suppose; scientifically, however, but one species has hitherto been determined, the *Anguilla Dieffenbachi Gray*. There are more than two dozen names in vogue among the Maoris for eels; and even after reducing the number for the different ages, which in the common parlance have frequently special names, we may still suppose, that they belong to a greater number of species. In the smaller brooks I gathered the Inangas of the natives, the whitebait of the colonists *(Eleotris)* of which thus far three species have been distinguished. Like our *Phoxinus* they are very plentiful in fresh-water lakes and even in the smallest brooks. Numerous sea-fish rove far up the rivers, where the water is only faintly brackish. Remarkable is the appearance of some sea-fish which seem to belong to the South Sea within a broad belt on both sides of the 40th paral. of latitude. *Thyrsites Atun*, which is chiefly caught in Simon's Bay at the Cape of Good Hope, and which we angled also in the waters girding St. Paul's Island, together with some species of *Cheilodactylus*, "Morue des Indes", is likewise found on the coasts of New Zealand. The brilliancy of colour of the New Zealand fishes is quite inferior to that of the species living in the Indian Ocean. The magnificently coloured Squamipennians are totally wanting, and the Julides number but few species displaying as beautiful a medley of colours as those of the other seas.

The most charming part of the fauna are the birds. The total number of species known amounts at present to 100 (England contains 273 species). Numerous species, however, exclusively peculiar to New Zealand, and the most remarkable ones there are rapidly dying out, and are already partly extinct.[1] One of the prettiest creatures is the Tui, Parson Bird of the colonists *(Prosthemadera Novæ Zelandiæ)*, which roves about in the lofty, leafy crowns of the forest-trees. "Larger than the blackbird and more elegant in shape, his plumage is lustrous black, irradiated with green hues and pencilled with silver grey, and he displays a white throat-tuft for his clerical bands. He can sing, but seldom will; and preserves his voice for mocking others. Darting from some low shrub to the topmost twig of the tallest tree, he commences roaring forth such a variety of strange noises with such changes of voice and volume of tone, as to claim the instant attention of the forest. Caught and caged, he is still the merry ventriloquist, mocks cocks and cats and attempts the baby. To add to his merits, he becomes a very fine eating in the season of the Poroporo berries."[2] The chief songster is the Kokorimoko *(Anthornis melanura)*. Of the *Certhiparus* species among the real warblers, likewise of the New Zealand thrush *(Turnagra crassirostris)*, and the starlings *Aplonis* and *Creadion*, I am not able to say, whether and how they sing. A striking exception appeared to me the New Zealand lark *(Alauda Novæ Zelandiæ)*, very common on all roads and hills, which I have, however, never heard utter a sound. A remarkably fine tenant of the forests is the large wood-pigeon Kuku *(Carpophaga Novæ Zelandiæ)*.

In the family of the parrots we met in New Zealand a very peculiar genus, that of the *Nestor.* They are characterized by an aquiline, far overlapping upper-beak. The brilliant hues of the parrot-family is bleached down in the chief representative of this genus, the Kaka — *(Nestor hypopolius,* synon. with *Nestor meridionalis* or *Australis),* — very numerous and common in all the woods, to a faint brown and grayish-green; only the exceedingly rare and larger species, *Nestor notabilis* and *Nestor Esslingii,* display livelier

---

[1] For a separate treatise on the remarkable wingless birds see the following chapter.

[2] Hursthouse, New Zealand I. p. 118.

colours, a greenish metallic hue, and under the wings red, yellow and blue. A fourth species, *N. productus*, is known to exist on Philip's Island, a small isle near Norfolk Island in the North of New Zealand. The several *Platycerus* species, Kakariki of the Maoris, are parrots with brilliant colours in green, blue and red. A perfectly anomalous form, on the other hand, is the yellowish-green owl- or night-parrot, Kakapo of the natives *(Strigaps habroptilus)*. It lives in crevices of the ground under tree-roots or in rocks and comes out only at night to pick the berries of the Tutu shrub *(Coriaria sarmentosa)* and to grub fern-roots. Although it can fly, it seems to use its wings very seldom. It always lives with its mate. The natives used to chase it with dogs or to catch it in snares. Thus it has been totally exterminated on the North Island; and now it is confined to the remotest Alpine valleys, on the South and West Coasts of South Island; yet it is still quite frequent in those parts.

Another famous bird of chase with the natives is the Weka *(Ocydromus Australis)*, or the wood-hen, belonging to the class of rails, which have already become quite scarce upon North Island. In the grassy plains and forests of the Southern Alps, however, they are still found in considerable numbers. It is a thievish bird, greedy after every thing that glistens; it frequently carries off spoons, forks and the like; but it also breaks into hen-coops, and picks and sucks the eggs. Among the swamp-fowls there are especially some herons, an oyster-catcher, the New Zealand plover, and the beautiful Pukeko *(Sultana)* to be mentioned. These together with wild ducks, including the splendid Paradise duck, and several species of cormorants enliven the numerous water-channels of the river estuaries, which are unapproachable on account of their extensive marshy bottoms; on the banks, beneath the dense foliage of the overhanging trees they find everywhere safe hiding places. [1] On

---

[1] The specimen of *Notornis Mantelli* caught by seal-hunters in 1850 on Dusky Bay, South Island, and preserved in the British Museum, London, has, as far as I know, hitherto remained unique; and it appears to me, that this family of birds is now totally extinct. It is nearest related to the Pukeko, of the size of a turkey,

the southern extremity of New Zealand there are two small pin-guins; while the coasts round about are teeming with albatrosses, storm-petrels, sea-gulls, and sea-swallows. Of the yellow headed Australian Sula we found numerous bevies swimming about outside the entrance to the harbour of Auckland.

The number of molluscas found on the coasts of New Zealand is very considerable. In all there have been described about 344 species, belonging to 123 genera, and every research furnishes new species. *Strombus, Triton, Murex, Fusus, Voluta* number species both considerable and highly prized; *Voluta magnifica* is the largest species of the latter genus. Of *Struthiolaria*, belonging exclusively to the Australian seas, there are three species known. The *Cyprœa aurora*, living in the South of the Pacific Ocean, so highly prized by the savages, and still dearly paid for by zoological collectors, is found also here. Of the numerous top-shells, there are three species of the genus *Imperator*, — which likewise belong only to the South Sea, — besides some beautiful, but rare species of *Turbo*. *Haliotis Iris* ("Mutton fish" of the settlers) is here found in colossal specimens. The nipple-shells *(Patella)* are also very abundant. The extensive cliffs of the inlets near Auckland, which cut deeply into the land like so many rivers are covered with savoury oysters, which in time of low water can be knocked off very easily. Of Brachiopodes four species are known; the pretty red coloured *Tere-bratella rubicunda, cruenta, Bourhdi* and *suffusa*.

Less numerous are the land- and freshwater-molluscas. Exclusive of some very fine species, such as *Helix Busbyi, Bulimus Shongii, B. Novoseelandicus* and the lately discovered magnificent *Helix Hochstetteri Pfr.*[1] from the Alps of South Island, the land-

---

its plumage magnificent and resplendent with the most beautiful metallic glare. The natives upon North Island named it Moho; upon South Island, Tukahe. The beautiful Paradise duck *(Casarca variegata)* I saw frequently in the highland valleys of South Island near Nelson, and always in pairs. — In the mountains of the Province Otago recently a large owl has been discovered, which digs holes in the ground. Dr. Haast also has observed this owl several times in the Alps at night.

[1] Dr[s]. L. Pfeiffer and W. Dunker have described the new species brought home by me, in the Malakozoological Journals (Vol. VIII. pp. 146—154).

shells are mostly but small and insignificant; they live very clan-
destinely and have but recently been brought to light by the closer
researches of science. The
rivers which I had an opportunity to examine, are inhabited by small and very
peculiar species of *Hydrobia*
and by two kinds of *Unio*,
large quantities of which are
caught by the natives with
whom they constitute a chief
article of food.

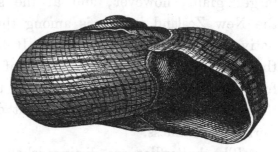

Helix Hochstetteri Pfeiffer, a new land-snail from
the South-Island, in life size.

Of crustaceæ, the number of species described are 56; of
insects 265 species, belonging to 215 genera. Among them 179
species *Coleoptera, 11 Neuroptera, 18 Hymenoptera, 13 Homoptera,
11 Hemiptera, 55 Lepidoptera, 57 Diptera,* and *21 Arachnida.* The
Orthoptera of New Zealand are characterized by this particular,
that the majority of the species of all orders possess either no organs of flight, or but very stunted ones; which peculiarity was
pointed out already by Erichson in speaking of the fauna of Tasmania; which, however, occurs to a far less degree in Australia.

The largest beetle, which I collected myself in the woods on

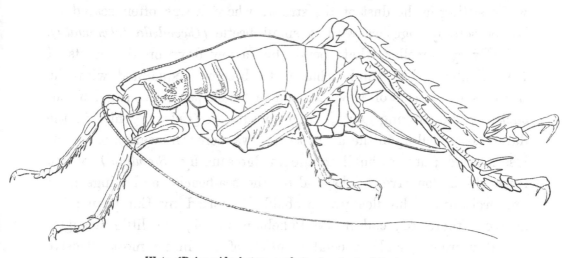

Weta (Deinacrida heteracantha), female in life-size.

the Waikato, is the *Prionoplus reticularis White*, a goat-chafer swarming at night. This, when full grown, is 1¹/₂ inch long. A much larger giant, however, and at the same time the oddest one of the New Zealand insects is among the Orthopteræ the *Weta* of the natives *(Deinacrida heteracantha)*. It lives in rotten wood and under the bark of trees, and the length of a large full grown specimen from the ends of its hind-legs to the tips of its feelers is 14 inches, the body measuring 2¹/₂ inches; but despite its hideous looks it is perfectly harmless.

The butterflies are distinguished neither by size nor by richness of colours. Night-butterflies are more frequent than day-butterflies; and among the former it is especially the family of moths, which is most extensively represented both as to the number of species and that of individuals. Among the few day-butterflies, which most easily strike the eye of the traveller, there are no strange forms; but some few Europeans, such as our "painted lady" Very common was in Auckland *Leptosoma annulatum Bod.* *Libellæ* (dragon-flies), — although their species are comparatively few, — nevertheless exist in large numbers in the swamps and stagnant waters about Auckland. Of the three known species of Cicadæ I observed one frequently in Auckland; it was met with in every street; everywhere it was heard chirping its shrill notes, even while sitting in the dust of the street, where it was often scared up by passers-by together with a small beetle *(Cicendella tuberculata)*.

Every traveller, that spends but a few days on the coasts of New Zealand, has ample chance to become acquainted with the troublesome insects of New Zealand in the shape of two small blood-suckers, the stinging gnats, vulgo mosquitoes *(Culex)*, which in the damp forest fall upon the unsuspecting wanderer in swarms of countless myriads; and a small midge, vulgo sand-fly *(Simulium)*, which lives chiefly on river-banks and on the sea-beach, and stings most unmercifully. The flea was probably imported by Europeans; the natives, therefore, call it the Pakeha-nohinohi, the little stranger. I will moreover make special mention of an insect most offensive because of its noisome smell. In Sydney already I met with a

Blatta (vulgo "cock-roach") of which I convinced myself, that like the chinches it can really squirt at pleasure a corrosive fluid, the penetrating smell of which is intolerable. In Auckland it is dreaded so much, that in wood-houses, vaults, and damp places special care is taken, not to come into contact with the vermin, since in that case it infects every thing for days at a time with the most terrible stench imaginable. Because of this property which is especially characteristic of the wood-bugs, the colonists call it wood-bug. Of the real wood-bugs, however, there are likewise some species. This Blatta is doubtless the same as the insect named by the natives Kikararu, which was erroneously taken for a bug. It is a new species, described by Mr. Brunner of Vienna, as *Polyzosteria Novæ Zelandiæ*.

It might not be improper to mention here also the Aweto or Hotete the large (nigt-butterfly) caterpillar, from the head of which a parasitical fungus, *Sphæria Robertsii*, grows out; hence the name "Vegetating Caterpillar" among the colonists. A large portion of such caterpillars die of it, while burying themselves in the ground for the purpose of changing into a chrysalis. A peculiarity of this fungus is this, that the stem bearing the seed-spurs as its end, rises nearly exclusively in the neck of the caterpillar between the head and the first ring of the body. Of hundreds of specimens that I examined, there was only a single one, the fungus of which had grown out of the aft-end of the caterpillar. The natives eat this vegetating caterpillar.

The known species of *Crustaceæ* in New Zealand were increased through the Novara collections by twelve new ones. The Brachyura are most numerous, and crabs are found everywhere on the seashore. The Bernhard crabs, although not very numerous, are still not utterly wanting. Of the Macrura I may mention especially *Paranephrops tenuicomis Dana*, which I found to be quite numerous in all the rivers; the natives call it Koura, which

The Vegetating Caterpillar, Sphæria Robertsii, 1/2 of life-size.

term, however, is applied to various crabs, even to the lobsters fished from the sea. Of Stomapodes, Isopodes, Myriapodes various kinds have been collected and recorded, without, however, having been subjected to a more minute examination. The same must be said respecting the remaining classes of the lower animals. The Radiaria and worms of the sea are as yet but little known; neither are the Spongiæ, of which New Zealand possesses numerous as well as interesting species, known to any extent.

CHAPTER IX.

Kiwi and Moa, the wingless Birds of New Zealand.

The Struthionidæ family. — Species now living. — Aepiornis of Madagascar. — The Dodo of Mauritius. — Discovery of the Kiwi (Apteryx) in New Zealand. — Three, perhaps four, different species of Kiwi. — Discovery of Moa bones. — *Dinornis, Palapteryx.* — Eggs. — Moa stones. — New discoveries in the caves of the Aorere Valley. — Complete skeleton of *Palapteryx ingens.* — Distribution of the Moas. — Different species upon the North and South Island. — Whether still living or extinct? — Causes of their dying out. — The giant-birds of New Zealand once the chief game of the natives. — Cannibalism the consequence of the extermination of the Moas. — Struggle of life. *Appendix.* Comparative table of the size of Struthionidæ.

The family of the ostrich-like birds (Struthionidæ) can boast not only of the most marvellous forms, wholly different from the common type of birds, but also of by far the largest representatives of the "feathery tribe". Therefore they are also called giant-birds *(Proceri)*. They are birds with short, rudimentary wings, which are totally unfit for flying. The muscular power, which nature has to dispose of in this case, would not have sufficed to keep the bulk of such birds in the air. Hence the bones are almost without air-cells, the breastbone is a convex plate without ridge, the muscles of the breast are thin; their plumage is loose and flabby; the feathers light and shaggy, resembling hair. Instead, however, the muscles of the upper and lower thighs are of unusual strength and thickness, the feet are long and most fully developed for running, with two or three toes having a callous sole.

The types of this remarkable family of birds all belong to the southern hemisphere. Only the African ostrich, the original home of which is probably also to be looked for South of the Equator, has in course of time spread into the northern hemisphere. They are, as it were, the pachyderms among the birds, and it is especially upon the limited territories of the islands of the southern hemisphere, — which are too small to sustain large mammalia, — that they take their place in every respect; but they die out, wherever they come in contact with man.

But few traces of them have as yet been found in the older strata of the earth, which might enable us to infer the existence of this bird-family in the periods of the earth previous to the appearance of man. All that can be adduced in proof of this, are tracks found in the New Red Sandstone of Connecticut, North America. If those so-called Ornithichnites are really impressions made by birds, they, indeed, betoken the existence of birds of a colossal size, the paces of which measured five feet, and which by their weight pressed the mud up from the ground, as though elephants had been wading about in it. Recently Professor Owen has described the remains of a fossil bird *(Gastornis Parisiensis Hébert)* from eocen strata of Paris. This excepted, whatever is known of giant birds, belongs to the present world, although many a species has long ago succumbed in its struggle with man, and vanished again from the stage of actual life.

The number of species living is very small. In all there are only about 12 species known; two, perhaps three species of ostrich in Africa, three cassuary species[1] in southern Asia, two Emu's *(Dromaeus)* in Australia, an East and a West Australian, three species *Rhea* in South America, and three or four species of Kiwi *(Apteryx)* in New Zealand. Among all these the African ostrich, 6 to 7 feet high, is known to be by far the largest and most numerous species.

But greater than the number of living species is the number

---

[1] Among them Casuarinus Bennetti, the Mooruk of the natives of New Britain, discovered in 1858.

of extinct species, which used to inhabit the islands from Madagascar to New Zealand even within the memory of man; and it
is among these that we become acquainted with by far the largest
representatives of the family of giant birds.

Marco Polo already, in the famous account of his travels, lo-
cates the giant bird Ruc of the myth upon Madagascar, and relates
that the Great Khan of the Tartars having heard of this bird at
the far off borders of the celestial empire, sent forthwith messengers
to Madagascar. They really brought a feather back with them,
9 spans long, and 2 palms in circumference, at which His Majesty expressed his unfeigned delight. People laughed at this tale,
as a fable, and like so many other relations made by Marco
Polo on real facts, it was declared vain swaggering talk; — until
tidings came establishing the fact, that very recently a gigantic
bird was, and is still existing in Madagascar. This happened
thus: Natives of Madagascar had come to Mauritius to buy rum;
the vessels they had brought with them to hold the liquor were
egg-shells, eight times as large as ostrich-eggs, or 135 times the
size of a hen-egg; eggs containing 2 gallons. They related that those
eggs were now and then found among the reeds, and that the
bird also was occasionally seen. This was not believed either until
the Museum at Paris in 1851 received such an egg from a landslip
in Madagascar, measuring $2^3/_4$ feet in circumference, and holding
$2^1/_2$ litres; it was in a state as though it had been laid but very
recently. Now Marco Polo's fabulous Ruc has become the *Aepiornis
maximus* of Madagascar. Yet that colossal egg, the casts of which
are exhibited in almost every Museum in Europe, besides some
fragments of bones in the British Museum, is all, that has hitherto been obtained of this bird. Whether it still lives, is uncer-
tain. The natives assert to this day, that in the thickest forest,
there still exists a giant bird; but that it is very rarely seen.

East of Madagascar, upon the Mascarene Islands (Bourbon,
Mauritius, Rodriguez), — from the bones collected by Mr. Bartlett
upon Rodriguez in 1855, there are three species known, the Dronte
or Dodo *(Didus ineptus)*, the Solitaire *(Pezophaps)* and a new, much

larger bird. All are extinct; but concerning the former two, it has been proved, that they lived in great numbers on those islands till within the 16. and 17. centuries. Even as late as 1638 a live Dodo was exhibited in England, the skin of which was afterwards transferred to the famous Museum of John Tradescant. At a revision of this Museum in 1775 by the trustees, the damaged specimens were condemned among the rubbish, and unfortunately also the Dronte; the head and feet were all that was saved of it, and these parts, the only remnants of the extraordinary animal, are now exhibited as a great curiosity in the Asmolean Museum in Oxford. [1]

Yet, whatever had been heard, known and collected of ostrich-like birds, was far surpassed by the discovery of the Moas and Kiwis of New Zealand. There, both forms were found united, dwarf forms, such as they had never been known heretofore, and giant forms, such as they had been merely sketched by fancy; of New Zealand alone there are now already nearly as many species known as of all the rest of the globe.

In 1812 the first skin of a Kiwi was brought to England. The Zoologists were at a loss, what to make of the strange bird. It was named by Dr. Shaw *Apteryx australis,* the wingless Australian bird; it next passed into the collection of the late Lord Derby, but not until after many years, in 1833 — at that time unique — was it described by Mr. Yarrell. Thus a dwarf form had become known, a bird, not larger than a hen, without wings and without tail, four toes on its foot, with a long bill resembling that of a snipe, the body covered with long brown feathers resembling hair. The skins of this bird brought to Europe, were sold for 200 and 300 francs a piece; they were considered the greatest rarities, the more so, as it was believed that the bird was almost extinct. It has, however, been exterminated only in the inhabited parts of New Zealand, while it is a fact, that in the primitive forests of the mountain regions there are up to this day great numbers of

[1] Besides this, there is a breast-bone in Paris, a skull in Prague, a beak in Kopenhagen and a foot in London.

Kiwi (Apteryx) and Moa (Palapteryx),
the wingless birds of New Zealand.

them living; of course, disappearing rapidly even there, as man advances to subdue the land. [1] There appears to be evidence of the present existence of three or four species of the genus Apteryx in New Zealand. [2]

The above mentioned *Apteryx Australis* was the first species made known to science. The original specimen in 1812 was obtained by Captain Barklay of the ship "Providence", and is stated to have come from Dusky Bay in the Province of Otago, South Island. Some time afterwards a second specimen from the same locality was procured by Dr. Mantell and examined more closely by Mr. Bartlett. This specimen became the property of the British Museum. All the other specimens exhibited in the European collections as *Apteryx australis* come from the North Island, and belong to the species described by Mr. Bartlett as *Apteryx Mantelli*. In fact, this very common species is so closely allied to the *Apt. australis* as to render it very desirable, that additional specimens of the later should be obtained and a rigid comparison instituted between the two.

The *Apteryx Mantelli* [3] is, as far we are now informed, confined to the North Island. This bird differs from the original *Apteryx australis* of Dr. Shaw, in its smaller size, its darker and more rufous colour, its longer tarsus, which is scutellated in front, its shorter toes and claws, which are horn-coloured; its smaller wings, which have much stronger and thicker quills; and also in having long straggling hairs on the face. In the northern districts of the Northern Island this species of *Apteryx* appears to have become quite extinct. But in the island called Houtourou, or Little Barrier Island, a small island, completely wooded, rising about

---

[1] Report on the Present State of our Knowledge of the Species of Apteryx living in New Zealand. By Philip Ludley Sclater, M. A., Ph. D., J. R. S., and Dr. F. v. Hochstetter. Read at the Meeting of the British Association, Sept. 1861.
[2] Apteryx Australis, Shaw, Nat. Misc. XXIV. pl. 1057, 1058, and Gen. Zool. XIII. p. 71.
Apteryx Australis, Bartlett, Proc. Zool. Soc. 1850, p. 275.
„ „ Yarrell, Trans. Zool. Soc. I. p. 71., pl. 10.
[3] Apteryx Australis, Gould, Birds of Australia XI. pl. 2.
„ Mantelli, Bartlett, Proc. Zool. Soc. 1847, p. 93.

1000 feet above the sea level, and only accessible when the sea is quite calm, which is situated in the Gulf of Hauraki, near Auckland, it is said to be still tolerably common. In the inhabited portions of the southern districts of the Northern Island also, it has become nearly exterminated by men, dogs, and wild cats, and is only to be found here in the more inaccessible and less populous mountain-chains, that is in the wooded mountains between Cape Palliser and East Cape. It is therefore not so easily obtained as we might suppose. Dieffenbach already mentions that during an 18 months' stay in New Zealand (1840—1841), despite the rewards he promised the natives everywhere, he succeeded in obtaining but one skin, and that in Mongonui Harbour, North of the Bay of Islands, from a European settler. I fared no better. I travelled through many a district on North Island, where according to the statements of the natives the bird still exists and is occasionally caught, but despite all my efforts was unable to procure a single specimen.[1] Of the attempts hitherto made to bring the peculiar bird alive to Europe, only one, to my knowledge, has proved successful.[2] In the Zoological Garden in London there has been since 1852 a live hen-kiwi, which is fed with mutton and worms. Its daily rations are half a pound of mutton, and it has already laid a number of barren eggs. The bird weighing not more than four pounds and a-half, lays an egg weighing $14^1/_4$ oz. and of an astonishing size. Taking this as a criterion, it is to be supposed, that the New Zealand Moas laid eggs as colossal as the famous egg of the giant-bird of Madagascar.

There is as yet no second species of Kiwi known to exist on North Island. But the natives speak of sorts of Kiwi, which they

[1] The skins which the Zoologists of the Novara Expedition brought with them, they are indebted to our excellant German friend in Auckland, Dr. Fischer, who but very recently forwarded some additional specimens to Vienna. One of them had been presented to the captain on board alive but unfortunately died during the voyage. In 1862 Mr. Buller of Wellington obtained two specimens of *Apt. Mantelli* at the sources of Wanganui; *"at no little expense"* he writes in his letter.

[2] Very recently the Zoological Society of London has received a pair of living Kiwis.

distinguish as Kiwi-nui (large Kiwi) and Kiwi-iti (small Kiwi). The Kiwi-nui is said to be found in the Tuhua district, West of Lake Taupo, and is in my opinion *Apteryx Mantelli*. Kiwi-iti may possibly be *Apteryx Owenii*, though I can give no certain information on this subject.

Apteryx Owenii,[1] the third species, somewhat smaller than the former two species and with grayish plumage, was first described by Mr. Gould in 1847, from a specimen procured by Mr. F. Strange and believed to have been obtained from South Island. The four specimens in the British Museum most certainly come from the South; and during my stay in the Province of Nelson, I had an opportunity to convince myself with my own eyes, that this species is still quite frequent in the spurs of the Southern Alps on Cook Strait. Some natives I met in Collingwood on Golden Bay, upon a promise of £5 agreed to go out kiwi-hunting for me; and, in fact, after only three days they brought me two living specimens of *Apteryx Owenii*, male and female, which they had caught close by the sources of the Rocky and Slate rivers, tributaries of the Aorere river, at a height of 3000 feet above the level of the sea. I kept them for several weeks alive in a hen-coop in Nelson, until one fine morning the male made its escape; the female I brought with me preserved in spirits.[2] When Mr. Skeet in 1861 examined the mountains between the Takaka and Buller rivers in the Province of Nelson, he found in the grassy ridges on the East side of Owen river such numbers of these Kiwis, that with the aid of two dogs he could catch 15 to 20 of them every night. He and his men lived on kiwi-meat, and the range they called kiwi-range. Also in the Wairau ranges East of Blind Bay (Prov. Marlborough) Kiwis are said to be still quite frequent, and as far as my individual experience goes, it is here also the species *Apteryx Owenii*, which consequently is the common Kiwi of the northern portion of South Island.

[1] *Apteryx Owenii Gould*, P. Z. S. 1847, p. 94.
    „   „   „   Birds of Aust. VI. pl. 3.
[2] I presented the specimen to our celebrated anatomist, Prof. Hyrtl, for an anatomical examination.

Besides *Apteryx Owenii*, however, a second larger species lives on the Middle Island, of which, although no examples have yet reached Europe, the existence is nevertheless quite certain. The natives distinguish the species not as a *Kiwi*, but as a *Roa*, because it is larger than *A. Owenii* (Roa meaning long or tall). The existence of such a bird in the South Island has long since been affirmed, and though no specimen of it has yet reached Europe, it has been registered as *Apteryx maxima*.[1] What I am able to state on this subject, is the following: Mr. John Rochfort, Provincial Surveyor in Nelson, who returned from an expedition to the western coast of the province while I was staying at Nelson, in his report, which appeared in the Nelson Examiner, of August 24th, 1859, describes this species, which is said to be by no means uncommon in the Paparoa ranges, between the Grey and Buller rivers, in the following terms: a Kiwi about the size of a turkey, very powerful, having spurs on his feet, which, when attacked by a dog, defends himself so well as frequently to come off victorious. My friend, Julius Haast, writes to me in a letter, dated July, 1860, from ten miles above the mouth of the river Buller, on the mountains of the Buller chain, which at the height of from 3000 to 4000 feet, were at that time, it being winter in New Zealand, slightly covered with snow, that the tracks of a large Kiwi of the size of a turkey were very common in the snow, and that at night he had often heard the singular cry of this bird, but that as he had no dog with him he had not succeeded in getting a specimen of it.

Only very little is known of the mode of living of the Apteryx. They are night-birds, hiding themselves in day-time under the rootstocks of forest-trees, and going in search of food exclusively in the night-time. They feed upon insects, grubs, worms and the seeds of various plants.[2] They live in pairs. The hen lays but one egg, which is hatched, as the natives say, by the male and female

---

[1] "The Fireman", Gould, Birds of Australia, sub. tab. 3, Vol. VI. *Apteryx maxima*, Bp. Compt. Rend. Acad. Sc.
[2] Especially *Astellia Banksii*, *Elæocarpus* (Hinau) and *Hamelinia veratroides*.

alternately. The male is larger than the female and has a longer bill. They can run and hop very fast. The female Apteryx, which I kept for several days alive in my room in Nelson, hopped very readily over objects two or three feet high. Dogs and cats are, next to man, the most dangerous enemies of the bird. The natives, by imitating its call, — at night, of course, — know how to call it up to them, and to confound it by a sudden glare of torch-light, so that they can either catch it with the hand or kill it with a stick. Dogs also are used in kiwi-hunting.

The Kiwi, however, is only the last and rather insignificant representative of the family of wingless birds that inhabited New Zealand in bygone ages. By the term "Moa"[1] the natives signify a family of birds, that we know merely from bones and skeletons, a family of real giant-birds compared with the little Apterygides.

Missionaries were the first that heared from the natives of those gigantic birds, against which the ancestors of the present Maoris had been engaged in fearful struggles. The natives even pointed out a Totara tree on Lake Rotorua as the place, where their ancestors slew the last Moa, and in order to corroborate the truth of their narrative they showed large bones, which they found scattered on the banks of rivers, on the sea-coast, in swamps and limestone-caves, as the remains of those extinct giant-birds.

In 1839, Mr. Rule brought to England a fragment of a thigh bone of a Moa, from which Professor Richard Owen drew up a wonderfully correct idea of the bird. Almost at the same time the Rev. Mr. Colenso described in the Tasmanian Journal Moa-bones as the remains of gigantic birds. These facts excited interest and caused fresh researches, in consequence of which the Missionary, the Rev. W. Williams in 1842 sent several chests full of such bones, — which had been gathered on North Island in the coast districts about Poverty Bay and Hawkes' Bay, — to Dr. Buckland. Dr. Buckland

---

[1] *Moa* or *Toa* throughout Polynesia, is the word applied to domestic fowls, originating perhaps from the Malay word *mua,* a kind of peasants. The Maoris have no special term for the domestic fowl; *tikaokao* is the cock, and *heihei* the hen; the former probably an (onomatopoëtical) imitation of the crowing of the cock; the latter a corruption of the English word *hen.*

presented the treasures to the Museum of the Royal College of Surgeons, and Prof. Owen constructed out of them the gigantic legs of *Dinornis giganteus*, which are one of the greatest curiosities of said Museum, legs over 5 feet high, which intimate a bird of at least 9½ feet in height. [1] This is by far the most colossal from all the birds known. The *tibia*, the shin-bone alone, measures 2 feet 10 inches.

Upon South Island it was Mr. Percy Earl and Dr. Mackellar, who made collections at the mouth of the Waikouaiti, North of the Otago peninsula. But by far the most copious harvest was that gathered by Mr. Walther Mantell in the years 1847—1850 upon North and South Islands. He had collected more than 1000 separate bones and also fragments of eggs, which were bought by the British Museum, and furnished Prof. Owen the rich material for this celebrated works on the extinct families of *Dinornis* and *Palapteryx*. In this collection there was the famous skeleton of the elephant-footed Moa *(Dinornis elephantopus)* from Ruamoa, three miles South of Oamaru Point (First Rocky Head), Province Otago, a species, which while it fell far short of the height of *Dinornis giganteus*, — measuring hardly over 5 feet, — was distinguished by an extraordinarily massive construction of the bones, and, as Mr. Owen says and indicates by the nomenclature, of all birds represents most the type of the pachyderms. Very appropriately, therefore, this skeleton has been placed in the British Museum by the side of the gigantic elephant *Mastodon ohioticus*.

Colonel Wakefield, Dr. Thomson and many others have also made up collections partly on North, and partly on South Island, [2] and according to Prof. Owen there are already 12 to 14 different species of Moas known. [3] Most of them have three toes like the

---

[1] From the leg-bones Owen calculated the height of another species, *Dinornis robustus*, at 10 feet 6 inches. But Dr. Thomson judges the height of those birds to have been 13 to 14 feet. (Edinb. New Philos. Journal V. LVI. p. 277.)

[2] The large collection of Moa bones, which Sir George Grey had placed in the Governor's house in Auckland, was unfortunately lost in the conflagration of said building in 1848.

[3] The names mentioned in Prof. Owen's treatises, are: *Dinornis giganteus*,

Australian Emu. These Prof. Owen classes with the genus *Dinornis*, the four-toed species with the genus *Palapteryx*. From smaller bones, which were found, the genus *Aptornis* was established. However, the whole family of those wingless birds seems to have been very variable, since nearly every individual found, varied not only in size, but also in the number and proportion of the bones (especially of the vertebræ). I tis, therefore, very doubtful, whether all the species, distinguished by Prof. Owen, are good species.

Besides bones, there were also fragments of egg-shells found on North and South Islands, indicating eggs of a size much larger than ostrich-eggs, but not quite equal in size to the egg of *Aepiornis maximus*, and of a thin shell with linear furrows. In 1865, Mr. J. C. Stevens, Natural History Agent in London, received from New Zealand an almost perfect egg of Dinornis. The egg is about ten inches in length and seven inches in breadth, the shell being of a dirty brownish colour, and about $1/12^{th}$ of an inch in thickness. According to the Wellington papers, the egg was discovered in digging the foundation of a house at Kaikoras (Prov. Marlborough) enclosed in a small mound, supposed to be a native burying place, as a human skeleton, buried in a sitting posture, was found within the grave holding the egg in its hands. This interesting relic was offered for public sale on November 24th. A *bona fide* bid of £115 was actually made, but it was bought in at £125.[1] Besides bones and eggs, little heaps of small rounded stones are very frequently found, generally chalcedony, carnelions, opals, and achates, which are designated by the natives as "Moa stones". They are sometimes found together with Moa skeletons, partly also in places, where there are no traces of Moa bones. It is probably correct to suppose that those stones come from the stomach of the birds, which like the ostrich and the Australian Emu were in the habit of swallowing little stones to assist digestion, ejecting them again from time to time, in order to swallow others less rounded.

---

robustus, crassus, elephantopus, struthioides, casuarinus, rheides, didiformis, curtus, gracilis, Palapteryx ingens, dromioides, geranoides, Aptornis otidiformis.

[1] Geological Magazine, 1865. p. 576.

I recollect with much pleasure the grand impression, the sight of those Moa bones made upon me, when for the first time I entered the halls of the famous North gallery of the British Museum. It was a few weeks before the departure of the Novara. Among the islands of the South Sea, which we were to visit, the name of New Zealand was also registered. Ever since that time I cherished the hope, that I might be able to fetch from New Zealand similar treasures; and happily I did not fail in my exspectations, although I suffered much disappointment at the beginning of my researches.

Upon North Island I had scoured every district, that had been noted for the occurrence of Moa bones, I had ransacked all the so-called Moa caves, but all in vain. The Moa enthusiasts, that had been there before me, had carried off the last fragment of a Moa bone, and the Maoris on having discovered, that they could make some money by it, had gathered whatever there was still to be found, and sold it to European amateurs at enormous prices. The only relic I at least found out, was in the possession of a chief in the Tuhua district, who produced from the dust and rubbish of his raupo-hut an old bone, which he had hidden for a long time, and with which he parted only after lengthy negotiations. It was the pelvis of a small species. In addition I procured a smoked leg-bone — likewise of a small species — which from all appearances had served as a club for a long time.

Upon South Island I had better luck, and that in the very last months of my stay in New Zealand. It was upon the goldfields of Nelson on the Aorere river that I heard from diggers of a cave very recently discovered, in which the almost perfect skeleton of a colossal bird had been found, and in which, as the report went, there were still numbers of bones so strong, as to require the utmost effort to break and shatter them. I was conducted to the cave, and after a short search I had the pleasure of exhuming some fragments of bones from the loam at the bottom of the cave. I at once ordered a thorough search of the cave, leaving it to my friend and fellow-traveller, Dr. Julius Haast, and a young English surveyor, Mr. Maling, to do their best for Moa digging. My services being required on the goldfields and coalfields of the district I could not indulge in the pleasure, of forming one of the party of the Moa diggers. In the town of Collingwood, we had appointed a rendezvous after three days; at the expiration of which my friends came in triumphant, conducting oxen, decked with flowers and heavily laden with Moa bones, amid the concurse of the whole population of Collingwood. Dr. Haast had scoured three caves.[1] They occur in a tertiary limestone on the right bank of the Aorere river, about eight miles above its mouth near Washbourne Flat, a small gold-digger colony. In the northern-most cave, Stafford's Cave, — through which a rivulet flows, forming at its outlet the Doctor's Creek, — nothing was found. The more surprising was the result of the diggings in the other two caves, which Haast named

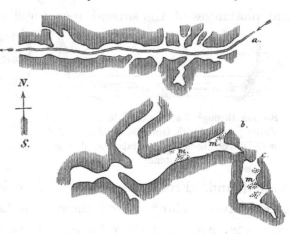

Caves with Moa-bones in the Aorere Valley.
a. Staffords-Cave; b. Hochstetter's Cave; c. Moa-Cave; m. places, where Moa-bones were found.

Hochstetter's Cave (the same, that I had visited myself) and Moa Cave. The bones lay partly quite on the surface, covered only by a few inches of loam, partly under stalactite incrustations. Very remarkable is the fact, established during the diggings in the Moa

[1] On the damp ceiling of those caves "glow-worms" are found to live, small grubs, one inch in length, enveloped in a slimy mass, which radiate from behind a phosphoric light similar to our glow-worms. A second tenant of the caves is an insect belonging to the *Homoptera*, resembling the Weta, with long feelers, hopping like a locust.

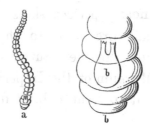

The Cave glow-worm.
a. Natural size. b. The phosphorescent part magnified.

Cave, that the remains of *Dinornis didiformis* presenting a very recent appearance were always on the top, while the bones of *Dinornis elephantopus* were dug from a deeper stratum, sometimes from under stalactite three feet thick, and in a half-fossil state like mammoth bones, so that it almost seems as though the different species of those colossal birds had not all lived contemporaneous. Nor were the bones of the various individual birds piled up pell-mell so that they might be supposed to have been carried piecemeal together; but the skeletons lay there whole, each bone in its proper place, the phalanges of the several toes together, next the legs, than the pelvis, the ribs and the breast-bone, finally the vertebral column with the skull and the bill; even the rings of the bronchial artery were in their proper place; and where the stomach had been, the "Moa stones" were found.

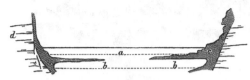

Section through the Moa-Cave in the Aorere Valley; a. Bed with Din. didiformis. b. Bed with Din. elephantopus. c. Stalactite. d. Limestone.

From these facts it is evident that the birds died in those caves, which served them as a hiding-place. Unfortunately, however, several of those bones were so rotten that they crumbled to pieces on being taken out, and notwithstanding the utmost caution used in handling them, the skeletons could not be preserved perfect.

The collections, which my friends brought to Collingwood, contained bones of ten different individual birds, belonging to six or seven species, among them also Kiwi bones. [1] This collection was increased moreover by the addition of the skeleton first found, — a nearly perfect skeleton of *Palapteryx ingens*, — which the finders had presented to the Nelson Museum. The trustees of said Museum, in their turn, destined it as a most valuable present to the Museum of the Imperial Geological Institution at Vienna, where it is at present exhibited. My friend Dr. G. Jæger has devoted

---

[1] The Provincial Government of Nelson after my departure ordered new diggings to be made in the caves of the Aorere Valley, which yielded results equally favourable.

himself to the difficult task of restoring the original skeleton and to multiplying it by plaster-casts.

The bones of this skeleton of *Palapteryx ingens*, — a species of which only rudiments were known heretofore, — are those of a young not quite fully-grown individual, since the last rib-bearing vertebra is not solidly connected with the pelvis. The height of the restored skeleton in plaster-cast, as it stands, is 6½ feet to the top of the head. This is the average height of an ostrich. A grown specimen, however, might have been taller by about ⅙th of the whole. The original bones required nearly all a more or less thorough repair before a cast could be made of them; and various wanting parts, such as the femura, had first to be modelled after corresponding parts of other individuals. The pelvis especially was in a very deficient state of preservation and has been imitated for the most part from the pelvis of *Dinornis didiformis*, which I had brought with me from North Island. Of the skull also there was only a fragment left. However, I was fortunate enough to find in the same cave, from which the respec-

Kiwi-like Moa.
Palapteryx ingens. Front of the skeleton.

tive bones had been dug, a very well preserved skull of a quite recent appearance, which probably belonged to another, older specimen, but, as was seen on comparing it with said fragment, doubtless to a specimen of the same species. Even the little bones in the auditory passage and the bony shell of the nose are preserved in this skull. Only the lower jaw had to be supplemented from fragments. This skull served as a model in the construction of the skeleton.

Without entering upon osteological details, I remark here, that the massive structure of the posterior extremities and the open pelvis, — the *ossa pubis* are not grown together, — as well as the number of toes are the most striking peculiarities, distinguishing this skeleton from that of an ostrich. *Palapteryx ingens* had a fourth hind-toe like the Kiwi, and is thereby distinguished from the Dinornis species. On the other hand, the fore-extremities are very rudimentary, so that even wings such as the ostrich has, are entirely out of question. The front-edge of the breast-bone has two small impressions, fitting to rudimentary bones, scarcely two inches long; shoulder-blade and wing-bones were probably wholly wanting.

The construction of the plaster-model was a difficult task, requiring much patience and mechanical skill. It is erected without a visible support; the iron stags passing through the leg-bones, and thus giving the skeleton that position, which the living bird had naturally to take so as to balance the bulk of its body upon its feet. The center of gravity in the body lies in the middle of the breast; therefore the hip-joint cannot

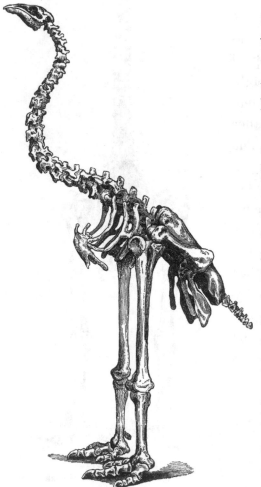

Kiwi-like Moa.
Palapteryx ingens. Side-view of the skeleton.

be the axis for the equilibrium of the bird, but this axis must pass through the centre of the body and can only rest in the knee-joints, which must be so placed, that the line connecting them, passes through the middle of the breast. If this be the case, the femura must not take a vertical or slanting position in the direction upward from the knee-joint to the hip-joint; but they must be made so as to incline slightly from the knee-joint towards the hip-joint. By this arrangement, the skeleton, of course, loses part of the imposing height, that might be given it by a more vertical position of the femura.

The whole work deserves to be designated as a master-piece, highly creditable to Dr. Jæger, and to Mr. Magniani, the artist who was engaged in executing it; a result, to which we are the more induced to tender our unfeigned acknowlegments, as by this complete plaster-cast of the skeleton of a New Zealand Moa, a lasting monument has been secured to the Novara Expedition in the numerous Museums at home and abroad.

We now arrive at the most interesting question: how, where and when did the Moas live, and what are the causes of their being extinct?

From the localities of Moa bones, hitherto discovered,[1] it appears first, that those birds were distributed over North Island, as well as South Island. Yet, as the Apteryx species of the two islands are different, so also the Moa species of North Island seem to be different from those found on South Island. Cook Strait, now separating the two islands, may have proved to these birds, which could neither fly nor swim, an unsurmountable obstacle, preventing them from migrating from one island to the other. New Zealand was perhaps a large continent when the Moas were first created. And if we suppose this or at least that the two islands were formerly contiguous to each other, we of course suppose also, that the separation took place so long a time ago, that the originally identical species, after the separation of both islands, may have been changed in course of time into the present varieties or species.

See Chapter III. p. 64.

According to Prof. Owen, the birds of South Island present stouter proportions, a compact, rather bulky frame of body, such as *Dinornis robustus, elephantopus, crassus,* and *Palapteryx ingens,* while those of North Island are distinguished by more slender and lengthy forms, like the *Dinornis giganteus* and *gracilis.*

These various species inhabited the plains and valleys and had their hiding-places in forests and caves. Their food doubtless consisted of vegetables, especially fern-roots, which they dug up with their powerful feet and claws. To assist the process of digestion, they swallowed small pebbles. According to native tradition, Moas were decked with gaudy plumage; and the present New Zealanders describe a cochin-china fowl as what they conceive to have been the shape and the appearance of Moas. The formation of the skull leads us to infer, that they were stupid, clumsy birds, which we must not suppose to have been swift runners like the ostrich, but sluggish diggers of the ground, the nature and habits of which demanded no larger scope, than such as the limited territory of New Zealand presented.

Most Moa bones still contain 10 to 30 per cent of organic (gelatinous) substance, and are not even in the state called semi-fossil.[1] Only the bones that had been lying deeper in the loam of the caves, are in a half fossil state, similar to the mammoth-bones *(Elephas primigenius)* found in Europe in post-tertiary deposits. The almost undecomposed state of the bones, their occurrence in the sands of the sea-shore, in swamps, forests, river beds, and in limestone caves, together with bones of animals still living in New Zealand,[2] all this proves beyond the shadow of a doubt, that those birds belong to the recent period of the earth, and that their age can only be counted by hundreds, instead of thousands of years. From the traditions of the natives it appears, that great numbers of Moas were still living upon the islands at the time when they were first populated, and that the last of those birds

[1] Fresh ostrich-bones usually contain $\frac{1}{3}$ organic and $\frac{2}{3}$ inorganic substances.
[2] Together with Moa bones in several places were found bones of Apteryx Notornis, Nestor, Pinguin and Albatros; likewise dog and seal-bones.

probably disappeared from the surface of the earth but a few generations ago. [1] It is even rumoured in the colony and is certainly not utterly impossible, that in unaccessible solitudes there might still be some few living strugglers of that giant-family "the last of the Mohicans". [2] However, I am not inclined to believe the stories of the natives, that Heretaunga in the vicinity of Ahuriri on the East Coast of North Island or Whakapunake on Poverty Bay is the haunt of the last living Moas; and I likewise discredit the assertions of American sailors and seal-hunters, who pretend to have seen monster-birds of 14 or 16 and even of 20 feet in height stalking to and fro on Cloudy Bay and on the inhospitable southwestern shores of South Island. And certainly it is a remarkable fact, that in those extensive, wholly uninhabited regions of the Southern Alps, which within the last years have been explored, no reliable traces [3] could be found anywhere. It is therefore my opinion, that all the larger species are wholly extinct, and that the above mentioned Roa-roa (*Apteryx maxima*) is probably the largest living representative of the former giant-family.

To the question about the causes of the dying out of those

[1] Berthold Seemann ("Viti" p. 383) says: 'Toa' is the Fijian form of the word 'Moa', applied to domestic fowls, and by the Maoris to the most gigantic extinct birds (*Dinornis*) disentombed in New Zealand. The Polynesian term for birds that fly about freely in the air is Manu or Manumanu; and the fact that the New Zealanders did not choose one of these, but the one implying domesticity and want of free locomotion in the air, would seem a proof that the New Zealand Moas were actually seen alive by the Maoris about their premises, as stated in their traditions, and have only become extinct in comparatively recent times.

[2] Of *Notornis Mantelli*, since the specimen caught in 1850 on Dusky Bay, South Island, not the slightest trace has been found anywhere.

[3] What the Nelson Examiner of January 12, 1861, relates of enormous tracks in the mountains on Blind Bay, appears to my judgment a joke rather than a serious assertion: "In June, while Messrs. Brunner and Maling, of the Survey Office, were surveying on the ranges between Riwaka and Takaka, they observed one morning the foot prints of what appeared to be a very large bird, whose track, however, was lost among the scrub and rocks. The foot-prints were 14 inches in length, with a spread of 11 inches at the points of the three toes. Similar foot-prints were seen on a subsequent morning, and as the country is full of limestone-caves, it is thought that a solitary Moa may yet be in existence. But no other trace of this *rara avis* has as yet been discovered."

gigantic birds, we must necessarily connect the question about the causes of the final extermination of other large animals of the present period. It is in the "struggle of life", that we are to seek the clue to the solution of this problem.

There are many facts, showing that in the struggle for existence, man acts the main part; that man has already swept quite a number of species from the surface of the earth, and that it is chiefly the largest animals that first succumb. We may even say, that all the larger animals are gradually being exterminated excepting those, which as domestic animals save their existence merely by their absolute dependence on man. The reasons for this are quite obvious. The animal is either useful or noxious to man. If it be a large animal, its useful or noxious qualities are the greater; and in both cases man will strive to kill the beast, either in order to secure to himself the benefits of it, or to avert the great damage. This struggle of extermination will last a longer or shorter time, according to the number of individuals engaged, or, — since in the case of large animals, it can be only comparatively small upon a given space, — in proportion to the greater or smaller area of distribution of the animals in question. The huge animals once populating the forests of Europe, furnish a great many examples and proofs, too well known to require any further explanation. I will mention only two facts to show, how rapidly often the struggle is brought to a close with species having only a very limited range of distribution, how little there remains of such animals exterminated by the hand of man, and how fast every thing relating thereto is forgotten.

About the middle of the eighteenth century, during Behring's second voyage, in 1741, the Zoologist Steller discovered on the coast of Behring's Island near Kamtshatka a colossal seacow, the *Rhytina Stelleri*, great numbers of which lived on that coast. Its body weighed 80 cwts.; and the savonry meat and the lard being great inducements to give it chase, as early as 1768 already the last specimen is said to have been killed. Consequently twenty-seven short years sufficed to sweep the last trace of that animal from the earth. Notwithstanding the general and diligent search for it, and the considerable rewards offered, not a vestige has since been discovered of this animal; man has destroyed it; a jaw and a skull-fragment preserved in the Petersburg Museum are the only relics left us, and but for the description of the zoologist Steller who was shipwrecked on the coast of Kamshatka, we should be utterly in the dark as to the animal in question. [1] But if we were to ask the seal-hunters and whale-fishers on those coasts, they would certainly reply that they had never heard of such an animal.

Almost as quickly was the Dronte or Dodo exterminated. The sailors of the Dutch Admiral Wybrand of Warwyck, whose vessel was stranded upon the coast of Mauritius, in 1598, commenced the war of extermination. Although they were utterly disgusted with the meat, — they, therefore, called those birds Walgh-birds, i. e. loathsome birds, — and although the whole crew, half-famished as they were, could not consume over two birds at one meal, yet the stupid, clumsy creatures were killed by scores. Already in 1607, the merchant Paulus van Soldt reported that the number of those birds was greatly decreasing on the coast; his crew also, subsisted for 23 days on nothing but drontes and some few tortoises. In 1681 the bird is mentioned for the last time, and the researches instituted on the spot by Bory St. Vincent in the beginning of the present century have shown, that upon that island the memory of those marvellous birds had entirely disappeared even from tradition.

Nor is it to be doubted, that the extermination of the gigantic birds of New Zealand was chiefly accomplished by the hand of man. In briefly retracing the past to the times when New Zealand was not yet trodden by the foot of man, we must assume, that at that time the large Dinornis and Apteryx species, whose bones we find to-day, lived in great numbers upon open fern-land, subsisting on the roots of *Pteris esculenta*. Dr. J. Haast notices

[1] But very recently, — a year or two ago, — some more complete remains were exhumed.

also the occurrence of bones of the Dinornis in the moraines of the glaciers of South Island, and observes that the present Alpine flora furnished a large quantity of nutritious food quite capable of sustaining the life even of so large a creature; and as the fruits of these plants seem at present to serve no evident purpose in the economy of nature, he argued the former existence of an adequate amount of animal life, to prevent an excessive development of vegetation. This part was played by the Dinornis.

Those huge birds were then the only large animal beings that populated New Zealand; for of indigenous mammalia, except a little rat, there is nothing known. The first immigrants, [1] who throughout the whole length and breadth of the extensive forests found nothing for man to subsist on, except the native rat and some small birds, obtained from the giant-birds the necessary supplies of meat, enabling them to increase in course of time to a whole nation numbering hundreds of thousands. But for those colossal birds, it would be indeed utterly impossible to comprehend, how 200,000 or 300,000 human beings could have lived in New Zealand, a country which even in its vegetable world offered nothing for subsistence, except fern-roots.

That such was really the case is sufficiently proven in the traditions of the natives. Ngahue, one of the discoverers of New Zealand, — so tradition says, — describes the land as the haunt of colossal birds. There are yet some Maori poems extant, in which the father gives his son instructions how to behave in the contests with the Moas, how to hunt and kill them. [2] The feasts

[1] According to what is said in Chap. X., there are no other points to go by for fixing the time of the first immigration to New Zealand. Nor are we better informed as to whether another tribe had inhabited the islands previous to the Maoris. In 1862, on the Pararua road near Wellington, peculiar utensils were found, under the roots of a Totara trunk wholly different from those commonly used by the Maoris of to-day: especially round sinkers for fishing-lines, of crystalline lime with a hole in the middle; also cutting-tools of a peculiar quartz-stone, which is only found in the Wairau, on South Island. These facts have been of late adduced as proofs for the existence of a tribe, that inhabited South Island previous to the Maoris, and thence had come over to Wellington Harbour.

[2] The birds being unable to swim, were driven towards a river or a lake, or towards

are described, which were wont to be instituted after a successful chase. Mr. Cormack as well as Mr. Mantell have found the bones on both the North and the South Islands in great number in the vicinity of camping-grounds and fire-places of the natives. Mounds were found full of such bones, in which after great feasts the remnants of the meals were promiscuously interred. The flesh and eggs were eaten; the feathers were employed as ornament for the hair; the skulls were used for holding tattooingpowder; the bones were converted into fish-hooks, and the colossal eggs were buried with the dead as provision during their long last journey to the lower regions.

Consequently those huge birds were in former times the principal game of the natives, and were probably altogether exterminated in the course of a few centuries. They succumbed, — the larger the species, the sooner, — to the same fate, that is gradually sweeping the Kiwi, the Kakapo and the rat Kiore[1] in a similar manner, and before our eyes, from the face of the land.[2]

the sea-coast, until they could escape no further, and thus slain. That also small heated stones were thrown into their way, which they swallowed and of which they died, is probable a mere Maori fable.

[1] This indigenous rat was so scarce already at the time of the arrival of the first Europeans, that a chief, on observing the large European rats on board one of the vessels, entreated the captain, to let those rats run ashore, and thus enable the raising of some new and larger game.

[2] Dr. Thomson believes, that the Moas have become extinct since the middle of the 17th century. Meurant, a seal-hunter, according to a communication of the Rev. Mr. Taylor (New Zealand Magaz., April 1850), asserts his having seen Moa bones with the flesh on in Molyneux Harbour, South Island, as late as 1823. At any rate, natural phenomena such as volcanic eruptions, conflagrations of woods and heaths are likewise very probable to have contributed to the diminishing of the Moa family. In the swamp near Waikouaiti in South Island, Moa feet and legs have been found in an erect position, and the extraordinary number of Moa bones found in swamps is probably to be explained in this manner, that large flocks of those birds driven by fire or by men, got lost in the swamps and perished there. Dr. Haast very recently had the good fortune to make a most extraordinary discovery of that kind. A swamp near the Glenmark home station (Province Canterbury) has long been celebrated for the quantity of Moa bones that have been found there. Dr. Haast found no less than twenty-five skeletons of the *Dinornis elephantopus* and *Dinornis crassus*, of different ages. The bones were in excellent preservation and perfect condition. They retain the usual proportion of animal matter, and have

But what next? The Maoris had increased to a very numerous people, the Moas were exterminated; whence were the natives thenceforth to get their animal food? This question leads us to the cause and the origin of the terrible cannibalism, that held its sway of terror over New Zealand, when towards the close of the last century the first Europeans landed on its coast. What else is there that could induce a human being to devour his own kindred, but want and starvation? There is no other reasonable way of explanation for an act, which is so abhorrent to nature; that it occurs even with animals only in exceptional cases when want compels them to take to it as a last resource. It was not barbarity, not savage cruelty, not monstrous heathenism, that drove the uncivilized man of the South Sea so far, as to drink his fellow's blood, and eat his flesh; the cannibalism of the South Sea Islanders is to be accounted for in no other way, than the cannibalism of the civilized European, when, shipwrecked, and on the point of starvation, he lays hold of his ill-fated fellow. Cannibalism also, is but one of the manifold forms of the struggle of life.

It is thus alone that we are able to explain, why the history of the past century of New Zealand is but a terrible tale of war and carnage and horrid cannibalism, and why this unnatural state was put an end to within less than twenty years, when by the importation of swine and potatoes on the part of Europeans new means of subsistence had been placed within the natives' reach.

Cannibalism has ceased, as it began; but not so the struggle for existence. This has again assumed a new form. From the struggle with the animal world the native, as the stronger, had come out triumphant. But now the tawny South Sea Islander has to wrestle for his existence with "the pale faces", and there is no doubt, for whom the dooming die is cast in this contest. I am

undergone no mineral change. It is evident from these and similar discoveries that the birds of the *elephantopus* and *crassus* species congregated together in flocks, while the more monstrous specimen known as *Dinornis giganteus* must have been a comparatively solitary bird, as the bones of this class are scarce, and never found in any numbers in one spot.

speaking here not only of the open, bloody war between the natives and the English, but of the struggle for existence, as it is carried out between man and man in all those innumerable circumstances, which are adduced as reasons, why in all parts of the world, in America, in Australia, upon Tasmania and at the Cape of Good Hope, as well as in New Zealand, the natives are continually growing fewer and gradually dying out.

In the vegetable and animal worlds, and among mankind this struggle is carried through according to unchangeable laws; — among mankind not only between tribes of different races, but in the same manner between natives of the same race, between states and states, between families and families, between individuals and individuals.

What may be a consolation amid this ever lasting struggle, is that we know it to be a law of nature, on which the development of all creatures depends, that this struggle is not only a destructive one, but in the same measure a preserving and creating one. None but the weaker, the inferior, perish; the stronger and nobler element remains victorious. Thus every progress in the world depends on this struggle for existence, and as far as man is concerned in it, we may above all be consoled by the fact, that it is not physical force, which decides the issue, but moral power and mental superiority!

Appendix.

Synoptical table, comparing the size of the foot, and the vertical height of several Moa-species with the African Ostrich, the Australian Emu, and the New Zealand Kiwi.

Name of Species.	Femur		Tibia		Metatarsus		Vertical height in feet.[1]	Remarks.
	Length in Inches and Lines.	Least circumfrence of the shaft.	Lenght	Least circumference of the shaft.	Lenght.	Least circumference of the shaft.		
Dinorn. giganteus	16″ 0‴	7″ 3‴	35″ 0‴	6″ 6‴	18″ 6‴	5″ 6‴	9′ to 10′	From North Island
Dinorn. robustus	14 2	7 10	32 3	6 9	15 9	5 3	8 to 9	From South Island
Palapteryx ingens	12 7	6 5	30 0	5 7	12 4	4 9	6 to 7	From South Island according to the skeleton in Vienna.
Din. struthioides	11 0	4 2	25 0	5 0	12 0	4 3	6	From North Island
Din. elephantopus	13 0	7 9	24 0	6 5	9 3	6 6	5	From South Island according to the skeleton exhibited in the British Museum.
Din. crassus	11 10	6 0	19 6	4 10	8 8	4 6		
Palapt. dromioides	9 6	4 0	21 0	4 0	10 5	3 9	4½	From South Island
Din. didiformis	a) 8 0	4 0	16 3	4 1	7 0	3 6	4½—5	From North Island
	b) 10 0	3 7	15 0	3 6	7 2	3 0	3½—4	a) From North Island b) From South Island according to a skeleton in the Novara Museum in Vienna.
Struthio camelus (Afric. Ostrich)	a) 11 0	5 3	18 6	4 3	16 0	3 7	6—7	
	b) 11 6	5 3	21 2	3 7	18 7	2 10		b) According to a skeleton in the Imp. Cabinet of Natural History at Vienna.
Dromæus Novæ Hollandiæ (Emu)	9 0	3 7	16 10	3 4	15 0	3 0	5	
Apteryx Mantelli (Kiwi).	a) 3 9	1 0	5 3	1 3	3 3	1 0	1—1½	a) From North Island
	b) 3 9	1 2	5 2	1 0	2 8	1 1		b) According to a skeleton in the Imp. Cabinet of Natural History in Vienna.

[1] The vertical height varies according to the position of the neck and foot. The figures given are the average for a natural, erect position.

CHAPTER X.

The Maoris.

Two races on the Islands of the Pacific Ocean. — Melanese and Malay-Polynesian race. — Difference between Micronesians and Polynesians proper. — The Maoris true Polynesians. Traditions and Mythology. — The Ika a Maui. — The legend of the migration from Hawaiki not historical, but mythical. — The Maui-myths are myths of the Sun. — South Island populated from North Island. — The Maeroes and Ngatimamoes not aborigines but wild Maoris. — The Maoris at the time of the discovery of New Zealand. — Cannibalism. — Origin of cannibalism in New Zealand. — The Maoris of the present day. — Injurious influences of European civilization and colonization upon the natives. — The Maoris are dying out. — Census of 1858. — Appendix, the story of Te Uira, chief of the Ngatimamoes.

Two human races differing widely in physical and mental qualities, in language, manners and customs, inhabit the islands scattered over the Pacific Ocean. One race of very dark complexion, almost black, of ungainly make, of an extremely low grade of mental faculties, savage and for the most part incapable of civilization, occupies the southwestern part, comprising New Guinea, New Ireland, Louisiade, Solomon Islands, Nitendy, the New Hebrides, New Caledonia with the Loyalty Islands and the Archipelago of the Fejee Islands. They are generally designated as Melanesians or Papuas. Their tribes are nearest akin to the aborigines of Australia and Tasmania, and like these seem to be the remnants of a very ancient race possibly the oldest branch of the human family, once occupying larger territories, but for a long time past, and up to this present age, oppressed and supplanted by more civilized, and more gifted nations.

The second race, of a lighter complexion, in all the various shades of brown, of an admirably regular make, ranks much higher. A characteristic feature in them is their partiality for the sea. They are chiefly distributed over the islands extending from the Pelew group in the West as far as Easter Island East. Among their numerous places of abode there are only two territories of a somewhat more extensive area, the Sandwich Islands to the North and New Zealand in the South. They are nearest akin to the

Aborigines of Australia, man and woman, from Murray-River (Colony of Victoria).

numerous Malay tribes inhabiting the peninsula of Malacca, Borneo, Celebes, the Sunda Isles, the Moluccas and the Philippines, and are generally designated as the Malay Polynesian race. But this race has been again subdivided into Mikronesians and Polynesians proper. The former comprise the western portion of that range of islands from the Kingsmill group to the Pelew Islands; the Polynesians the eastern half from the Samoa and Tonga Islands as far as Easter Island, together with the Sandwich Islands and New Zealand. The former have a complexion somewhat darker than the latter; their several tribes speak different languages; and in their stature, their manners and habits they are more akin to the Malay race than to the real Polynesians. The latter, compared with the Mikronesians, are a separate, far more characteristic, and exclusive family of their own.

Nothwithstanding the great distance separating the several tribes, they speak essentially the same tongue, although divided into many dialects. They have all one and the same legend concerning the creation of the world by Maui; and the same system of Tapu (or Tabu), the custom of canonizing persons and things, and declaring it sacrilege to touch them; they are also alike in complexion and structure of the body. The representatives of this race on Society Islands (Tahiti), and Friendly Islands (Tongatabu) are noted for their handsome exterior as well as engaging manners. They bear a striking resemblance to the Caucasian race in features

Natives of New Zealand.

and mental endowments; they entertain commercial intercourse with Europeans, they easily and readily adopt European manners, and consequently rank much higher in civilization than the Melanesians and Mikronesians.

To these Polynesians proper the natives of New Zealand belong; they are moreover the most important family of the Polynesian race both as to number, and mental and physical faculties

It may be owing to the temperate climate of New Zealand, its extensive area in comparison with the other islands, to the manifold features of its soil, and above all to the indispensability of labour in a country ill adapted by nature for a life of ease and idyllic pleasure, that the qualities of the great gipsy-race of the South Sea, the Polynesians, upon New Zealand have been developed to the highest degree they can attain.

The New Zealanders style themselves Maoris, and consider all the other oceanic races as far beneath them. Whether the word maori may be derived from the root uri, dark, or considered synonymous with moor (negro) I leave to others to determine. With the natives, maori simply signifies indigenous or peculiar to New Zealand.[1] Tangata maori is the native in contradistinction to Tangata pakeha, the stranger; wai maori signifies common or fresh water, and wai pakeha the spirituous beverages of the Europeans.

The origin of the Maoris is shrouded in profound obscurity. Mysterious legends and traditions seem to convey vague historical reminiscences. The history of the creation of New Zealand is given in the form of a fisherman's legend, corresponding well with the notions of a people, whose abode is surrounded by the sea, and whose chief occupation was fishing. The natives call the Northern Island "Te Ika a Maui", the fish of Maui.[2] Maui according to their

---

[1] The same word occurs in the language of the natives of other Polynesian Islands. Upon Mangarewa and Hawaii *maoi* signifies native, indigenous, and upon Tahiti *rai mauri* means fresh water. C. Schirren (die Wandersagen der Neuseeländer p. 48) has justly proposed applying the word as comprising all the Polynesian tribes. At all events all the other Polynesian tribes belong to one and the same race with the Maoris of New Zealand, which, therefore, may be properly named after the most populous and most important tribe, the Maori race.

[2] The Maori name for South Island is "Te Wahi Punamu". It owes this name to a mineral found there, to the Nephrite, or Jade of the mineralogists; by the colonists called "green-stone", by the natives "punamu". This stone is greatly prized by the natives, who make ear-rings, neck-laces and battle-axes *(Mere)* of it, and numerous varieties of it are distinguished by hardness, colour and transparency. It is found on the West Coast of South Island among the boulders of the sea shore and the shingle of the rivers. The natives used to make expeditions to South Island for the purpose of gathering punamu, and hence the name "tewahi punamu", literally signifying the place or the land of green-stone, may have been used to denote the the whole of South Island. The usual, but incorrect, way of spelling the word is

traditions is a hero, as it were, the Hercules of their mythology, who achieved great many wonderful exploits. He was their first instructor in boat and house building, the inventor of the art of twisting flax into cords and snares; he killed the sea-monster Tunarua; he is lord of fire and water, and likewise of air and sky; among the deities and spirites he is so to say the national god. To Maui is due the honour of fishing up the land out of the ocean; hence its name the fish of Maui. His hook on this occasion was the jaw-bone of one of his ancestors. The fish was hardly above water, before the brothers of Maui fell upon it, to cut it up. Thence originate the mountains and valleys and all the irregularities on the surface of the country.

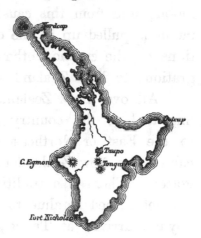

Te Ika a Maui.

Strange to say, the outlines of North Island actually resemble the form of a fish, and the natives even designate the localities corresponding with the respective parts of the fish. The southern portion is the head, the northwestern part the tail; Cape Egmont the back-fin, East Cape the lower fin. Wanganui a-te-ra (Port Nicholson on Cook Strait), they say, is the salt-water eye of the fish, Wairarapa (a fresh-water lake near Wellington), the sweet-water eye. Rongorongo (the North Coast of Port Nicholson) is the upper gill, Te Rimurapa (the South Coast) the lower gill; the active volcano Tongariro in the centre of the island, and Lake Taupo at its foot are, according to their notions, the stomach and the belly of the fish. Certainly an interesting proof

Te wai punamu, which signifies "green-stone water". The Maori name for Stewart's Island is Rakiura, compounded of raki, dry, and ura, fine weather, or brillant sunrise and sunset. Perhaps the natives thus designated the small, southern-most isle, because to the inhabitants of North Island the wind, which brings fine, clear weather with a bright morning and evening-sky, blows from the South. Strange to say, however, the natives have no general name for the whole of New Zealand; they employ the European word New Zealand, which in the Maori pronunciation becomes *Nuitireni*. Sometimes they say also Nuitereni or Niutereni.

of the correct knowledge the natives had arrived at regarding the form of the island, long before its outlines were represented on a map.

The legend of Maui's fishing exploit is only one of the many traditions about Maui which constitute the Maui mythos, as it is disseminated over the Tonga, Samoa, Tahiti, and Hawaii groups. According to an other tradition a pigeon, into which Maui put his spirit, flew to heaven with a line in its beak and assisted in elevating the land above the water; and Maui tied the sun to the earth, with the ropes which have since become the sun's rays. As Maui could not prevent the sun going down, he tied it to the moon, and from this cause it results, that when the sun sets, the moon is pulled up at the other side of the earth. To these traditions on the creation others about the discovery of, and the immigration, to New Zealand are attached.

All over New Zealand, the natives state that their ancestors migrated to the country from a place called Hawaiki, situated to the East or Northeast of New Zealand.[1] The motives which caused the New Zealanders to migrate from Hawaiki, are not forgotten. There is a tradition that a civil war in Hawaiki caused a chief named Ngahue to flee from the country, who after a long voyage arrived at Tuhua, an island on the East Coast of North Island. Thinking himself pursued by the enemy, he continued his flight as far as Aotearoa.[2] In order to be perfectly safe he proceeded along the coast as far as Arahura (on the West Coast of South Island), where he discovered the highly prized punamu-stone. On his return voyage he touched Wairere, Tauranga, Whangaparoa, points on the East Coast of North Island, and thence he sailed straightway to Hawaiki, bearing the tidings of the glorious country inhabited only by the gigantic bird Moa, and abounding

[1] Allusion is also made by the natives in their traditions on this subject to distant and a larger Hawaiki, and a nearer or smaller Hawaiki.

[2] Aotearoa is a second mythical Maori name for North Island, which is sometimes applied to the whole New Zealand group. Aotea in the first place is the name of one of the canoes, which, as the legend goes, came from Hawaiki, and is equivalent to glaring light, sun-light; roa signifies long, big, large, hence aotea-roa big glaring light.

with green-stone. Strife had not ceased, when Ngahue returned to Hawaiki, and the weaker party, in order to save their lives, determined to migrate to Aotearoa, the newly discovered land.

Other traditions make Kupe the Columbus of the country. [1] His younger brother Hoturapa had carried off his wife Kura Marotina, and Kupe put to sea in the canoe Mataorua in pursuit of the fugitives. After a long voyage he landed at Wanganui a-te-ra (Port Nicholson on Cook Strait), he proceeded through Cook Strait, arrived at Patea; but not finding a single soul anywhere he returned to Hawaiki, where several canoes were now built and fitted out for a voyage to the new country. The canoes were named: Arawa (said to have been a double canoe), Tainui, Aotea, Mataatua, Takitumu, Kurahaupo (alias Kura aupo), Orouta, Pangatoru, Tokomaru, Motumotu ahi, Te Rangi ua mutu, Whaka ringa ringa. Each of these canoes has its own tradition, and its own wandering heros like the Argonauts of old. The people seem to have preserved these first records of their history so faithfully and scrupulously, that in relating these traditions they display a vivid recollection of the very particulars of the voyage, of the names and adventures of the individual leaders, of the places where they first landed and settled, as well as of the various plants and animals which were brought from Hawaiki at the time. [2] The ancestral records of the different tribes are most carefully traced down to the present generation, and to this day the natives show to the traveller the supposed remains of the canoes of their ancestors. [3]

---

[1] See Taylor, Te Ika a Mani, p. 116.

[2] According to these traditions the *Kumara* or sweet potato (*Convolvulus Batata*), the *taro* (*Arum esculentum*), the calabash-plant *Hue* (*Lagenaria vulgaris*), the Karaki tree (*Corynocarpus lævigata*), the rat *Kiore*, the *Pukeko* (*Porphyrio*), and the green parrot *Kakariki* are said to have been imported from Hawaiki. In fact, all these plants and animals still existing in New Zealand deviate strikingly from the natural character of the flora and fauna of the country, and are evidently imported from the tropics.

[3] A rock with its peak projecting from the sandy downs of Kawhia Harbour is pointed by the dwellers on that harbour as the remains of the Tainui-canoe, from the intimates of which they date their origin. In Maketu (Eastcoast of North Island), — so tradition says, — the canoe Arawa lay "high and dry"; but it was burnt, and now there is nothing to be seen but the stone-anchor. Likewise at the

All this seems to justify the supposition that the traditions of the migration of the Maoris contain historical reminiscences; and much pain has been taken to ascertain from the genealogical trees, from the number of ancestors registered, the approximative time of the first immigration. Most writers on New Zealand are of opinion, that the ancestral records prove about eighteen to twenty generations, and that consequently the arrival of the first settlers from Hawaiki took place 500 or 600 years ago, somewhere about the year 1300 A. C., and that their number may have amounted to about 800 souls. As to the situation of Hawaiki, the inquirers differ in opinion, because there are several islands in the Pacific Ocean, to which the term Hawaiki bears a strong resemblance; first of all Hawai, the well-known island of the Sandwich group. Mr. H. Hale, however, the ethnologist of the United States Exploring Expedition under Captain Wilkes, is of opinion, that Sawaii [1] in the Samoa group (Navigator Islands) had a far stronger claim to be the Hawaiki of the tradition, and that from this central island all the other Polynesian Islands were peopled. He traces the first immigration to New Zealand and Tahiti as far back as 3000 years, whereas the Hawaii of the Sandwich Islands is said not to have been populated before about the year 450 A. C. Hence according to one hypothesis New Zealand was first populated from the Sandwich Islands 1300 A. C., and according to the other, from the Samoa Islands about the year 1300 B. C.

mouth of the Mokau the stone-anchor of the canoe Tokomaru is to be seen in the cliff Punga-a-Matori.

[1] Sawaii and Hawaii are identical. The Maoris so pronounce the H at the beginning of a word, as hardly to be distinguished from S or Sh. I merely refer here to the different ways of spelling some well-known New Zealand names, Hongi or Shongi; Hauraki or Shauraki; Hokianga or Shokianga. The dialect of the Samoa Islanders deviates so little from the Maori language, that people from Sawaii, who occasionally come to New Zealand with whale-fishers understand the Maori without difficulty. There is moreover upon Sawaii a place, Mata Atua, of quite the same name as that of one of the canoes mentioned in the tradition. All these circumstances tend to support Hall's views. Some other names might, however, be adduced as synonymous with Hawaiki, such as *Avai*, with the adjective *poere* = "dark night", a spot on the eastern peninsula of Tahiti; *Habai*, an island North of Tonga; *Hunga tonga habai*, also in the Tonga-group.

But we may well ask, are these migration-legends not as mythical as those on the creation of New Zealand? In my opinion, Mr. C. Schirren has adopted the only correct method of interpreting those traditions. In a very able treatise,[1] in which the traditions of the New Zealanders are most ingeniously analyzed, Schirren shows that we must not search for historical truth in those traditions, as we thereby involve ourselves in a labyrinth of mythical fancies, out of which only the thread of mythical analysis can point the way to light. I cannot refrain from quoting here the principal conclusions at which Schirren has arrived; especially because the work of that learned gentleman seems to be hardly known by English writers on New Zealand, and because I believe that an erroneous opinion generally prevailing, is thoroughly disproved in it. Hawaiki etymologically means lying beneath. According to Schirren, it is not originally the name of an island, and has not a geographical, but a mythical signification. It denotes the lower regions, the realms of the dead.[2] In this sense, according to the Polynesians, Hawaiki is the beginning and the end, the place whence their fathers came, and to which the souls of the departed return.[3] But if thus the pretended home is stripped of its claims to reality, the migration legends will also prove to be not facts, but fables. Just as Maui, the God of the lower regions, and at

---

[1] C. Schirren, Die Wandersagen der Neu-Seeländer und der Mauimythos. Riga 1856.

[2] Hawaiki in this sense is synonymous with Rarotonga and Raiatea.

[3] *Reinga,* on the North Cape of North Island, according to the notions of the Maoris, denotes the earthly portal, the entrance to the realms of the dead. On the margin of the cliff there is a cave. Through it the spirits descend; then they ascend a hill and finally on the spirits path, *Rerenga wairu,* they arrive at a lake. A canoe conveys them over to Hawaiki. Fleet as flitting shadows, ever escaping the empty grasp, they glide along towards their final home. At night especially, after heavy battles, the dwellers on North Cape hear the flight of the spirits as they rustle through the air. Chiefs ascend to heaven first, there leaving their left eye as a twinkling star; thence they proceed to Reinga. The path of the spirits is the same for all. An old Pohutukaua-tree (*Metrosideros tomentosa*) sends its branches down. These are the ladder for the dead. The Maoris fear, that if a white man were to cut those branches through, the road to eternity would be destroyed for ever, and the island annihilated.

the same time the first man, lord of water, air and sky, raised the earth out of Hawaiki, so also all the first immigrants hailed from Hawaiki. Maui, the god, is the prototype of the migrating heroes, whom we may regard as humanized gods or deified men. Notwithstanding the change of scenes and names, there is a certain uniformity, a repetition of stereotype personal transactions, such as abductions, persecutions, open feuds and treacherous stratagems in all the stories of migration. The adventures and incidents of the heroes are traceable to natural phenomena, and their great number to the number of the different New Zealand tribes. Each tribe endeavours to trace his descent to one of those mythical heroes, and thereby to establish their special claim to this or that country. The traditions of the migration of the New Zealanders are nothing but versions of the Maui mythos. Thus every single link is removed that might aid in tracing the descent of the New Zealanders to the immigration from this or that South Sea Island, and the only question is whether traces may not be found in the Maui myths, from which conclusions may be drawn as to the origin of the Maoris.

Schirren proves that the Maui myths, varying in individual features, but identical in the main, exist throughout Polynesia. Maui, as god of the atmosphere and lord of the deep, as god of the creation in heaven and on earth, is identical with the cosmogonic supreme deities of other Polynesian Islands; [1] Maui represents the national deity of all the Polynesian tribes. Hence the Maui mythos proves, as irrefutably as the conformity of the language, the original unity of the Polynesian race. As this race is now dispersed over islands far remote from each other, migrations must have of course taken place. Concerning this there can be no doubt, and indeed to this very day the Polynesians — true gipsies of the sea — are distinguished by an innate love of wandering. [2]

---

[1] For example with Tangaroa (Tangaloa in Tonga, Taaroa in Tahiti), Tiki and Atea.

[2] It is a wellknown fact, how willing the inhabitants of Tahiti and New Zealand have been, ever since Cook's times, to accompany sea-farers on their voyages.

However, when, and in what direction those migrations took place, and which was the original starting point of the whole race, are problems for the solution of which we are wholly without even the slightest link of evidence.

America or Asia have been thought of; the former existence of a continent in the South Sea has been also suggested, which may have served as the natural highway for the migrations of nations in olden times, but being vent to pieces by violent disturbances of the earth, has only left its highest points projecting from the water, forming the numerous islands of the South Sea. But in both directions, to the East and West, the connecting links are wanting. It is worthy of remark that neither in the social habits of the Polynesians, nor in the original mode of their government, is the slightest trace of influence from abroad, or of an intermixture with other nations to be observed. In vain we search for foreign elements in their language, for a connection of the Maori tongue with the Malay language;[1] nor are foreign ingredients recognizable in the Maui mythos. Schirren, therefore, traces the whole cycle of traditions back to the worship of elementary spirits, such as must be common to all nations on their earliest conceptions of the existence of God, an original suggestion arising independently in every nation, and indeed within the bosom of every individual man; to a simple worship of the sun, dressed in the concrete garb of legends and fables created by the brilliant and vivid fancy of an original and childlike people. To the child of nature, the sun, the earth, the ocean, the air are not mere elements; to him they are personifications, with whom he is in constant intercourse, whose favour he entreats, whose anger he is scrupulously anxious to avert. Thus the mythology of the Poly-

[1] Dr. Fr. Müller (Reise der österreichischen Fregatte Novara, linguistischer Theil 1867) remarks that in the Polynese languages the character of a primitive language is preserved, and that the southern part of the asiatic continent together with the circumjacent islands are to be looked upon as the home of a primitive population, which spread over the South Sea and out of which the different races developed themselves, which we now distinguish as the Malay, Melanese, Micronese and Polynese race.

nesian tribes so fully corresponds with their nature, their manners and their language, that the same cannot be interpreted as the issue of a remoter and older system, but must be considered as original to them. The Polynesians, we must conclude, had their own sphere of creation, and science has at yet not been able to reduce the different spheres of creation to one common centre.

Equally obscure as the origin of the Maoris is their history, after they had once established themselves in New Zealand. It is only certain, that North Island was first populated, and that South Island was colonized by degrees from the North.[1] Previous to the immigration of the Maoris, New Zealand seems to have been totally uninhabited. The inhabitants of North Island, it is true, tell of savages with long hair, long fingers and nails, who eat their food raw, and who are supposed to live in the most inaccessible ravines and forests of the Tararua range. These savages who are supposed to be the last of the

A cannibal of olden times.

[1] In Cook's times the whole population was estimated at 100,000, in 1859 it only amounted to about 56,000, and of this number 53,000 fall to North Island and only 2283 to South Island. Only within the latest period, in 1838, the Chatham Islands, also, were populated from New Zealand in consequence of the emigration of a portion of the Ngatiawa tribe, who by war were forced to leave their home on Cook Strait, and who were conveyed over to those islands by a European Captain. In 1859 the number of Chatham Island Maoris was estimated at 500.

aborigines of New Zealand, are called Maeros. In like manner the natives of South Island tell of the Ngatimamoes, as savages living in the mountains. However both the Maeros and Ngatimamoes are probably degenerated Maori tribes, which have been driven back into the mountains.[1] The first historical account of the natives of New Zealand we have from Tasman. He describes them as a brown-skinned race of gigantic stature, their jet-black hair tucked up in a bob behind, after the Japanese fashion, and in it a large white feather. They hailed Tasman with loud voices and when a boat was sent on shore, they furiously attacked the crew, killed three men and mortally wounded a fourth. This first hostility at once branded the New Zealanders as a most savage and barbarous race; and when subsequently other navigators disclosed their blood-thirsty cannibalism, the Maoris were regarded in Europe as monsters of human beings with absolute abhorrence. The name "Massacre Bay," by which Tasman cursed the spot where he had anchored, was the first European name on the coasts of New Zealand!

In how different a light does all that appear now-a-days! European towns and settlements are to be seen thriving and flourishing on the distant coasts of New Zealand; "Massacre Bay" is changed into "Golden Bay", and for many years European settlers have lived in peace with the cannibals of old. We know now their language, their manners and customs, and we have found them to be a people, whose qualities remind us of the ancient Germans, as Tacitus describes them; whose dauntless courage in their struggles against European immigration and civilization excites our admiration, and in whose fate we take a lively interest.

The European settlers found the New Zealanders in a state of civilization, scarcely to be expected from so-called savages. The Maoris lived together in villages; their huts, constructed of wood and reed-work, were ornamented with ingenious wood-carvings and painted with gay-coloured arabesques, and we are justly astonished on reflecting, that those wood-works were executed only

---

[1] See Appendix to Ch. X.

with chisels and axes of stone, and that in order to obtain a single board a whole log had to be worked off. They were in the habit of tattooing their faces and bodies. The villages were protected by ditches and palisades and surrounded by extensive plantations, on which sweet potatoes, taro, and melons were cultivated. Besides agriculture, fishing and hunting were the chief occupation of the people, by which they made their living. Fishes, molluscs, birds, rats, dogs, seals, fern-roots *(Pteris esculenta)*, and woodberries, besides the produce of the fields, were the staple articles of life. The Maoris were exceedingly well skilled in preparing the fibre of the flaxplant *(Phormium tenax)* peculiar to their country, and in plaiting and weaving mats and garments out of it. Kaitaka, Korowai, Wakawae, Kotikoti etc., are names of the different mats made out of flax. For dyeing the flax they used various kinds of bark and roots, and as ornament the feathers of sea- and land-birds. From dog-skins were manufactured also cloaks of a high value. The natives had no written language; but the numerous legends, fables, songs and proverbs of the people were transmitted by oral tradition from generation to generation. They knew precisely every plant, bird, insect of the country they inhabited, and knew how to designate them by special names; and even the various kinds of rock they distinguished with a keen talent of observation.

The religious belief of the New Zealanders was that which belongs to the infancy of a race. Their religion was a kind of polytheism, a worship of elementary spirits and deified ancestors; yet without idols and temples. They believed in a future state of existence, and that there was a spirit within their bodies, which never died.

They were divided into a great number of separate tribes, which were ruled by families of chiefs after a patriarchal system. Moreover there existed closely defined classes in six ranks, from that of the first priest and chief down to the slave. The chiefs lived in polygamy. The property was common to the whole tribe. But the several tribes were engaged in continual feuds with each other, generating an uncommonly warlike national spirit. Bravery and cunning, command of temper, revenging injuries and hereditary feuds, suffering torture without complaint were considered the principal qualities of chiefs virtue.

The education of the youth reminds us of Spartan discipline. The boy belonged to the tribe rather than to the father. Corporal punishment was of rare occurrence. The boy was not to become cowardly and submissive, but brave and independent. The youth grew up amid games, dances and wrestling. Boys had to learn the art of catching birds by creeping up to them, of angling fish, of setting traps and snares for rats. The son of the chief had to learn the traditions, laws and rites of the people; he had to be orator and poet, statesman and warrior, farmer and seaman, hunter and fisherman, — all in one person, if he wished to occupy in future years a position suited to his rank, and to do honour to his name.

Only one gloomy feature spread over the life of that people, casting into shadow all the other bright sides; it was cannibalism. At the time of the discovery of New Zealand by Europeans it was more prevalent there than anywhere else, so that the Maoris, especially when even Europeans became their victims, were considered the real type of cannibals. And yet, this was not always the case. From the traditions of the people it appears, that cannibalism did not exist upon New Zealand till long after the immigration of the Maoris, and it almost seems, as though at the very time of the discovery of New Zealand the anthropophagism had reached its *maximum*.[1] But its origin is as mysterious, as all the former history of the people. In speculating on the origin of cannibalism in New Zealand, it is requisite to remember that by the increasing population on the islands the resources of animal

---

[1] The chiefs of the Bay of Islands answered the missionaries, as these were lamenting about the monstrous custom of man-eating: "Large fishes eat up the small ones; dogs eat men; men dogs; dogs each other; birds each other; one God another." The women, however, were allowed the luxury of human flesh only in exceptional cases.

food became scarcer by and bye, so that about new hunting-grounds, good arable land, and productive fishing-places, hostilities arose which afterwards led to open war. By these wars the spirit of the people grew wild, agricultural labours were neglected, want came upon them, and starvation, superstition, revenge, hatred and other motives may in time of war have led to the first cases of cannibalism. As the wars continued, the want of animal food, in consequence of the gradual extermination of the beasts and birds, which composed the principal game,[1] became more and more sensible, and what at first had occurred only in the utmost need, and at the highest pitch of passion as an exceptional case, became gradually a horrid custom, which did not cease, until, by the introduction of more productive resources, want and misery were removed, and the principal cause of the bloody wars eradicated.

This change took place with the introduction of pigs, potatoes and grain by the seafarers at the end of the last century. In addition to this came the beneficial influences of Christianity, soothing and softening the savage mind, and thus it is that already in 1843 the last real case of cannibalism upon New Zealand is recorded in history. True, there are still many alive, who in their youth tasted human flesh, but to the younger generation already all and every recollection thereof sounds like a fable.[2] An aged chief who was travelling with a young Maori, on passing an old Pah, remembered the days of yore and related to his young friend: "Behold! here we caught and slew your father, and yonder we cooked and ate him!" The young man listened to the story, as though it did not concern him in the least; both slept cosily

[1] First among them were the gigantic wingless Moas, *Dinornis* and *Palapteryx*, which seem to have been exterminated already about the middle of the 17. century; then the New Zealand rat Kiore, which disappeared within the present century, finally Kiwi (*Apteryx*), Weka (*Ocydromus*) and Kakapo (*Strygops*), which are likewise all exterminated in the vicinity of settlements, and are only still to be found in uninhabited or inaccessible mountain regions. Quadrupeds, we know, there were none upon New Zealand except the above said rat and the dog (comp. Chapter VIII. and IX.).

[2] During the last war, the Maoris have relapsed into Cannibalism and heathenism.

and socially together in the same tent, ate out of the same pot, and were good friends. — Thus the memory of older time is now like that of a dream to the Maori of the present day.

With the exception of a few aged chiefs the Maoris are all converted to Christianity.[1] Being trained in excellent missionary schools with native teachers and preachers, the most of them know how to read and write, and sometimes display an astonishing proficiency in the knowledge of geography and history. While agriculture and the raising of cattle are their chief occupation, they participate also in commerce and trade, and especially a large portion of coast-navigation is in the hands of the natives, who have acquired an extensive reputation as well-skilled and dauntless seafarers. Richly endowed by nature with intellectual and physical powers, of a lively temper, energetic and open minded, and with natural wit, the Maori is fully aware of his progress in moral improvement and culture;[2] yet he is not capable of attaining the full height of a Christian civilized life; and it is from this very incompleteness, that his race is doomed to gradual extinction.[3]

The incapability of the Maoris of attaining the full height of European erudition and morality is probably seen most strikingly in their indifference to the English language, and in their views of Christianity. Since the English language has made such rapid progress in other countries owning to British sway, it is very singular, that it has been so much neglected upon New Zealand. Much as the Maoris have adopted of the habits, manners and customs of the English, their language thus far has remained almost entirely strange to them. While in other British colonies the na-

[1] About one half were converted by the missionaries of the Anglican High Church; the other half by Wesleyans and Roman Catholic Missionaries.

[2] A chief describes in a letter the former state of affairs in the following words: "We did wrong to each other, made war upon each other, ate upon one another, and exterminated each other."

[3] The frontispiece of this work gives us an idea of the half-civilized state in the very exterior of a still living chief. He wears European shirt and neck-lace, over it his Maori mantle, in one hand a gun, in the other a Maori weapon, the *mere* of nephrite. The Albatros feathers of old in their head-dress are supplanted by those of a peacock.

tives are made to learn the language of their lords, upon New Zealand the Englishman is compelled to study the Maori tongue.

It is wholly unjustifiable to impute the fault thereof to the missionaries, because they, instead of introducing the English language into their schools from the very beginning, had given themselves the unnecessary trouble, of first compiling a Maori grammer. The few Maoris, who understand English, have learned it in missionary schools. The cause is a deeper one. Perhaps there is nothing more adapted to prove the limits, set by nature herself to the natives' capacity for civilization, than the remarkable fact that even though they understand English, and are able to read and write it, they never acquire a plain English pronunciation. A Maori once remarked to me quite correctly, that he believed, his ear would catch the English well enough, but he could never get it out of his mouth again. This is proved by numerous words and phrases, which have passed into the Maori language, but in a form, in which the original word is sometimes hardly recognised. Of New Zealand, for example, they make Nuitireni; Victoria the Queen of England, in the mouth of the Maori becomes Wikitoria te Kuini o Ingarangi; Governor is Kawana; Auckland Akarano; Christchurch they call Karaitihati; Gold is Kaura; doctor te Rata; and my name they changed to Hokiteta. In a similar manner all the English names of places, persons, months etc., have been changed to suit the Maori tongue. Because they are unable to pronounce the language correctly, from its seeming to them so very difficult and hard, they believe, that the same must be the case with the Europeans. I once heard a Maori argue: "We can see very well, what trouble you have in your schools to teach your children English; we need not teach our children Maori, they acquire it quite by nature."

Of Christianity the Maoris have adopted only the outward form. Instead of their old pagan rites and ceremonies, they now have Christian forms; biblical history is to the Maori only a new edition of traditions, which he exchanges for, and perhaps also mixes up with, his own ancient traditions; a great many had themselves baptized merely for the sake of the material advantages. [1]

With regard to outward observance, therefore, the Maoris are the best and strictest Christians. At regular intervals the little bell tolls in their villages the summons to morning and evening prayers, and in the strict observance of the Sabbath they are ahead even of their teachers. The English troops at one time are said to have gained an advantage over their enemies during the last war by storming a Pah on Sunday, while the natives were assembled for divine service, the latter not dreaming, that Christian soldiers would think of such a thing as waging war on the Sabbath-day. That the natives do not allow the Europeans to travel on Sunday, refusing them even any aid or assistance whatever on that day, I have experienced myself. The liturgy, the old and new testaments, are translated into the Maori language, and it is remarkable, what an intimate knowledge of the Bible many Maoris display. The old family and village names are changed into biblical names, and on every occasion biblical sentences are heard quoted. But how much of the deeper, moral and spiritual substance of Christianity has entered the newly converted pagans? To answer this question satisfactorily might prove rather a difficult task. Maori Christianity is only an external creed, which has become the fashion. He prays regularly, but lives and works irregularly; even the missionaries confess, that religion alone can not save this people, but that they need labour and a regular mode of living. Progress in the civilization of a people is positive gain and real progress only when at the same time the vital powers are improved; when the civilizing influence contributes also to the moral and physical improvement of the people. It is but too easy to adduce proofs, that just the contrary is the case with the Maoris. I will only mention a few facts such as I find them recorded in my diary.

It seems natural to think, that the introduction of the plough, the trashing-machine, or the construction of mills and the like

---

[1] This holds good especially of the slaves, who were far more ready to adopt Christianity, than the chiefs.

caused the natives incalculable advantages, and that thereby no disadvantages whatever could be connected either for the individual or for the people collectively; and yet the benefits arising therefrom are of a very dubious kind. Formerly they used to work twenty to thirty upon an acre at a time; now the plough is busy, and the twenty or thirty sit about the acre laughing and joking, eating and smoking, — thinking, the Europeans had invented all such things only to keep them from working. A native, for instance, who had erected a mill at a great expense, expecting to be able to grind a large quantity of flour, and to sell it with great profit to the Pakehas, has found afterwards that he has reckoned without his host. The natives of the vicinity have heard of the mill, they go to visit the rich miller; according to the wonted communism they consider his flour as theirs also, and ere long the supply of flour is consumed. The mill, which the miller is not inclined to work, that others might reap the benefit, stands still. Money and labour are uselessly expended; and the enterpriser, instead of having become richer by his mill, is poorer than before.

As with agriculture, so it is with navigation. The natives on Tauranga Harbour, for example, worked and saved for a number of years, to get money enough together, that they might buy a schooner and be able to say, "we are ship-owners and captains as well as the Pakehas." By the sale of wheat and potatoes they really made up the sum of £800, and at last, thirty or fourty of them together owned a fine schooner. But what happened now? "We have worked very long," they said, "now let us rest a while," and thus the ship was in the water, but the natives on the shore. Once or twice perhaps they made a trip to Auckland; but they had beforehand sold so much of their produce, that now they had nothing left to sell or to ship; like children they got tired of their plaything; the schooner lies idle; it is owned by forty, hence by nobody; none of them are willing to undertake any repairs; the vessel goes to ruin, and again money and labour are lost. Had those thirty or forty kept to their old canoes, they would have done better.

As regards dress, also, the European influence has proved injurious rather than beneficial. The original dress of the Maoris consisted of cloaks and mats of different forms and sizes. They were ingeniously wrought of New Zealand flax or made of dog-skins, and were very durable. They formed an excellent protection against rain and cold. The women manufactured those mats; but now only a few old women still understand this useful art; the younger generation knows nothing of it. And what has now taken the place of the Maori mat? Let us not judge by some chiefs, who are seen parading about in the cities in suits of black and with Parisian hats, making all possible efforts to keep up their proud carriage in that unused dress and in boots which are either too tight or too wide; or by the ridiculous rigging of some Maori fair ones, who represent the demimonde of their race; but look at the natives in their villages and pahs, when they are among their own, and at home. There the woolen cover, the blanket is the only fashion. When these blankets first came into fashion, the Maoris were perfectly crazy to give up every thing in order to procure those rags, which after very short use look abominably dirty. Money they had none; therefore they gave potatoes, pigs, or the far superior flax and dog-skin mantles away, and now they sit before their huts, scantily wrapped in old ragged, tattered, dirty blankets; and by those very blankets they have become proletarians in reality, not only in appearance. It is certainly most true, that in this imperfect, miserable dress of the Maoris we must find one of the various causes of the chest-complaints and rheumatic diseases, to which of late they are subject. Not less it has been imputed to the potatoes, that they have only contributed to the physical degeneracy of the race, since they have become almost the sole and exclusive article of food of the natives. They have taken the place of the original, generally far more nutritious, and by virtue of their variety also much more salubrious, victuals, and have with the Maoris the same deleterious effect, as with our poor mountaineers, who are exclusively limited to this food.

Formerly the hospitality of the Maoris was praised; traits of generosity and highmindedness were related: to-day little is heard of such instances. The younger generation, — "the young Maori School," — cares for hardly any thing but the money of the Pakeha. One example may serve for many. A settler wished to be rowed across the Waikato river. The natives demanded for this trifling service not less than £5. After lengthy and tedious negotiations he succeeded in reducing their demands to 15 shillings. He, however, reported the case to Auckland, and the Government felt induced to take measures against such extortions. On his return, the same gentleman offered the natives 7½ shillings to be carried across. In the middle of the river, however, the canoe capsized in consequence of an intentional motion of one of the natives, and the settler, who with the heavy bundle on his back could not swim ashore, was compelled to cling to the canoe and beg for help. "Why did you write to Auckland?" asked the natives. "Because I had a right to do so, and because I felt insulted at your extravagant charge," was the reply. "Well then," said the Maoris, "you will recollect, that we asked only 7½ shillings to carry you across; if now we are to help you, and keep you from drowning, you have to pay another 7½ shillings." And in this manner they made again their 15 shillings.

Numbers of the saddest proofs of the deterioration of the manners and the character of the natives in their intercourse with Europeans, are furnished in the cities by the class of "town Maoris." Too proud or too lazy to engage their services to Europeans, and by regular work to earn an honest living, they loaf and lounge about the streets and taverns, — morally and physically bankrupt, a loathsome burden to the Europeans, and an abomination to their own countrymen.

Thus despite the many advantages it has brought to the natives, the European civilization and colonization acts upon them after all like an insidious poison, consuming the inmost marrow of their life; like a poison which not only whalers and sandal-wood traders import in the shape of plagues and cutaneous diseases, but which every European brings with him. The naivity of manners disappears before the formalities of civilization. The hospitable savage becomes a calculating and deliberating trader; our dress renders him stiff and awkward, and our food makes him unhealthy. Compared with the fresh and full vigour, with which the Anglo-Saxon race is spreading and increasing, the Maori is the weaker party, and thus he is the loser in the endless "struggle for existence."

In conversations about the natives I frequently heard the opinion expressed, not only by missionaries and Government officers, but also by independent colonists, that in the colonization of New Zealand special pains were taken, to make up, in the treatment of the Maoris, for the sins committed against the aborigines of Australia and Tasmania. And, in fact, the history of New Zealand furnish plenty of cases of noble-hearted men, who with truly selfsacrificing charity and devotion espoused the cause of the natives, leaving nothing undone to make those uncouth, but highly gifted savages Christians and civilized men. And it is equally true, that also the English Government, with regard to the Maoris, has pursued a policy quite different from that shown in most of the other colonies.[1] Yet, if we inquire after the result of those philanthropic efforts, we shall have to acknowledge, that the result is scarcely any better than there, where the European in the treatment of helpless natives has enforced the right of the stronger in the most merciless, most brutal, and sometimes horribly bloody manner. — The Maoris too are dying out.

The official document which gives the latest statistical accounts concerning the Maori population on both islands,[2] calculates at the same time, from the rate of decrease of the native population, the approaching time, when it will have altogether vanished from the

---

[1] I am by no means a pessimist, who sees in all that merely an egoistical act of necessity or of shrewds, calculating prudence, because the Maoris are an intelligent, energetic race. An English writer on New Zealand says literally: "I have long since come to the conclusion, that the modern Englishman is as cruel and unprincipled a scoundrel as the world has ever seen. — In simple truth, we pay the Maori large sums for his land, because he is an acute and powerful savage, we swindle the Australian out of his birthright, because he is simple and helpless."

[2] Observations on the State of the Aboriginal Inhabitants of New Zealand by F. D. Denton. Auckland 1859.

face of the earth. The decrease of the population, as far as it could be ascertained, amounted within the last 14 or 15 years to 19 and 20 percent.

Should the decrease continue at this rate, then the Maori population will number:

In the year 1872 only	. . . . . .	45,164
,, ,, ,, 1886 ,,	. . . . . .	36,363
,, ,, ,, 1900 ,,	. . . . . .	29,325
,, ,, ,, 1914 ,,	. . . . . .	23,630
,, ,, ,, 1928 ,,	. . . . . .	19,041
,, ,, ,, 1942 ,,	. . . . . .	15,343
,, ,, ,, 1956 ,,	. . . . . .	12,364

and we come to the conclusion, that about the year 2000 the native race will have quite died out, while the European population at the present rate of increase will have risen from 84,000 inhabitants in the year 1860, to half a million in the year 2000.

The Maoris themselves are fully aware of this, and look forward with a fatalistic resignation to the destiny of the final extinction of their race. They themselves say: "as clover killed the fern,[1] and the European dog the Maori dog; as the Maori rat was destroyed by the Pakeha rat, so our people also will be gradually supplanted and exterminated by the Europeans."

Appendix.

The story of Te Uira, chief of the Ngatimamoes.[2]

The most ancient tribe, that inhabited South Island, is said to have borne the name of Waitaha. This tribe was exterminated by the Ngatimamoes, who had come over from the Wanganui, North Island. The Ngati-

---

[1] In consequence of the most bloody war of the last years this proportion has become much less favourable for the Maoris, whilst the European population of the South Island was increased by immigration in a proportion quite unexpected, in consequence of the gold discoveries.

[2] Communicated by Dr. J. Haast.

mamoes lived on the shores of Cook Strait and subsisted principally on the gigantic Moas. They were subsequently joined by the Ngatitaras, who likewise came from North Island; and when their friends, the Ngatikuris, heard of the charming beauty of the new country, and of the excellent eel-fishery there, they also emigrated from North Island to South Island. Now hostilities arose. The Ngatikuris joined their kindred tribe, the Ngatitaras, assumed the name of Ngatitahu, and waged war with the Ngatimamoes. The latter in the course of long and bloody wars were driven more and more to the South and from the coast into the interior of the island, to take up their abode among the inhospitable Southern Alps.

About a century ago — thus related Taitai a chief from the West Coast of South Island, who died 1861 — the Ngatimamoes were driven back as far as Jackson's Bay (lat. 44⁰), whilst their Pahs at the mouth of the Mawhera (now Grey river) fell into the hands of the Ngatitahus. At that time the Ngatimamoes were led by a famous warrior, who on account of his quick and sure execution was called Te Uira, "the lightning". They were in possession of a precious "mere punamu" (battle axe of nephrite), named by them Taonga or Tonga, which they regarded as the last symbol of their tribe, and which they held in high esteem, like the banner round which a regiment rallies amid the din of battle. The Ngatitahus had endeavoured for a long time, but without success, to get possession of Te Uira and his precious mere; at last they accomplished this by a stratagem, and conveyed their captive to their Pah on the Arahura river. Here he was tied to a tree and was destined to be tortured to death. But in order to share the pleasure of seeing Te Uira die with their friends from Mawhera Pah, the execution was postponed till after their expected arrival. Meanwhile Te Uira managed to escape. Possessed of great strength of body, he burst his fetters and fled from his pursuers into the depth of the bush. He was thus again free, but before returning to his kindred and friends, he determined to try, whether he could not recover also the mere punamu; for he was ashamed to return home without that jewel of his tribe. At night-fall he stole up to the enemy's Pah, and watched for an opportunity to accomplish his design. This soon presented itself. One of the Maoris left the watch-fire to refresh himself by walking to and fro in the cool night-air. Swift as a tiger Te Uira bounded upon him, strangled him, and after having donned the cloak of his victim to disguise himself, he calmly walked up to the fire and sat down among his enemies. They were just entertaining themselves with the wonderful mere, handing it around and expressing their unfeigned admiration for the same. Te Uira, who knew, that the man he had just killed, and whose cloak now served him as disguise, stuttered, imitating the organic defect asked for the mere. No sooner did he feel the wonted weapon in his hand, than he dealt a blow to the left and another to the right, striking down the two men, between whom he sat; and with one bound he was out of their midst and speeding towards the woods, without his enemies being able to overtake him. When they saw that Te Uira had escaped from them, they endeavoured to cut off at least his retreat to his Pah, and to this end proceeded forthwith along the coast to Jackson's Bay. They found the Pah well fortified and were not able to take it. But on the second day they saw a fire blazing on the top of a steep rock to the rear of the Pah. This was a preconcerted signal of the Ngatimamoes to retreat into the interior of the country to a place previously fixed on. At dead of night and with the least possible noise they left the beleaguered Pah and retired into the wilderness, taking with them the only token of their former greatness, their celebrated mere punamu. Since that time they have never been heard of; but there is a rumour current, that in the interior of the Province Otago, in the wild, unexplored mountainous regions between Lake Wanaka and Milford Harbour there are still some few living, — the last of the Ngatimamoes. The Maoris of the coast represent them as savages, and both, natives and European settlers, pretend to have seen such wild Maoris, who are said to be extremely timid.

CHAPTER XI.

The Isthmus of Auckland.

Situation. — Waitemata and Manukau. — The City of Auckland. — The town of One-hunga. — Geological features of the Isthmus. — The extinct volcanoes. — 63 points of eruption. — Tuff-cones and tuff-craters. — Cinder-cones. — Combinations of both. — Lava-streams. — Manukau lava-field. — Waitemata lava-field. — Lava-streams of different age. — Mount Welling-ton. — Lava-cones. — Rangitoto. — The Auckland volcanoes of very recent date. — The Isthmus as it was and as it is. — Past, Present and Future.

The southern portion of the North Island is connected with the long-stretched northwestern peninsula by a small isthmus upon the 37th parallel South latitude. On the East-side the sea penetrates through the isle-studded Hauraki Gulf far into the land, washing in its southwestern ramifications, in the Waitemata River, the North-side of the isthmus. On the western coast, the weather-side of the island, the ocean has forced a narrow entrance through hard volcanic conglomerates, thence assuming wider dimensions, and forming the extensive basin of the Manukau Harbour on the southern shore of the isthmus. The land between both seas has a breadth of only five to six miles, and at two places, where from the Waitemata River in a southerly direction narrow creeks are cutting deep towards the Manukau Basin, it dwindles down to one single mile, forming "portages," which the natives have used since olden times for the purpose of dragging their canoes from one side of the sea to the other,[1] and which have roused among the colonists

[1] The western or Whau portage is one mile wide, and at its highest point 111 feet high. The eastern, the Tamaki portage near Otahuhu, South of Mount Richmond, is only 3900 feet long and 66 feet high.

the idea of connecting both harbours by means of a canal. While the Waitemata River forms the most central harbour of the numerous harbours on the Eastcoast, the Manukau Basin is on the Westcoast the only harbour, which is accessible to larger vessels. [1]

No second point upon the North Island possesses such an extraordinary facility for inland communication in all directions by means of the harbours, estuaries and rivers, all of which may be navigated to near their sources, either by boats or canoes, and between most of which very short portages intervene. In a northerly direction the Waitemata Creek extends to within a few miles from the Kaipara Harbour. Upon the branches of this extensive estuary and upon the Wairoa River, navigable to a large extent of its course, the water-route leads through regions abounding in timberwood, and through the most luxuriant Kauri forests far to the North. In a south-easterly direction, there is a passage between the islands of the Hauraki Gulf to the mouths of the rivers Piako and Waiho (the New Zealand Thames), and upon the latter river far into the interior of the country. In a southerly direction, there lies between the end of Waiüku Creek, a side-branch of the Manukau Harbour, and the Awaroa Creek a portage of only 1½ miles; this creek flows into the Waikato, the principal river of the North Island, navigable for 100 miles up and leading through the most fertile and smiling tracts of land into the very heart of the country.

Such are the natural advantages of the spot, which in 1840 was fixed upon by Captain Hobson as the site for the seat of the Government of New Zealand, and as the proper spot for a large and prosperous agricultural and commercial settlement. Experience appears to have justified the wisdom of this choice. The city of

---

[1] The passage into the harbour in front of which extensive sandbanks are spread out, is not practicable during bad weather. The dangers, however, have been considerably reduced by accurate surveys, which have proved the existence of a middle channel and a southern channel; likewise by the pilot station erected at the entrance. Nevertheless within the last years the harbour has been used only by the mail-steamers and by small coasting vessels.

Auckland,[1] extending the rows of its houses along the charming shores of the Waitemata on the Northside of the isthmus, will prosper and flourish, and secure its rank as a metropolis of the North Island, even after the seat of Government is removed to Wellington, and the centre of gravity of the colonial development seams to shift more and more from North to South.

In the year 1860, Auckland numbered 10,000 inhabitants, and probably to the same number amounts the population of the Auckland district. Though the city with its mostly timber-built houses has not quite lost all traces of its late origin; yet it continues to improve from year to year by the construction of large stone-buildings from the porous basalt-lava ("scoriæ-stone") of the surrounding volcanic cones, yielding an excellent building material. The circumference of the town and suburbs, planned on a large scale, is at present already a considerable one. From East to West, including the suburbs, Auckland has a frontage on the water of a mile and a-half and extends from North to South to the distance of about one mile. The centre of the city is situated on a ridge, between Mechanics' Bay East and Commercial Bay West, shelving off abruptly to the harbour at Britomart Point. Upon this central ridge, on the extremity of the head land, Britomart Fort and Barracks are built, next the Metropolitan Church of St. Paul, the rows of houses in Princes' Street, the Governor's Mansion, the Albert Barracks, and overlooking town and harbour the Windmill is seen with Mount Eden in the back ground. East of this central line, round about Official and Mechanics' Bay, are the Government buildings and offices, and the detached cottage-like houses of military and civil officers, of clergymen and missionaries; West of it on Commercial Bay is the mercantile quarter, forming a mass of houses closely packed together. Freeman's Bay, to the westward of Commercial Bay, is occupied chiefly by saw-pits, brick-kilns, and boat-builders' yards. The site of the city, the variously jutting hills, and the intervening bays remind us of Sydney with its "coves." The Harbour of Auckland being very shallow on the

---

[1] To its position Auckland owes the surname: "Corinth of the South."

town-side, the construction of piers extending far into the water was a matter of necessity. On Mechanics' Bay, the Official or Wynyards Pier affords a convenient boat landing-place at all times of tide; on Commercial Bay, the Commercial Pier or Queen-street wharf more than a quarter of a mile long, is, in fact, one of the most reputable works in the young colony, and of incalculable benefit for the commerce of Auckland; alongside of it, coasting vessels are able to land and to take in their cargoes, and in the continuation of it runs Queen-street, the main commercial street of the town. Whoever has not been accustomed to city life on too extravagant a scale, will scarcely miss anything in social life. Considering its size, Auckland possesses the elements of considerable society. The most cheering and encouraging beginnings have been made in every thing; even a botanical garden and a museum of natural history have been already founded; and numerous societies and institutes for benevolent, scientific, agricultural, horticultural and other purposes have been formed. [1]

Auckland is the starting point of the two principal road-lines on North Island; the Great South Road leading to Mangatawhiri on the Waikato; and the Great North Road, which will lead overland to the Bay of Islands. A third, macadamised road is cros-

[1] Auckland numbered in 1860 twelve churches and other places of public service: St. Paul's Church, St. Matthew's Church, St. Barnabas' Church (in Parnell), St. Patrick's Church (catholic), Church of the immaculate conception of Mary (catholic), Presbyterian Church, Wesleyan Chapel, Primitive Methodist Chapel, Independent Chapel, Second Independent Congregation, Baptist Congregation, and a jewish synagogue; — ten public schools, four female seminaries, four Maori schools, and the following public institutions and societies: Mechanics' Institute, Choral Society, Chamber of Commerce, British and Foreign Bible Society, Auckland Museum, Auckland Dispensary, Young Men's Christian Association, St. Andrew's Society, Hibernian Benevolent Society, Auckland Land Association, Auckland Medical and Surgical Society, Acclimatisation Society, Agricultural and Horticultural Society, Auckland Homeopathic Hospital and Dispensary, Auckland Bethel Union, and several Masonic Lopges. Auckland has also three Banks, several Insurance Companies and six public news-papers. The principal papers are: the "New Zealander" and "Southern Cross"; besides, there are published in Auckland the "New Zealand Gazette", and the "Auckland Government Gazette"; moreover the "Examiner" and "Auckland Register", and a Maori paper, the "Maori Messenger", or Te Karere Maori.

View of Mount Eden near Auckland, from the Domain

sing the Isthmus to the town of Onehunga, situated on the shores of Manukau Harbour, a distance of five miles.

Onehunga, originally a settlement of civil and military pensioners who from the Government had received a small house and one acre of land each, has already risen to the rank of a town, which, being the chief trading-place of the natives, is gaining more and more importance and, in consequence of its pleasant situation and charming environs, has become the favourite residence of business-men who, having their business-establishment in Auckland, prefer to live in or close by Onehunga. Along the road between the two towns farms and cottages are seen scattered about. The land, however, is not exclusively in the hands of farmers; merchants, civil and military officers also invest their savings in landed property. Charming country-houses surrounded with lovely gardens grace the country all over the Isthmus; while at the crossings of the principal roads already villages have sprung up, such as New Market, Mount St. John Village, Epsom, Panmure, and farther on Otahuhu and Howik. It is not to be wondered therefore, that in the course of time the land in and about Auckland has enormously risen in value. [1]

The Isthmus of Auckland is one of the most remarkable volcanic districts of the earth. It is characterised by a large number of extinct volcanic cones with craters in a more or less distinct state of preservation, and with lava-streams forming extensive stone fields at the foot of the hills, or with tuff-craters surrounding, like an artificial wall, the cones of eruption piled up of scoriæ and volcanic ashes. These cones are promiscuously scattered over the Isthmus and the neighbouring shores of the Waitemata and Manukau. The

---

[1] In Auckland from £10 to 12 are paid for one foot front in the principal streets for ware house and store purposes; at a late sale of real estate two miles from Auckland at the foot of Mount John, at the crossing of the Onehunga and Great South Road £6 to 7 were paid per foot, and £1100 for one acre. The usual price of an acre of cultivated land is £15 to 30. With such prices it will be easy to account for the fact, that in Auckland the project has been carried out, to fill up the shallow dent of Commercial Bay for the purpose of gaining by it level house lots and sufficient room for a slighty street along the sea-shore.

volcanic action seems to have made itself a new way nearly at every eruption, and has thus splintered into a number of small cones, while by always keeping one and the same channel, it might perhaps have formed one mighty volcanic mountain. On the geological map of the Isthmus I have traced, upon a rectangle twenty miles long and twelve miles wide, or within a radius of only ten miles from Auckland, not less than 63 separate points of eruption. They are volcanoes on the smallest scale, cones of only 300 to 600 feet above the level of the sea. The Rangitoto, the highest among them, rising at the entrance to Auckland Harbour as it were the Vesuvius of Waitemata Bay, measures 920 feet. Nevertheless, they are perfect models of volcanic cones and craters presenting a rich field for observation and ample material for the discussion of the question of the formation of volcanic cones and craters. A full description with all details I have given in the Geology of New Zealand (Scientific Publications of the Novara Expedition, Geological Part, Vol. I.); here I may be allowed to quote the principal results.

The volcanic cones of the Isthmus are rising on the basis of tertiary sandstone and shale, the horizontal strata of which are laid bare in numerous sections on the precipitous bluffs and steep banks of the Waitemata and Manukau Harbours. Fossils are an extremely rare occurrence in those strata. Only now and then, near the water's edge, thin layers of lignite or drift-wood, changed to brown coal, are seen, and on the Northside of the entrance to the Orakei Bay I discovered glauconitic strata with fossil species of *Pecten, Nucula, Cardium, Turbo, Nerita,* and replete with fossil Bryozoes and Foraminiferæ. These strata were completely broken through by the volcanic action from below, and an examination of the points of eruption proves, first of all, that volcanic action has repeatedly exhibited itself at one and the same place.

The first outbursts — as a closer examination shows — were probably submarine; they took place at the bottom of a shallow, muddy bay little exposed to waves and wind, and consisted of flowing mud mixed with loose masses, such as fragments of sandstone and shale, lava-debris, cinders and scoriæ (lapilli), which now form beds of volcanic agglomerate or tuff. The eruptions occurred, no doubt, at intervals; for in this manner alone can the fact be accounted for, that the ejected material has been deposited round the point of eruption in layers one above the other, forming low hills gradually rising and with a circular basin or dish-shaped crater in the middle. A cross-section presents clearly the different layers, which usually slope inwards towards the bottom of the crater, as well as outwards down the sides. The hills formed by these first eruptions we may designate as tuff-cones, or in as much as they enclose circular craters, as tuff-craters.

Tuff-cone.

Lake Pupuke on the Northshore, Orakei Bay East of Auckland, Gedde's Basin (Hopua) near Onehunga, the Waimagoia Basin near Panmure, the Kohuora Hills, South of Otahuhu, and a good many others are striking examples of such tuff-craters. Like the lake-craters ("Maare") of the Eifel, the crater-basins are sometimes very deep and full of water, — the fresh-water Lake Pupuke has a depth of 28 fathoms, [1] — sometimes flat and dry, or covered only with swamp and turf-moors. Where they lie close to the sea, the latter has generally forced an entrance on one side or the other, breaking down the circumvallation, and thenceforth passing in and out from the crater-basin. Where there are several such cones close together, as at Onehunga, and in the vicinity of Otahuhu, it is very often a difficult matter, to designate the individual craters, because a space bordered by a number of craters, very easily assumes for itself the form of a crater.

The excellence of the soil of Onehunga and Otahuhu is owing to the abundance of tuff-cones. Nearly each one of them harbours

[1] According to Captain Burgess, pilot of the Waitemata Harbour. The natives, however, sustain the peculiar notion or legend, that the opposite volcanic cone, the Rangitoto, was taken from the deep abyss of this lake. However, it is very probable, as Dr. Fisher thinks, that the lake is connected by a submarine channel with Rangitoto, which may be the source of the water of the lake.

the house or premises of a farmer; and it is curious to observe how the shrewder amongst the settlers, without any geological knowledge, and without dreaming even that they were building their houses at the very brink of a crater, have long since picked all those tuff-craters for themselves. The meadows and clover-fields upon them are proudly waving in softest verdure, while upon the sterile clay-soil nothing but ferns and manuka-bushes are strutting.

With the beginning of the volcanic action, by which the tuff-cones were formed, a slow and gradual upheaving of the whole Isthmus seems to have taken place, so that the latter eruptions were supra-marine. In this second period the volcanic action caused the ejection of glowing masses of scoriæ and cinders ("lapilli"), and of fiery fluid drops of lava ("volcanic bombs"), which from the rotatory motion through the air assumed the pear or lemon-shaped form peculiar to them; and finally great out-flowings of lava-streams took place, which were rolling their glowing waves, like rivers of fire, through the valleys. At that period the Auckland volcanoes were "burning mountains" in the true sense of the word; they were piling up their steep-rising scoria or cinder-cones; and, where re-peated and frequent out-flowings of lava from the same crater were taking place, there also lava-cones like the Rangitoto formed them-selves.

It was not on all points of eruption, that cinders and lava came bursting forth, but at several points the first formation of the simple tuff-crater remained, and the volcanic forces afterwards opened new channels. But where the new eruptions followed the old channel, there we observe in the middle of the flat tuff-crater, the outer slope of which hardly ever rises at an angle larger than 5 to 10 degrees, a steep cinder-cone with a decli-vity of 30 to 35 degrees, and a deep, funnel-shaped crater at the top. The cinder-cone very often has entirely filled up and even covered up

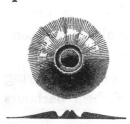

Tuff and Cinder-cone.

the tuff-crater, as is, for example, the case with the Takapuna, the North Head of the Auckland Harbour; or it rises in the shape

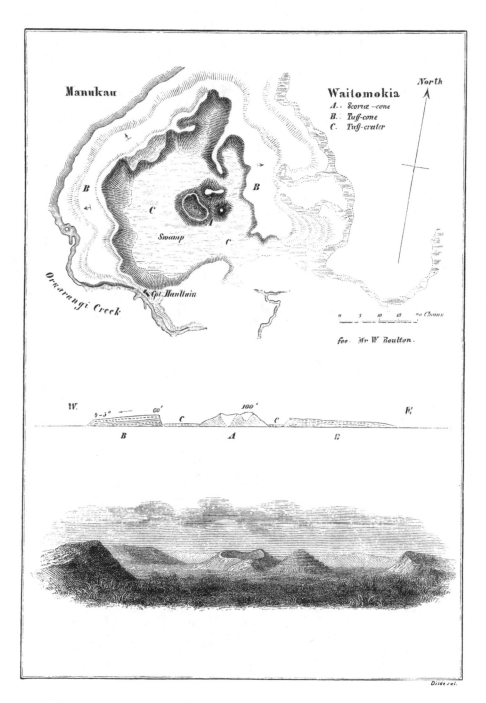

Waitomokia, an extinct volcano on Manukau Harbour, South of Onehunga.

of an island, in the middle of the larger tuff-crater, and is surrounded either by water or swamp, as at Mt. Richmond, Fort Richards and several points South and Southwest from Otahuhu. In fact, examples of every gradation may be seen — from the simple tuff-crater without any cinder-cone, to those which are entirely filled up by the cinder-cones. Especially interesting are those which may be said to represent the middle state; perhaps the most perfect specimen of this kind is the Waitomokia Southwest of Otahuhu. No doubt, such forms may partly be explained also by a later sinking down, by a sagging of the cinder-cone within its tuff-enclosure, by which even cones, that formerly had towered high above the tuff-crater, sunk in to their topmost points, some of them perhaps disappearing entirely. The destroying influences of water and atmosphere have also wrought a change of the original forms. This may be the case especially with the remarkable point of eruption within the very precincts of the city of Auckland, upon the half-destroyed tuff- and cinder-cone of which the central parts of the city are built. I will give a more detailed description of this point. The Wesleyan Church, Mechanics' Institute, the Auckland Hotel and other buildings are built upon a kind of terrace, higher by about 40 feet than Queen-street; behind the terrace at a steep angle and almost with a semicircular slope the hill rises, upon which the barracks stand, and on the northern and eastern slope of which the Governor's house, St. Paul's Church and the Princes'-street are situated. I consider the terrace to be the central point of eruption, or the remains of a sunk cinder-cone. The foundations of the buildings are resting there on more or less compact masses of cinder and basaltic lava, blocks of which are everywhere in the vicinity protruding from the ground. The steep, almost semicircular declivity, 200 feet high, however, is to be considered as the eastern half of a large tuff-crater, the western half of which, beyond Queen-street, is scarcely to be recognized by a thin layer of volcanic tuff, which has been almost entirely decomposed into yellow clay. Queen-street intersects the former tuff-crater in the direction from North to South. Near Odd Fellow's Hall I saw the layers of scoriæ and cinders in new cuttings on both sides of the street. Near the barracks loose masses of scoriæ and cinders are obtained as metalling-material from shafts 12 to 16 feet deep; and I was told that on the occasion of special trials nothing but scoriæ and cinders have been found at a depth of 340 feet, so that it almost seems as though, just beneath the barracks there were another centre of eruption. Farther below, near St. Paul's Church, on Shortland Crescent, and in the ditches of Fort Britomart we meet again decomposed tuff-strata; likewise at the Clipp house and in the yard of the Victoria Hotel. The little valley, however, leading through these tuff-strata to Wynyard Pier, the natives call very strangely Wai ariki (warm water), as though a warm spring had been flowing there in olden times.

The scoria and cinder-cones, although not adapted to agricultural purposes are nevertheless of practical value. They furnish an excellent material for macadamizing roads, which can be easily obtained; and it is to this material, that the Isthmus of Auckland owes its beautiful metalling roads. The metalling-quarries are opened everywhere at points contiguous to the road, as on the foot of Mount Eden, on One Tree Hill, Mount Wellington and others.

The lava of the Auckland volcanoes consists of a scoriaceous basalt, containing small grains of Olivin. The lava-beds of Mount Eden are on various places columnar. The porous mass, called "scoriæ-stone" is exceedingly well adapted for building-material, and is used for this purpose in Auckland in like manner, as in Melbourne quite a similar kind of basaltic lava, found in the vicinity of the latter city.

In the larger lava-streams, as at the "Three Kings," Mount Smart, Mount Wellington etc., caves are very frequently found, which are in reality nothing but the results of great bubbles in the lava — occasioned probably by the generation of gases and vapors, as the hot mass rolled onward over marshy ground. Where the roof of such caves broke down, there are deep holes, such as those near Onehunga, called the "grotto" and "pond," which have been mistaken for points of eruption or crater-sinkings.

The several craters differ very much as to the quantity of lava that has issued from them. Where the issue of lava has taken place only once, breaking through one side of the crater and flowing over the ring of the tuff-cone to the foot of the hill, there we see the volcanic system in a very simple perfection and distinctness, as at Pigeon Hill near Howick, at Green Hill and Taylor's Hill. At other points, however, there was a more copious out-flowing of lava; stream followed after stream, and the lava-streams of several craters uniting together, formed extensive lava-fields, upon which it is often a difficult matter to distinguish the streams belonging to the different craters. Thus the lava-streams of Mount Eden, Three Kings and Mount Albert are blended in the large Waitemata lava-field Southwest of Auckland. Those three mountains seem to have been active simultaneously; their streams, spreading round the basis of the cinder-cones, and rolling thence over the northwesterly slope of the Isthmus through former ravines and valleys towards the sea. Near the coast they met all together in a contracted valley, and there formed *one* large stream to the shore of the Waitemata, terminating on the long reef West of Sentinel Rock and opposite to Kauri Point on the Northshore. The idea had once been entertained of building a bridge from that point across the harbour.

Tuff-cone, Cinder-cone, and Lava-stream.

On the southeastern slope of the Isthmus the lava-streams of One Tree Hill, Mount Smart, and Mount Wellington form the large Manukau lava-field. Here, however, a remarkable difference of age appears in the streams of the various craters, proving distinctly that they were not active all at the same time. The lava of One Tree Hill is already entirely decomposed at the surface, it is covered with a fertile reddish-brown earth and beautiful grass and clover paddocks are laid down over those oldest lava-streams. The lava of Mount Smart, on the other hand, presents a stone-field very difficult to cultivate; and the comparatively newest lava-streams of Mt.

Wellington, together with the large stream, which flowed in a southwesterly direction as far as Onehunga, shows a surface as yet uncorroded by the action of the atmosphere or water. The lava presents a barren stone-field of black blocks of rock, between which only a few bushes have succeeded in taking root. The difference between the older and newer lava is very clearly shown upon the Great South Road. About one mile East of the "Harp Inn," the traveller will observe a sudden change in the colour of the road, which is most distinctly noticed after a rain. The red colour (from oxide of iron) changes suddenly to black, where the road leaves the older and more decomposed lava-streams of One Tree Hill and passes on to the new and undecomposed lava-stream of Mount Wellington.

Inferences thus drawn already from the state of decomposition of the lava-streams, are moreover proven by observations on the remarkable crater-system of Mount Wellington (Maunga Rei of the natives), which I needs must make special mention of, as being one of the most instructive points on this subject. Here the careful observer has ample opportunity of studying a whole system of craters and cones of different ages and different composition. The oldest member is a large tuff-crater intersected by the Panmure Road, and exhibiting most beautifully, in the northern cut of the road, the characteristic outward dip of its strata. In this tuff-crater arises a double cone[1] of cinder with two craters. The Northeast-side of this cone is now cut into by a quarry. Its old lava-streams are to be seen very much decomposed at the bottom of the tuff-crater. After a comparatively long period of quiescence, from the southern margin of the tuff-crater by new eruptions the large and very regular cinder-cone of Mount Wellington arose, from whose three craters large streams of basaltic lava flowed out in a westerly direction, extending North and South along the existing valleys of the country, one stream flowing

[1] I have denominated it as Purchas Hill in honour of my friend, Rev. Mr. Purchas, who assisted me in the exploration.

into the old tuff-crater and spreading round the base of the smaller crater-cones.

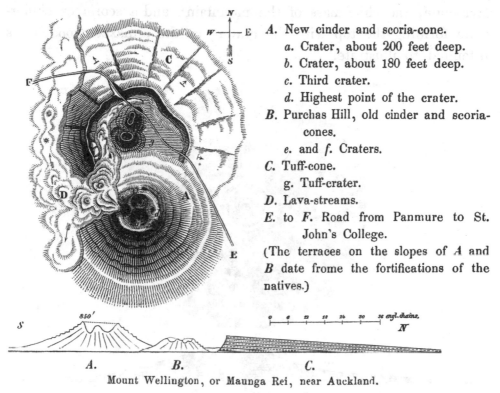

A. New cinder and scoria-cone.
  a. Crater, about 200 feet deep.
  b. Crater, about 180 feet deep.
  c. Third crater.
  d. Highest point of the crater.
B. Purchas Hill, old cinder and scoria-cones.
  e. and f. Craters.
C. Tuff-cone.
  g. Tuff-crater.
D. Lava-streams.
E. to F. Road from Panmure to St. John's College.
(The terraces on the slopes of A and B date frome the fortifications of the natives.)

Mount Wellington, or Maunga Rei, near Auckland.

Most abounding in lava, and in its last eruptions probably also the newest of all the Auckland volcanoes is the Rangitoto Mountain, 920 feet high, rising on the Eastside of the entrance to the Waitemata Harbour. This is at the same time the only point, where the lava-streams have built up a regular lava-cone, rising at an angle of 4 to 5 degrees, and bearing at its top two

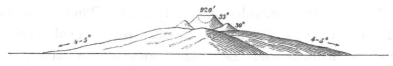

Rangitoto near Auckland.

steep cones of scoriæ and cinders, of which the upper one with a funnel-shaped crater appears set in the crater of the lower. Hence the characteristic profile of this mountain.

A complete volcanic system would accordingly consist of three parts: a tuff-cone, the basis and pedestal of the whole frame; a lava-cone, the chief-mass of the mountain; and a scoria or cinder-cone forming the top with the crater, as the annexed wood-cut is intended to illustrate.

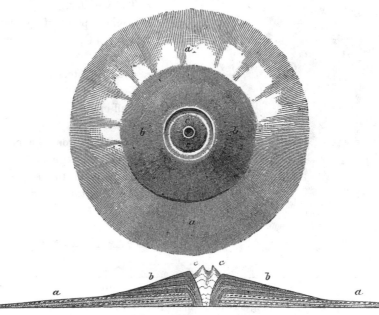

Volcanic Cone-formation.
a. Tuff-cone. *b.* Lava-cone. *c.* Cinder-cone.

In referring the Maori name Rangitoto, according to its literal signification "bloody sky," to fire-phenomena, such as a reflection of blazing lava-streams in the nocturnal sky, we would have to suppose, that the Auckland volcanoes had not ceased to be active till within the latest historical period. This, however, seems to me improbable, since there are nowhere traces to be found of solfataras, fumaroles or hot springs. That, however, the time of their activity belongs to a very recent geological period, is shown by the fact that the cinders everywhere occupy the top-most surface, and that the lava-streams have taken the course of the existing valleys. These valleys, consequently, were already formed, when the out-flowings of lava took place; and the surface of the country has not essentially changed its appearance since that time.

The question, how long the volcanic action might have lasted, and whether there is any probability of its returning, is of course not to be answered: yet, — if we take the example of Monte Nuovo near Naples, which in the month of September A. D. 1538, grew in two days and two nights to the size of a cone 400 feet high, — we may venture to say, that cones, such as Mt. Eden and Mt. Wellington, are likely to have sprung up in the course of a few days.

We have thus sketched the geological history of the volcanic cones in the vicinity of Auckland; moreover these take also a most remarkable part in the history of man. Now-a-days they are the ornament of a country richly cultivated by the industry of European settlers, who duly availed themselves of the fertile volcanic soil. Their summits present charming points from which the whole Isthmus can be viewed from sea to sea, and I love to linger a little while longer over the picture here presented to the eye, linking to it my thoughts about the "Past and Present" of this country, such as they suggested themselves to me, whenever from those heights I viewed the peculiar landscape.

The country is now almost bare of trees. Only on the hill-sides, in the craters or in a few gullies, there are some remnants

---

[1] Similar extinct volcanic cones with far-spread basaltic lava-streams, I met with in Australia, during my return voyage from New Zealand home, on excursions into the vicinity of Melbourne. There, however, the several cones are at a much greater distance from each other; their craters are in a less perfect state of preservation, and their lava occupies a far more extensive area. The porous basalt lava of the extinct volcanoes of Victoria is the principal building stone used in Melbourne. On the other hand all the features of the Auckland volcanoes seem to reappear in an equally typical manner in Western Victoria, at points described by Mr. James Bonwick (in a little work: "Western Victoria, its geography, geology, and social condition, Geelong 1857"), such as Mount Leura, Lake Purrumbete, Mount Noorat, Mount Gambier, Tower Hill, and many others. I am in possession of a view of Tower Hill or Koroit, sketched by a German artist, Mr. Gerhard in Melbourne, and lithographed in London, which is quite the counterfeit of the Waitomokia Crater near Otahuhu, only on a larger scale. Likewise the region of the *Kata-ke-kaumene* in Lydia, described by W. J. Hamilton ("Travels in Asia Minor and Armenia," German by Otto Schomburgk, 1843), seems to be studded with similar extinct volcanic cones. European countries abounding in craters, such as the Campi Phlegræi near Naples, the Auvergne, and the Eifel present far fewer points of comparison.

of the forest-vegetation, formerly covering the Isthmus. To judge from the Kauri gum found upon the Isthmus, the queen of the New Zealand forest, the Kauri pine, had likewise a share in those forests. On the road to Onehunga, at the foot of Mount Eden,

Cabbage-Tree, on the Road from Auckland to Onehunga.

there stands an isolated "cabbage tree" (Ti of the natives; *Cordyline australis*), nearly 30 feet high, with ramified branches and a crown of luxuriant growth, — a true representative of the original vegetation, and a magnificent specimen of its kind, fully deserving the indulgence bestowed upon it. A tree quite peculiar to the volcanic cones is *Brachyglois repanda*, by the natives called Pukapuka, meaning book or paper-tree because of the white lining of its leaves. On the high bluffs of the Waitemata, there are some few scattered Pohutukaua trees *(Metrosideros tomentosa)* the last remains of the beautiful vegetation, that once decked the shores of the harbour. About Christmas these trees are full of charming purple-blossoms; the settler decorates his church and dwelling with its lovely branches, and calls the tree "Christmas-tree." As to the rest of the landscape, it presents only shrubs.

Nearly every vestige of the former wilderness has disappeared from the Isthmus. The former vegetation has been mostly supplanted by European domesticated plants, and the weeds always accompanying the latter are mingling with the remnants of the indigenous flora. Roads are intersecting in all directions the hilly country between the Waitemata and the Manukau. Cottages and farms are dotting the smiling landscape between Auckland and Onehunga. The premises are very substantially fenced in by solid basalt-walls and quick-set hedges; and wherever the nature of the soil or the structure of the ground admits of cultivation, there grass and clover paddocks, and gardens and fields have been laid out. Cattle are browsing in the meadows; omnibuses enliven the roads; here a farmer's family in the one-horse dog-cart; there ladies and gentlemen mounted on horse-back: — a picture full of fresh-ness and vigour and of gay and merry life, as in the happy, idyllic spots and cherished haunts of our native home.

Like mirrors artificially enframed, the ponds, set in the old circular tuff-craters, are glistening afar. The sea is dashing far into the land, as though water and land had as yet not sti-pulated for their proper limits. Towards North arises the majestic Rangitoto from the waters of the Waitemata; and opposite to it, the smaller cones of the Northshore. Sailing vessels are passing in and out through the channel, and boats are racing over the harbour. On the other side of the Isthmus, on the waters of the Manukau, the mail-steamer, with its long whirling columns of smoke, is bringing us letters and taking along our greetings to "the loved ones at home." On surveying all this, how should the stranger realize, that he is in New Zealand at the Antipodes.

Only yonder in the distant horizon, towards West and South, where sombre shadows are hovering over lofty mountain-ranges, there are still traces to remind us of virgin forest and of primeval wilderness. Yet, the curly wreath of smoke ascending there, is a proof, that even there the son of man has fixed his abode. There are the first settlers, pioneering for generations to come. A small log-house is standing in the midst of the dusky bush, it is the scanty shelter of a family, that has come many a thou-sand miles far o'er the deep, to found a new homestead in a new country. The father is in the bush; trunk after trunk, is falling prostrate under the powerful strokes of the woodman's merciless axe; the mother at home is preparing the frugal meal in the iron pot suspended by a chain over the merrily flickering chimney-fire. Children are playing in front of the sylvan hut, radiant with health, and with their cheeks flushed with the forest-breeze; a faithful dog, chickens and pigs are their playmates. "It is hard work, indeed" the industrious house-wife is perhaps chatting with her husband on his return from the combat with those antiquated wood-giants, "a life full of trouble and privation; no physician, no drug-store, no church in the neighbourhood; nor even a friend, to talk about the dear, old home; yet, what we see before us and all around us, is ours, we may call it ours, and the Giver of All, I trust, will grant us His help for the future." And so it is. From year to year improvements are going on; the bush dwindles away; crop succeeds crop; the log-house has been supplanted by a pleasant, commodious country-house, surrounded with blooming gardens and waving fields. Herds of well-fed cattle are grazing in the pastures; horses are skipping and plunging in the meadows; friends have settled in the neighbourhood; smooth lanes and neat paths are winding between hedges and through the woods from farm to farm. And close by the wayside stands a church; a tavern is there, and the first trading-shop too has already been opened. Where of late there stood but a scanty isolated log-house, there is now — it cannot be called a village, nor is it quite a town; it is — a fragment of a town. Town people are inhabiting it with town-wants and town-fashions; they have mail-communications and newspapers; horses and carriages, and are living like the lords and ladies in the "old country." Thus, in the evening of a busy life, the old ones are enjoying their plenty; their children have now advanced into the bush; father and mother have set them a good example, and a new vigorous generation, undaunted by obstacles, is taking with rapid strides possession of the land, once the native haunts of a race of men of another complexion, called savages, who also lived after a manner, but it was the manner of *their* fathers.

How very different is the fate of that race of men! They, too, had emigrated in olden times from distant isles, in order to enjoy a better, happier life in the new country. Perhaps they, too, found here, what they had anticipated, through a long series of generations. But their time is past; and like a dreary picture of the romantic medieval age their life appears by the side of the cheerful picture of to-day.

The Isthmus of Auckland was of yore the dwelling-place of a mighty tribe of Maoris, the scene of the peaceful occupations, of the festivals and games of a people, who, if barbarous, were not the less gifted; but also the scene of the bloodiest cannibal-massacres, in consequence of which that tribe gradually vanished from the earth. The Ngatiwatuas, who were living here, are said to have numbered but a few generations ago from 20,000 to 30,000, and those extinct volcanoes were at that time acting the part of mountain-forts like the castles of the middle ages. By their commanding position and the prospect far o'er the country, they were exceedingly well adapted for watch-towers and forts. As in Europe the ruins upon rocks and mountain heights are the gloomy mementoes of club-law, where "might alone made right;" so also the heights of New Zealand are peculiarly marked as the fastnesses and places of refuge of powerful and tyrannical warriors and chiefs. Their summits bore the well-fortified Pahs of the chiefs; and at the foot of the hills, the dwellings of the serfs ranged to a great distance, with the Kumara-fields which they had to till. The ruins of those dwelling-places at the foot of the heights are seen to this very day; most distinctly perhaps at the foot of Mt. Smart; nor the mountain-cones themselves bear less evident testimony to their former destination.

The slopes of the hills look, I might say, tattooed, like the faces of the old surviving warriors, who have been spared from the carnage of the cannibal age. They are terraced, that is to say, terraces are cut around the declivities, 10 to 12 feet high, which are visible

Upon the Isthmus of Auckland, of yore.
Maunga Wao (now Mount Edne) as ancient Maori Pah.

at a considerable distance. Upon those terraces double rows of stockades were planted in olden times, and deep holes dug, covered with branches, reed, and ferns, like wolf-traps, for the purpose of insnaring the assailing foes. Other pits less deep, connected by subterraneous passages from above and below, and having ingeniously concealed outlets, served the defenders of the fort as secret paths and hiding-places, or as ambuscades, from which they sallied forth upon the assailants; and in a third sort of holes in the ground they had their provisions stored away. The observer is justly struck with astonishment on seeing, how ingeniously and practically the Maoris had planned their forts, and what colossal works they were capable of executing with extremely rude and defective instruments of wood and stone; with wooden spades, with hammers, chisels and axes of stone, and with knives wrought of muscle-shells. Behind all those palisades and ditches encircling the slope of the mountain, high on the top lived the chief with his family and the nobles of his tribe.

Now-a-days the houses and huts are destroyed; the last vestige of the stockade has disappeared; the Maori-castle is in ruins. And as the crater on the top has remained as it were a scar of the fiery combat in the bowels of the earth; so the terraces with their deep holes and ditches are the scars that remind us of the bloody combat of nations long passed away. Heaps of seashells[1] are the remnants of the repasts of the savages. Fernweed, Manuka, and other indigenous plants, or the grass and clover of the European settler are covering with their soft vestment of luxuriant verdure the former scene of action of a valiant people past and gone, whose mighty deeds are now living only in songs and traditions. Of the tribe, once so numerous and powerful, there are but a few families inhabiting a small village on the Orakei Bay East of Auckland. The lava-caves at Three Kings, Mt. Smart, and Mt. Wellington are crammed with the skeletons of those unhappy victims, who perished, during the second decade of this century, in the murderous wars of the terrible Hongi. Upon one of the mountains, Mount Hobson, so called in honour of the first English Governor of New Zealand, I found one single, solitary inhabitant left, living beneath the scanty shelter of a tattered tent half underground, — an old deranged Maori woman, an out-cast upon that lonely spot according to the superstitious customs of her kindred, and doomed to die in dreary solitude where thousands of her tribe had died of old.

Such is the "Past and Present" of the Isthmus of Auckland.

---

[1] Mytilus, Venus, Ostrea, Turbo, Monodonta, Trochus, etc.

CHAPTER XII.

The North Shore.

When the inhabitants of crowded cities, disgusted with the continual smoke, dust, and busy bustle of town, and yearning after the free haunts of nature and the clear, open sky, avail themselves of the few days set aside for recreation, to expand once more the breast, contracted by continual sitting in the musty office or counting-house, and with wife and children to migrate into the country, — we find this very natural, the more so, when winter with its frigid cold and blustering snow-storms returns so soon, as in Middle Europe, confining us poor mortals within the four walls of home. But in Auckland, where, so to say, the town itself is in the country, where the mild and mellow climate — rainy days excepted — never prevents the happy inhabitants from passing their leisure-moments out in the beautiful garden around the house, and where the clear, sunny sky, not hidden by rows of

[1] Most of this chapter is from the pen of my friend Dr. Haast, who acted the historiographer of our adventures, while I was engaged in drawing the map. The woodcut, see Ch. I. p. 5, represents a view of the North Shore.

four or five-storied houses, smiles upon every street; — I say, on hearing somebody in Auckland speak of longing after country-life, we could hardly refrain from smiling at such an idea.

For some time past we had intended to go to the northern coast of the Waitemata Harbour, to the so-called "North Shore," for the purpose of paying a visit to the volcanic cones there, and to a remarkable lake lying in an ancient crater, of which we had heard so much. A friend of ours, however, wishing to accompany us, requested us to postpone our excursion for a short time. Knowing him to be an excellent guide and pleasant companion, we readily agreed to wait a few days longer, for the benefit of having him with us. We were, however, not a little surprised, when he informed us, that his wife and children were also to be of the party in order to enjoy the pleasures of country-life; and that he would, therefore, take two tents with him for the purpose of camping out and having the full benefit of a short stay in the country. It naturally made a somewhat ludicrous impression on us, to hear our worthy friend thus speak about the country, and the more so, since we had repeatedly admired his idyllic dwelling as the very ideal of a country-seat.

Let the reader fancy to himself, at one of the many small bays of the Waitemata, situated on a sheltered slope, a cottage-like house snugly nestled in the luxuriant shrubbery of its surrounding garden. The house is covered with passion-flowers, honey-suckle, and other beautiful creeping plants; the veranda in front of the house decked with fuchsias, whose charming bell-shaped blossoms have clothed roof and walls as with a purple robe; round about a large garden, its farthest end kissed by the rippling wave of the calm, blue sea. Boats and sails of every kind enliven the watery surface, which forms a part of the variously indented Waitemata Harbour. Beyond the harbour the North-coast with its volcanic cones is seen; above it the regular cone of the Rangitoto with its taper-points is looming high up to the azure sky, — in short a scene so charming, so pretty and picturesque, that we could never tire of admiring it. The house with this prospect is the very same idyllic dwelling of our friend, and the beautiful garden adjoining it is a perfect pattern of a New Zealand garden, a lovely spot of earth, on which one cannot help being contented and happy. Hedges, six to eight feet high, of monthly roses, fuchsias, and geraniums, their leaves and blossoms presenting a many-coloured tapestry, encompass the garden. The damp New Zealand climate secures to that luxuriant vegetation the charm of perfect freshness even in midsummer. And in the garden itself, what a variety of trees and shrubs and plants! All the plants of the temperate zone thrive here, and amongst them a great many forms reminding us of a warmer climate. The German oak-tree with its knotted branches towers by the side of the slender Norfolk-pine *(Araucaria)*; the blue gum-tree of Australia *(Eucalyptus)*, by the side of the weeping-willow and locust-tree. Interspersed between them are groups of oranges and lemons; the banana of India; the date-tree of North Africa; the granate-tree; the myrtle, and the fig-tree. Jessamines, bignonias and roses; heliotropes, coronellas, camelias and dahlias grace the flower-beds in rich abundance, forming upon them a gay, brilliant floral texture; while upon the verdant turf the agave of South America proudly rears its blooming shaft from between its juicy leaves. In truth, it is delightful, charming indeed, to walk among such beds and under such trees.

But *"le bonheur est dans l'inconstance!"* — Our friend had made up his mind to enjoy the country-air, and hence we rowed in his boat, — two Maoris managing the oars, — to the North Shore, a distance of about one league. We landed on the low sandy beach strewn with mussel-shells, and ere long the Maoris had two tents pitched, in which we made ourselves at home as well as the circumstances permitted. The larger tent was intended for our friend and his family, serving at the same time as common dining-room, while in the second tent we had just room enough to lie down to sleep at full length. The tents were stretched so close to the sea-beach, that at high-water they were almost touched by the waves. It was a clear sunny day, the scorching heat, however, was pleasantly moderated by the prevailing southwest wind.

We set out forthwith, to take a closer view of the place, which bids fair to become more and more a place of amusement for Auckland; which, however, as yet had but very little of the appearance of a fashionable summer-resort, although, as I have been informed, even the governor does not mind passing some weeks in the summer of every year with his family at this place; of course, in tents like all the rest. With the exception of some small huts and the pilot's house, there was no shelter to be found upon ꞏthe North Shore. But it seems to be a pleasure to a great many inhabitants of Auckland, to exchange for a short time the comforts and conveniences of a house for the simple living in a tent.

The North Shore is a peninsula, and was probably formerly an island. Only a narrow, slender strip of sand connects the peninsula with the mainland. Small as it is, measuring scarcely a mile at its widest part between the Waitemata and the East-branch of Shoal Bay, it nevertheless presents various points of attraction to the geologist. The western half consists of tertiary sandstone and shale, forming precipitous walls on the Auckland-side, in which near the high-water line small seams of lignite crop out. Farther East the tertiary strata cease, making room for a flat strand covered with muscle-shells. Those shells are piled up in heaps several feet high, and, there being no limestone in the vicinity of Auckland, they are burnt for lime. Behind the muscle-banks small extinct volcanic cones arise, the scoriæ and lavas of which extend farther West to the sea. The principal one of these cones is Mount Victoria, formerly called Takarunga, a crater-cone nearly 300 feet high, upon which a flagstaff has been erected.

In passing along the beach, we came to a kind of scaffold about 30 feet long. Our organs of smell betrayed to us at a considerable distance its object. A long row of fish, sharks and other kinds, were suspended from it to dry, tossed to and fro by the wind and promising the natives a favourite dish for the winter with a great deal of "haut gout." Fat pigs and lean dogs were running about; and farther on, there were some Maori huts. The old folks sitting at the door hailed us with their cordial "Tena-koe," while the black-eyed, half-naked children were staring at us with amazement, probably at a loss to know, what the two men with hammers in their hands were about. The plantings about the huts, consisting of potatoes, cabbage and other culinary vege-tables, were in a tolerably good state of cultivation, and surrounded with a wall of lava-blocks piled up one above the other four feet high, up which pretty lianas of luxuriant verdure were climbing.

Our object was, to visit and to examine the most easterly of the three cones, called Takapuna by the natives, and 216 feet high. It forms the North-head of Auckland Harbour, is of an almost hemispherical shape and on three sides washed by the sea, from which it rises in steep ascent. It is the most interesting of the North-shore hills.

Takapuna, the North head of Auckland Harbour.

The first eruptions were here evidently submarine, the basis of the hill round about being formed of regular layers 20 to 40 feet thick. These layers consist of volcanic ashes, scoriæ and lava-frag-ments baked into a solid breccia, and with an outward inclination at an angle of 12 degrees, so that at low-water, at the foot of the cliffs, which are 20 to 30, some even 40 feet high, the traveller can walk on the lower strata swept clean by the rolling surge, nearly all around the hill as on a roof inclined at an angle of 12 degrees. We made the attempt; of course we had at some places to climb along rock-walls upon projections scarcely half a foot wide, below which the foaming sea was dashing up its spray. At last we ar-rived at a point, where the surge had washed out a deep cave. Its walls were entirely lined with a salt-crust. Here large blocks of rock prevented our advancing farther, and we were obliged to climb our way back along the same difficult passage by which we had come. Above the regularly stratified tuff-cone arises with steeper inclination the scoriæ-cone. It is closed up at its top, but displays at its western declivity a flat indentation denoting the crater, from which in a westerly direction a small stream of lava had issued forth. Manuka bushes and ferns cover the slopes of the hill, between them the European high-taper rears its slender shaft. The stately plant with its bright leaves and its pure yellow blossoms looked quite strange amongst the gloomy, monotonous New Zealand vegetation. An observer, however ignorant of the geography of plants, would necessarily recognize the foreigner at first sight.

The Takapuna is very remarkable for the numerous lava-drops, or volcanic bombs, found on its surface, bombs of the most regular pear or lemon-shape with spiral points, which must have been formed by the rotation of the glowing masses thrown out in a fluid state. The interior structure of the lava-drops is dense and brittle like cooled cast-iron. They are found of all sizes: small like a lemon, and others 3 or 4 feet long with a thickness of 2 feet and a weight of several hundred pounds. Those bombs could have been only cast after the point of eruption had already risen above the level of the sea. This being quite a novelty to our Auckland friends, and it being moreover the intention to have some very characteristic specimens of such bombs deposited in the Auckland Museum, each of us took one of them on his shoulder, and thus heavily laden we returned to our tents.

The natives we met on our way, stopped in perfect amazement, and we heard one of them whisper to another the word "Kaura," i. e. gold, which was quite amusing to us. Of course, what could those children of nature know about volcanic bombs? They naturally reasoned, that

Volcanic bombs.

gold alone could induce us to bear those heavy clods in the sweat of our brows for a whole hour.

On arriving at our tents we found behind a bulwark constructed of lava-blocks a gaily blazing fire, over which the teakettle was suspended, and our Maoris were busy gathering oysters, which are found there in great quantities on the rocks of the beach. And in the tent the worthy spouse of our friend had an excellent dinner prepared, which we enjoyed with the best appetite possible. In vain, however, did I wait for the oysters. Being specially fond of this shellfish, I betook myself in person to the natives in order to see about the matter. I found them all busy opening the shells with rocks, and greedily devouring their contents. Three large stones, covered all over with the finest oysters, were yet steaming on the coals. The Maoris pointing to the stones, meant "kapai" (very good), and rolled one of the stones before me, after having duly tested the degree of heat which the shells had sustained. Of course, I did not wait to be asked too often, but helped myself freely. Oysters roasted à la Maori are indeed no bad dish. The lids were easily removed and the oysters roasted in their own liquor tasted daintily. After having most scrupulously cleared the stone holding about 25 oysters, I too said "kapai," and turned my steps again towards the pastry of our amiable hostess, who could not refrain from smiling mischievously at my gastronomical peregrinations.

Dinner over, we set out to ascend the flagstaff-hill or Mount Victoria. It is the highest volcanic hill on the North-shore, 283 feet high, and is called Takarunga by the natives. In former times the summit bore a pah, and it is from the fortifications of this pah that the terraces, cut along its slope 10 to 15 feet high, date, likewise a hole on the Northside of the hill about 20 feet wide and deep. The top is flat and truncated, it presents still a semicircular crater open towards Southeast, from which in the same direction several lava-streams issued forth as far as the sea, forming stony bars. The prospect from the top is truly charming. It opens a view over the whole Waitemata Harbour, and farther on, the Hauraki Gulf is visible with its islands and promontories, and the sea alive with sails of every kind. Beyond the sea a large Maori settlement is seen, belonging to a tribe emigrated from the Bay of Islands, who some years ago very readily paid here the

Government one pound sterling per acre, for the purpose of raising upon the fertile volcanic soil crops of maize, wheat, potatoes and other vegetables for the neighbouring Auckland market. The people are said to have attained to considerable wealth by their industry. Their vessels — among them several war-canoes with beautiful carvings in wood on the prow and stern, and two neat whaling-boats — lay upon the beach.

A War-Canoe of the beach.

Between Victoria Hill and Takapuna head there is a third small scoriæ-cone, about 100 feet high, with its crater in a tolerable state of preservation, which on the map I have styled "Heaphy Hill" in honour of Mr. C. Heaphy.

Towards evening we returned to our tents, and sat together for a good while in social confabulation. The roaring of the sea at our feet was like a grand cradle-song; our couch, however, though there was no scarcity of woollen blankets, would not suit us at all. A keen wind having arisen, our tent was most uncomfortably waving to and fro, threatening every moment to be upset. How easily we might have reached Auckland in the course of an hour, — its lights gleaming invitingly over to us across the waters, — thence to return in the morning and thus continue our excursions! However, our friend had invited us to his country-tour, and we were obliged to share also these pleasures connected with it. Despite wind and tempest we slept in peace the sleep of the just, and awoke in the morning at sun-rise, greatly refreshed from our night's repose. A bath in the waves sporting at our feet helped to make us wide awake; tea was soon made, and we set out to visit Lake Pupuke about five miles distant in a northerly direction.

Our path lay between the crater-cones we had visited the day before, across the slip of land between the North Shore peninsula and the channel on the North-coast of the Waitemata Harbour. Not a tree interrupted the monotony of the fern-land; only here and there a heifer was seen grazing, New Zealand field-larks were flattering about, and crickets and locusts were chirping their shrill notes. The soil consists of a stiff whitish clay ("pipe-clay"), being covered with dwarf Manuka *(Leptospermum)*, fern *(Pteris esculenta)* and a variety of small shrubs and tufts of grass. Nothing else seems to thrive in the sadly sterile soil. And yet that very soil bore in olden times luxuriant forest-trees.

This circumstance gives rise to various reflections; for the sterile pipe-clay soil occupying such extensive tracts in the vicinity of Auckland, and especially in the districts East and West of Auckland, in which nothing will grow, not even grass, is a real calamity. The question arises: is there no means to restore to the soil the productive power, which it must necessarily have possessed of old, when it produced those towering Kauri forests, traces of which are plainly seen in the Kauri-gum which the natives are everywhere digging from the surface of that soil? Experienced farmers must decide this question by direct experiments. At any rate, however, the method usually pursued by the colonists upon the fern-heaths, seems to be throughout a perverse one. If immediately after the burning down of the woods, grass and clover-seed had been sown into its ashes and humus, a heavy growth of grass might perhaps have preserved the humus-surface of the soil. But they burn again and again; the winds carry off the ashes; the rain is gradually washing the humus away, and at last nothing remains but the naked clay-soil, upon which only *Leptospermum* and *Pteris* are scantily thriving. And in consequence of the usual burning-system, even these plants are not allowed to grow strong and hardy, and gradually to reproduce humus; but year after year those bushes are set on fire and burnt down. The owners of the ground assert that it was done, because the cattle was fond of browsing the tender shoots that spring up after the conflagration. But these also are growing more and more scarce from year to year, and a system, perhaps correctly applied to very fertile alluvial soil, for the purpose of exterminating the exuberant growth of fern and preparing it for a crop of corn, is certain to crush upon this stiff clay soil even the last scanty trace of verdure. The method of burning out was originally a custom of the natives, who by burning cleared the forest, tilling it once after the burning, and then again went in search of new ground. Thus applied, the system is a correct one, but the repeated burning is an abuse. After the first burning luxuriant undergrowth is produced; after the second, tall flax and fern, finally dwarf fern and Manuka, and last of all the naked soil remains.

It was a fatiguing walk across a sadly waste plain, although on the right the view of the sea and the beautiful cone of the Rangitoto Mountain, on the left a glance at the deep incisions of Shoal Bay presented to the eye a great change of scenery.

It was not until we approached the lake and arrived upon volcanic soil, that both land and vegetation assumed a different character. Taller shrubbery intermixed with peach-trees made its appearance; a Maori hamlet with some twenty huts lay there, surrounded with fields and meadows, and the New Zealand flax-plant — its tall and luxuriant growth being always indicative of fertile soil — stood in powerful bushes by the way-side. We had arrived at the foot of a gently sloping hill.

Large fields, fenced meadows and a farm-house situated on the top between fruit and ornamental trees betrayed the industry and the labour of a farmer settled there. We ascended the slope and found ourselves suddenly at the brink of a nearly circular very steep basin, about one mile in diameter and four to five miles in circumference at the bottom of which, quite picturesque between wood-clad shores, lay the remarkable Lake Pupuke (alias Pupaki and Pupuki), the largest of the crater-lakes in the vicinity of Auckland. That settlement is beautifully styled "Flora See," and belongs to a worthy German physician, my friend Dr. C. Fisher in Auckland, who has established here extensive nurseries and vineyards, and expects to produce excellent wine within a few years. Lake Pupuke is a fresh-water lake of "unfathomable depth," as I was told. I therefore requested Captain Burgess, the pilot of Auckland Harbour, to measure the depth with his sounding apparatus. The greatest depth in the middle of the lake amounted to 28 fathoms, so that the bottom of the lake lies 140 to 150 feet below the level of the sea, from which the lake is separated only by a very narrow ridge, a part of the crater-frame. The lake fills the crater-basin of a gently sloping tuff-cone, rising only about 100 feet above the level of the sea and consisting of layers with a regular outward dip. On the steep inner crater-wall here and there basaltic dykes appear and on the Eastside of the cone larger masses of basaltic lava, dating from real lava-streams, are forming cliffs jutting far into the sea. In these lava-masses there are said to be caves full of human skeletons, memorials of the former outrages in the wars of the natives. Pupuke is the largest and deepest among the numerous tuff-craters in the vicinity of Auckland, and the question is justly asked, whence the lake, situated upon a low isthmus, derives its water-supply. Is it not natural to think of the opposite Rangitoto mountain, separated from Lake Pupuke only by a branch of the sea four miles wide, the highest among the lava and scoriæ-cones of Auckland? The natives assert that the Rangitoto was taken out of Lake Pupuke; but we say, that Lake Pupuke obtains its water from the extensive lava-fields of the Rangitoto by means of subterraneous or rather submarine channels.

Hundreds of wild-ducks were swimming on the lake; it is also said to be abounding in all kinds of fish, especially eels. On the shore we fished interesting fresh-water shells and fresh-water plants out of the water,[1] and the wood furnished us many a beau-

[1] Prof. Alexander Braun in Berlin, whom I sent specimens of those plants, writes to me as follows: "They belong to *Nitella hyalina* (Chara D. C.), first found at the Lake of Geneva, by and bye in various places of southern Europe, also in Belgium and, in somewhat deviating forms, in Middle Asia, the East Indies, northern Africa, at the Cape of Good Hope, and in the warmer North America, to which New Zealand has of late added its variety of *Novæ Zelandiæ*, differing from the European form by its smaller stature, larger seeds (sporangies) and very delicate points of its leaves.

tiful fern. At the sight of this charming lake we were vividly reminded of the "Laacher See" near the Rhine. The old celebrated cathedral on the Laacher See is represented here by the church and school-house of a deserted Roman Catholic Mission Station on the southern side of the lake.

Towards 2 o'clock we beat the retreat to our camping ground. The wind in the meantime had increased to a perfect gale, which blew into our faces with full blast from Southwest. On arriving at our tents we found our sleeping-tent upset by the storm, and the sea went so high, that we could not venture to return in our small boat. Fortunately, towards evening two of our acquaintances came from Auckland with a cutter, for the purpose of taking us back. We embarked behind the North Cape at a place quite protected from the wind, and although many a wave dashed over us, and our little craft leaned so much on one side, that we apprehended every moment its taking water, our sea-experienced friends brought us safely ashore at Auckland after a tiresome three-hour's cruise against the heavy gale.

Our worthy friend, however, determined to enjoy country life with his family, remained despite storm and tempest quietly in his tent on the North Shore, and returned the next day to Auckland, highly gratified with his short rustication.

CHAPTER XIII.

Round the Manukau Harbour and to the Mouth of Waikato River.

Onehunga. — Rev. Mr. Purchas. — The Manukau Basin. — Whau-Bay. — Volcanic rocks. — Magnetic iron-sand. — Taranaki-steel. — Huia Bay. — Saw-mills. — Maori path through the bush. — The pilot-station. — Character of the West coast. — The weather. — South head of the Harbour. — Kauri Point. — Schooner „Sea Belle". — Waiuku. — Lignite. — Caravan of Maori traders. — Te Rata Hokitata. — Collections. — Formation of sandstone banks from quicksand. — Mouth of the Waikato. — North and South head. Discovery of Belemnites. — Queen's road. — First view of Mount Egmont. — Maori mail. — Fossil ferns on the West coast. — Awaroa Creek. — From Waiuku to Mauku. — Farmhouses blessed with daughters. — Return to Auckland.

Captain Wing, the obliging pilot of Manukau Harbour, had offered me his excellent whale boat, for a cruise upon the waters of the Manukau, and promised to accompany me with my friends, the Rev. Mr. Purchas and Captain Ninnis of Onehunga. Our place of rendez-vous was Onehunga. I had made my appearance there two days previous, for the purpose of roaming, in company of Mr. Purchas, over the environs of the town. Before I begin to the describe the country, I may be allowed to dedicate a few lines to the cherished memory of that noble-hearted man. Endowed with an extraordinary talent of observation, and exerting himself with an untiring zeal and energy in the most different directions, Mr. Purchas has rendered great services to the colony by the share, justly due to him, in the discovery and management of the coalfields in the vicinity of Auckland, as well as by the invention of a new method of preparing from the leaves of *Phormium tenax*, the well-known New Zealand flax, so highly valued for its durability and tenacity. That worthy man attends with equal credit to his spiritual functions; and Onehunga is indebted to him for a beautiful school-house built of massive basalt-stone. In his agreeable company I spent many a pleasant day, and in the circle of his amiable family I passed many a happy evening. Whenever I went to visit the cheerful parsonage, homely situated in a small garden, I was always certain to meet with some little surprise; for the children also, after the example of their energetic parent, rivalled each other in industry for collecting curiosities, and had always something reserved for me, such as butterflies, beetles, or shells. Thus I am indebted to his family for many friendly services and also for valuable contributions to my collections.

Manukau Harbour[1] presents a very extensive sea-basin, hemmed in on the North by hills, on the South and East by low flats; its diameter is from 12 to 15 miles. It is crossed by three navigable channels: the Onehunga, Papakura, and Waiuku Channels leading to creeks of the same names, which recede far back into the country. The mail-steamers come to anchor in the Onehunga Channel at White Bluff, a few miles West of Onehunga on the North-side of the harbour. Papakura Creek leads to the immediate vicinity of the coalfields at Drury, and Waiuku Creek is the route to the Waikato country. At neap-tide the shallow mud and sand-banks between the channels become dry; at flood-tide, however, the water rises from 10 to 13 feet. Ebb and flow, and the prevailing Westwinds, cause the water to be in continual motion; the Manukau is, therefore, noted for being a boisterous water. As the harbour is frequented only by the mail-steamers and small coasting-cruisers, it is quite dull in comparison with Auckland Harbour; on the other hand, during calm weather, numerous canoes of the natives are to be seen passing between the Maori settlements along its shores.

[1] Manukau from Manuka *Leptospermum scoparium*, a shrub or a small tree, very frequent throughout the country.

Captain Wing having been detained by his pilot-duties at the departure of the mail-steamer, it was not until January 18, that we could begin our excursion. The weather was extremely favourable; there was a perfect calm. The fogs, which in the morning had covered, like a pall, that low flat region, had passed away by 7 o'clock, and the sun looked cheerfully from the cloudless sky upon the placid watery mirror of the Manukau Gulf. A better day we could not have desired.

We embarked from the Onehunga pier; five natives managed the oars, and with Captain Wing at the helm, we drove along the Northcoast of the harbour. The banks break off in steep declivities, presenting regular strata of sandstone and shale, and with here and there small seams of lignite at the water's edge, exactly as on the banks of the Waitemata. The strata lie mostly horizontal, local disturbances, such as at Matinga Rahi Point (Cape Horn), excepted. Opposite, on the southern shore, rises the volcanic cone Mangere, the volcanic island Puketutu (Weekes Island), and the smaller cones about Tumatoa Point.

In Whau Bay we had reached that point, where Whau Creek, a long narrow South-branch of the Waitemata cuts so deeply into the Auckland Isthmus, that the whole breadth between the two harbours amounts to but little over one mile. As the land is moreover very low — its highest point being only 111 feet above the level of the sea — the colonists have repeatedly entertained the project of connecting the two harbours by a canal; and, towards the execution of this plan, a benefit rather than disadvantage might arise from the fact, that ebb and flow do not take place at the same time at the East and West coasts; the flow being later on the West coast by three hours than on the East coast. From Whau Bay the coast, intersected by Little Muddy Creek, Waikumate and Big Muddy Creek, extends in a southwesterly direction to the long and keenly jutting peninsula Puponga. The land here begins to rise, and the mountain-ridges of the Titirangi Chain extend their wood-clad spurs as far as the coast.

On the North-side of the Puponga peninsula, in a small inlet, we went ashore, and encamped for dinner under the shade of a magnificent Pohutukaua tree *(Metrosideros tormentosa)*, the trunk of which measured 24 feet in circumference. In the back-ground rugged cliffs of a most remarkable appearance are seen. Tremendous blocks of volcanic rocks, of trachyte, andesite and basalt — blocks of 4 to 6 feet diameter, angular and sharp pointed, and of all colours, red, green, brown and black — are cemented together into a conglomerate or breccia, forming huge masses of solid rock, in which, nevertheless, a rude stratification may be observed. From some of those blocks I succeeded in knocking out neat crystals of pyroxene. Here we have the commencement of those strata of volcanic conglomerate and breccia which, more than 1000 feet thick, compose the Titirangi Chain, and from the Manukau Northhead stretch as far as close to the Kaipara Harbour, forming the rugged, rocky precipice of the Westcoast.[1] Proceeding along the beach of Puponga peninsula in a N. W. direction to Karangahape Bay, upon the banks of which, on a less favourable site, a township, called Cornwallis, has been laid out — of which, however, as yet, not a single house stands — the lower strata appear; first banks of a soft, rusty sandstone, which is sprinkled black with fine grains of magnetic iron; and farther on layers of sandstone and shale, such as the Isthmus of Auckland is composed of. One is able to convince oneself here, that the violent convulsions, which formed those colossal masses of volcanic conglomerates, are of a more recent date, than the formation of the tertiary strata on the Waitemata. Here also, in several places of the beach, black titaniferous iron-sand has been found of the same description as the iron-sand, which is an ingredient of the sand along the whole West-coast of the northern Island; which especially on the coast of Ta-ranaki covers the shore for miles, and which, according to ex-periments made with it in England, is said to produce most

---

[1] As far as these volcanic conglomerates on the sea-coast are exposed to the action of the sea-breeze, they form solid masses of rock, while farther inland they are totally decomposed into variegated clays.

excellent titanium steel. [1] The iron-sand here has evidently its origin in that rust-coloured, crumbling sandstone, which rises along the shore in banks measuring from 10 to 20 feet and is washed out by the breaking of the surge. The original mass, however, by the decomposition of which the grains of magnetic iron were mixed with the sandstone, must be a volcanic rock of prior date.

The peninsula of Puponga protects the interior portion of Manukau Harbour against the waves of the ocean, which penetrate between the "Heads" far into the outer portion of the harbour, and are broken on the Westside of the peninsula. We passed over the low ridge of the peninsula. The boat was awaiting us on the other side, and thence we crossed over to Huia Bay, where, close by the seashore, upon the dry sand of the beach, we pitched our tent, for the purpose of camping through the night.

January 19. It was a sleepless night, as in the evening, attracted by the glare of our light, numerous mosquitoes had found their way into our tent, which tormented us most unmercifully. Joyfully, therefore, we hailed returning day. A refreshing morning-breeze, fresh spring-water, and a cup of good coffee had to supply the want of the invigorating effects of healthy slumber; then we continued our expedition for the purpose of visiting the settlements situated farther in the background of the bay.

I was extremely surprised by the romantic character of the landscape now surrounding us: a wild mountainous region, covered with dusky woods; lofty, sharp-pointed peaks, steep and rugged precipices, and gloomy ravines, from which brooks and rivers of the clearest spring-water are continually issuing forth. Enterprising colonists have chosen this romantic wood-region for the establishing of saw-mills. The primeval forests furnish in the powerful trunks of the Kauri pine an excellent timber; brooks and rivers yield abundant water power for mills, and are at the same time used for floating the timber; and thus the busy life and work of a wood-cutters colony is developing itself here, as in the wild woods of

[1] Comp. p. 265.

California and Canada. The saw-mills in Huia Bay on the Manu-kau-side, and Henderson's Mill on the Waitemata-side are the principal establishments of this kind for the preparation of boards and all kinds of timber in the vicinity of Auckland. We paid a visit to one of the saw-mills, which is constructed and managed in the best manner possible; thence we followed a tramroad leading about three miles into the woods to the point, whence the wood-cutters were engaged in rolling the powerful Kauri logs down-hill upon the road; and by 11 o'clock we were back again on our camping ground.

From Captain Wing's house on the North-side of the entrance to the harbour, — which was to be our stopping-place for to-day, — we were distant only four or five miles, and we could easily reach it by water in course of an hour. We determined, however, on taking the more interesting route by land across the mountains. But we did not arrive at our destination until after a very trying march of five hours; nevertheless we were amply repaid for our hardships by the grandest and most imposing display of natural scenery, which on the way was were everywhere presented to our view. The passage we had chosen was a Maori path but rarely trodden, leading through the deepest ravines, and over the highest, steepest rocks, and which cannot be found without a very expert guide; for one would almost need the instinct of a wild beast to perceive in those wild haunts a track, which others have trodden before. Already from the heights near Onehunga I had observed, in the direction of the Manukau heads, several peaks of sugar-loaf shape, designated upon the maps as Omanawanui and Pukehuhu, and had been curious to know, of what kind of rock those singular peaks might be composed. To-day I had ample opportunity to convince myself with hands and feet, that they are very hard, pointed masses of volcanic conglomerate, like those upon the penin-sula of Puponga, shooting up as it were in taper-pointed masses; for it was over the topmost points of those very cones of rock, that we had to climb before arriving at Captain Wing's house, which is built high above the foaming sea upon a ledge of rocks like an albatross-nest. There we found late in the evening our hospitable quarters for the night. It was not until next morning that I was able to survey more closely the remarkable spot we had arrived at.

Entrance to Manukau Harbour.

January 20. The house was built by the Government as the station of the pilot for Manukau Harbour. It stands upon a rock 300 feet high, shelving off towards the sea so abruptly, that the passage down to the landing, at the time of my visit, when the projected stairs were not yet existing, was a really break-neck affair. The building material for its structure and furniture all had to be hauled up over the precipice by means of a crane, like up a tower. The house is splendidly furnished, and supplied with a measure of comfort, one would never expect to find in such a Robinsonian seclusion and upon a rock so inaccessible. This spot commands a wide view out to the sea. It has been asserted, that in perfectly clear weather even the snow-capped peak of the Taranaki mountain, at the entrance of Cook Strait, a distance of 140 sea-miles, is visible. From the signal station erected upon the opposite conical rock Paratutai (350 feet high), the signals are given to vessels arriving, for the purpose of guiding them safely through the channels between the dangerous sand-banks in front of the entrance.

After breakfast we climbed down the steep rock to the beach, and walked along the coast northward. The land shelves abruptly towards the sea over barren walls of rock, 400 to 500 feet high, along which mighty banks of rough, volcanic conglomerates and breccias, broken through by basaltic dykes, are laid bare. But a broad, flat sand-beach, upon which the waves roll sluggishly, and a row of sand-hills still separate the foot of that stonewall from the sea. Similar to this is the character of the coast on the whole West-side of the North Island, so that the sandy beach, for want of other convenient passages, is the natural road for the communication along the coast. It is only at a few projecting points, that the surge breaks directly against the rocks, which must then be climbed over, sometimes not without danger.

The whole sea-beach from Kaipara Harbour as far as Taranaki South, a distance of about 180 sea-miles, consists of fine sand abounding with titaniferous magnetic iron; and at places where wind and waves have distinctly separated the heavy black iron-sand from the lighter grains of quartz, the beach presents an appearance as if gunpowder had been spilled there. Already the earliest settlers on the Taranaki coast directed their attention to this peculiar black sand, which is strongly attracted by loadstone, like iron filings. Samples of it were repeatedly sent to England, and there submitted to chemical analysis. It was found that 100 parts contained 88.45 parts of black oxide of iron and 11.43 parts of oxide of titanium, a compound similar to the iron-sand of volcanic districts. An analysis by Mr. Moritz Freitag has yielded, protoxide of iron 27.53, sesquioxide of iron 66.12, oxide of titanium 6.17. For years past, therefore, it has been an object of speculation, to apply this iron-ore to technical purposes; and within the last years experiments have been made on a larger scale. The merit in this affair belongs to Captain Morshead, who went to New Zealand and brought several tons of the ore back to England for decisive experiments. These trials are said to have led to the most satisfactory results. The sand yields 61 per cent of iron, and steel of unusual hardness and tenacity can be produced from it. Messrs. Mosely in London have submitted the Taranaki-iron and titanium-steel to tests, and declare them to be of most superior quality. Should it be confirmed as a matter of fact, that the Taranaki steel surpasses by far all other sorts of steel in quality, we may well take it for granted, that the Taranaki iron-sand will prove of great value, if it is only possible to obtain large quantities of it sufficiently pure to be smelted, a thing I have reason to doubt.

Behind the sandy beach, basins of fresh-water are frequently found, and at the foot of the rocks deep caverns are seen washed out, in the background of which generally large masses of boulders are deposited. This would indicate a former period, when the surge washed the rocks themselves, and piled up those masses. Now, these caves are a safe camping-place at night for the cattle grazing upon the grass-covered sand-hills.

January 21. All night the storm was raging round the house, and when at break of day I looked from our lofty watch-tower over the mountains and the sea rolling beneath, it did not appear as though we should be able to undertake anything at all on that day. A violent N. E. wind chased heavy clouds from the sea upon the land; and the sand-banks at the entrance of the harbour, which during spring-tide or a calm sea are scarcely visible, were to-day plainly to be seen amid the wildly breaking surges. Towards 9 o'clock, however, the wind veered more to the West, and subsided; and we had after all a fine, clear day. We sailed across the strait to the South Head, the Mahanihani of the natives. What a remarkable difference between the two shores — North Head and South Head! On the North-side, ranges densely covered with forests, sharp pointed conical rocks, hard masses created to bid defiance for thousands of years to the impetuously dashing waves; upon the South-side nothing but loose sand, whirled up by the sporting winds to a height of 500 feet against the barren steep, which in some places free from sand presented only soft strata of clay and sandstone together with thin layers of lignite. This contrast in the composition of the two shores proves, that the entrance to Manukau Harbour owes its origin to a considerable dislocation in the coast-range.

Thence we rowed along the southern shore of the Manukau, and arrived in the afternoon close to Kauri Point at Mr. Graham's dwelling, where Mrs. Graham, in the absence of her husband, gave us a most hospitable welcome. As Captain Wing, in compliance with the duties of his office, was obliged to return to his station, our luggage was here put ashore. We were all under great obligations to the gallant captain for his valuable company and for the hospitable reception at his house.

The point we had now reached, offered very little for attraction. The steep declivity of the shore, about 100 feet high, as far as I had followed it up, presented, in unchanging monotony, clayey or sandy banks, alternating with strata of bituminous shales and lignite. The only remarkable feature of this spot is the presence of a thick bank of reddish-brown sandstone containing magnetic iron, the same as on the peninsula of Puponga. This bank can be traced along the whole South-side of the Manukau Basin. As to the rest, the region, almost treeless and little fertile as it is, presents a desolate and cheerless aspect, and I could well appreciate the complaints of good Mrs. G., who — after many sad blows of adversity, and after having finally, in Bass Strait between Australia and Van Diemensland, by a shipwreck lost all she had, saving nothing but the bare life — when cast upon this shore, was longing to exchange this dreary solitude for more genial regions. She exerted herself to the utmost of her efforts, to render the evening as pleasant as possible to her guests, and after we had retired, she even made her music-box play soft melodies to lull us into gentle repose.

January 22. The question, how from Kauri Point we were to reach Waiuku, a distance of about 18 miles by water, was most satisfactorily solved by a lucky accident. A small schooner of 20 tons, the "Sea Belle," which was plying between Onehunga and Waiuku, for the purpose of bringing from the latter place the produce of the natives, flour and flax, to the market of Onehunga, ran on her trip to Waiuku in the morning close to Kauri Point and took us on board. True; we had to confide ourselves to a vessel that was in a most miserable plight, and the master of which calculated besides upon the passengers performing sailor duties; for while the so-called captain sat continually at the pump, engaged in pumping out the water as it came streaming into the leaky boat, and while his only helpmate managed the helm, the weighing of the anchors and the setting of sails was left to the passengers. Meanwhile the wind was favourable; by 11 o'clock already we had reached our place of destination, and the "Sea Belle" lay before Waiuku; not, however, as one might suppose, at anchor, but, as the captain had deemed it the far simpler way of disposing of her, fast aground in the mud. There she is lying perhaps up to this day, rotten and decayed.

Waiuku, — the word is a compound of wai, water, and uku, white clay, — is a very appropriate name for the place. It is the name of the creek, which, in a southerly direction from Manukau Basin cuts deeply into the land; and the low banks of whose shores are formed of white clay and sand, below which, just at the water's edge, layers of peat-like lignite, in a thickness of several feet, crop out. These are the same post-tertiary strata, which reappear farther to the North West in the creeks at Drury and Papakura, and which compose the extensive flats on the East and South-side of the Manukau. Where the white clay forms the surface, there the land is sadly sterile. Luckily, however, the clay is covered to a large extent by basaltic conglomerate, the gradual decomposition of which furnishes a fertile, arable earth.

At the South-end of Waiuku Creek a cluster of houses, among which several mercantile shops and two taverns, presents the first start of the town of Waiuku; William's Hotel formed our headquarters. Though its environs can boast only few points of attraction, yet, the site of the place is an interesting one, lying as it does on the great thoroughfare of the natives from North to South, upon a kind of isthmus, which separates the Manukau Basin from the largest navigable river of the North Island, the Waikato. Between the upper end of Waiuku Creek, and Awaroa Creek[1] emptying into the Waikato River, there intervenes only a flat ridge 1½ miles wide, and not above 40 to 50 feet high, over which a brisk trade is continually carried on; because upon this road the natives bring their produces from the fertile Waikato valley, the granary of the North Island, to Manukau Harbour. It has, therefore, also been

---

[1] *Waikato* = streaming water; *Awaroa* = long water course.

considered profitable to have a canal dug for the purpose of connecting in a direct line the most important water route of the country with a harbour in the immediate neighbourhood of Auckland.

A walk from Waiuku to Purapura brought us to the camping-place of the natives on Awaroa Creek. There we met a large company, that had embarked upon a trading trip to Onehunga, and had just arrived from Waikato. It was really an interesting scene. The men, large, robust figures, with grave faces beautifully tattooed, were cozily smoking their pipes, while the women were busy peeling potatoes with the sharp-edged shells of the common Waikato Unio *(Unio Aucklandicus)*, and preparing their social repast. But as on such journeys nothing is left behind, there was also no lack of children, dogs and pigs playing about in unrestrained companionship. The provisions, which the natives carried with them, consisted of large quantities of dried Waikato-eel, and baskets of peaches, apples and potatoes. The articles for the market were flour, flax and Kauri-gum. We received a cordial invitation to partake of their meal, but contented ourselves with conversing with them instead. They were extremely talkative, and, like all Maoris, curious to learn our names and the object of our expedition. My name caused them no small difficulty, and it was really amusing to observe them, as they endeavoured to pronounce it in the most different manner, until at last one of them seemed to have found its true version. He called me Hokitata, and from the pleasure and satisfaction shown at it by all the rest, I could easily infer, that by it he had found the right word with the right meaning. My friend Purchas had previously informed them, that I had come from afar for the purpose of seeing their country, and that I was about to return soon. Then, of course, I could have no name more significant than *Hokitata*, literally meaning "about to return soon." This name, then, remained henceforth as my Maori-name upon all my excursions through New Zealand, with the single variation of *Hokiteta* in some districts, and the addition of *Te Rata*, the doctor.

Sometime afterwards, the whole company passed through

Waiuku. Men, women and children were pulling all together a large canoe, which they were dragging from Awaroa Creek to the Manukau; and one of them, in a red shirt, with a battle-axe (mere) in his hand, skipping about with the quaintest gesticulations, and singing and dancing, led the noisy procession.

January 23. — Sabbath day. — Waiuku at present belongs to the diocese of Onehunga, and having no pastor of its own, Rev. Mr. Purchas performed divine service to-day. To me the day of rest was a welcome one., as it gave me sufficient leisure to arrange my collections and to work at the map of the district. The evening we spent very pleasantly with the amiable family of Mr. Griffith. The pretty collections of beetles and plants made by the members of the family convinced me anew of the lively interest, taken by the colonists in everything that has reference to the natural history of New Zealand. The same observation I had afterwards occasion to make wherever I visited any one. Nearly in every house, in every family, I became acquainted with, there was somebody making collections. Here it was the husband, who had a collection of insects; there the wife, who pressed mosses and fern-leaves neatly between paper; or the sons and daughters, that gathered shells and seaweeds; and from their treasures I was always certain to have some new specimen or other, which I had hitherto not possessed, presented to me with a hearty welcome.

January 24. A most delightful morning. The fresh South-western breeze, in these parts always the harbinger of pleasant weather, had swept every cloud from the sky. After some fruitless endeavours to induce the natives to transport our luggage in their canoes to Waikato, upon paying them any reasonable demand, we found ourselves compelled to leave whatever we could possibly dispense with, behind at Waiuku, and, taking with us only the most necessary articles, to continue our journey to the mouth of the Waikato River. Mrs. Jenkins, to whose kindness I am indebted also for some fine specimens of *Helix Busbyii*, had furnished us with horses, which for the distance we had to go, were of great service to us.

We turned westward towards the sea-coast. Two miles from Waiuku we passed the small Maori village Tauwhare. Thence the road rises gradually to the height of the hills forming, 6 to 700 feet high, the coast-range. The higher the traveller ascends, the more his passage is obstructed by quicksand, till on the top he finds himself in a perfect desert of sand. I was not a little surprised to find, that the conical summits of the hills ascending steeply, sometimes at an angle of 45 degrees, and ending in sharp points, which here and there are seen to spring up from the ridge, and which had attracted my attention already from the distance consist of nothing but quicksand piled up by the wind. Nor was it less interesting to me to observe, how the sand is not only drifted together in loose heaps, which are subject to continual movement and change of form, but that, for considerable distances, the quicksand has, under the sole influence of atmospheric currents, deposited itself in regular strata almost in the same manner, as the drift-sand of creeks and rivers. According to the respective course of the wind and the surface of deposition, those strata assume different directions, and there are sections to be seen, where there

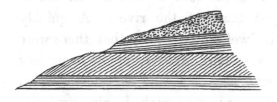

Section of strata of Sandstone,   Port Jackson, South-Head,
formed by quicksand.   Sydney-sandstone with cross stratification.

are quicksand-banks with a double stratification. They reminded me vividly of the cross stratification, such as the ancient palæozoic sandstone-banks at the Heads of Port Jackson near Sydney present them, the larger strata of which are composed of a number of minor layers placed obliquely to the general planes of stratification. While through the influence of rain and the atmospheric agents in general, a gradual decomposition of the magnetic iron, contained in the sand, into brown iron-ore (hydrated oxide of iron)

is produced, the layers of quicksand are gradually consolidated into
an iron-coloured sandstone, which differs from the sandstone deposited from water only in this, that in the former there are animal
and vegetable remains from both the land and the sea promiscuously
embedded in it; a circumstance, which, for the correct interpretation of a great many remarkable facts in older sandstone formations,
is decidedly worthy of notice. In this manner, a large portion of
the sandstone strata composing the coast-range between the Manukau
and the Waikato, seems to have been formed.

Section through the coast-range between Manukau Harbour and the West-Coast.

Passing through a small valley, we arrived by a rapidly descending path to a place on the coast, called Rukuwai after some
sandstone-cliffs, and hastened close by the white foaming surge
upon a sand-beach strewn with shells, in full speed towards the
mouth of the Waikato River. At about four o'clock we arrived at
the North Head upon the right bank of the river. A quickly
kindled fire informed the Maoris dwelling opposite, that there were
travellers waiting to be ferried over; and, indeed, ere long we saw
their canoes starting off to receive us. We sent our horses back,
and had ourselves rowed over at a place, which is already quite
out of the reach of the surge, and where the river measures about
half a mile in breadth. On the other side we found a Maori settlement
Maraetai, (i. e. near the salt-water), consisting of 7 or 8 huts, and but
a short distance from it, at the foot of a wood-clad ridge, at the
edge of a little dell, stands the house of the Rev. Mr. Maunsell,
a deserted missionary station. Since Mr. Maunsell has moved
farther up the river to Kohanga, the house has been utterly neglected, and seems to be but rarely inhabited. We, however, happened to find one of the missionary's sons there, and made ourselves
as comfortable as possible in the scantily furnished dwelling.

After a short rest we started off for the purpose of inspecting the mouth of the river. Despite a breadth of half a mile at its mouth, the Waikato does not succeed in making the same grand and imposing impression here as farther up, where it flows between green, wood-clad mountains, and around islands with a most luxuriant vegetation. The mouth is dammed up by a sand-bar, over which the sea is continually breaking heavily. Only during a perfect calm is it advisable and practicable for little boats. to venture out and back again. Inside the bar the depth of the water is as much as five and six fathoms. The tide comes and goes with a velocity of 4 miles, and makes itself felt to within 10 or 12 miles up the river. It is remarkable, that at the mouth of the Waikato there is not an estuary similar to that of the Manukau, Kaipara, and Hokianga in the North; or as at Waingaroa, Aotea and Kawhia in the South. With regard to this point I have repeatedly heard the opinion expressed, that the Waikato River had formerly emptied into the Manukau Gulf, and that its present mouth is comparatively of recent date. Yet, I cannot corroborate this opinion; I believe myself right in assuming, that the Waikato River also had in former periods a similar estuary, and that the extensive swamps, beginning two miles above the mouth, and now partly covered with bush, through which the Awaroa Creek is meandering, are parts of that former estuary, which the river has gradually almost filled up with masses of sand, mud, and pumice stone, which it always carries along. The bed of the river also, between the North and South-side of the mouth, has in the lapse of time changed its course in consequence of alluvial deposits and quicksand. At present, the river channel is situated on the North-side, which, being almost entirely destitute of vegetation, presents the dreary aspect of sand-hills rising successively higher and higher, one behind the other, the gray colour of which is interrupted only by accumulations of shells. On the western corner of North Head the surge is continually washing out sea-shells, on the opposite side fresh-water shells from the Waikato. The quicksand extends to a great distance both up the river and inland.

Where nothing is to be seen now but a sandy desert, there, it is said, stood generations ago a Maori village with luxuriant Kumara plantations. The natives tell of an extraordinarily high tide, and of a violent hurricane, that have wrought such changes, and turned the river from its bed. Up to this day, a broad alluvial plain, strewn with fragments of pumice stone, and a lagoon in the middle thereof, called Totomoaku, a favourite resort of wild ducks, indicate the river-bed of old at the foot of the steep, rocky South Head.

E.   W.

Section along the Waikato South head.
1. Strata with Belemnites. 2. Tertiary sand-stone. 3. Sand-hills.

To the geologist, the South Head is a locality of considerable interest. Where the sand-hills cease, there a steep, rocky coast commences, extending towards Okariha Point, and washed by the surge at high tide. At neap tide, a narrow strip of sand-beach is laid dry, which presents a convenient path nearly as far as Okariha Point. The rocks consist of marl and sandstone, the strata being piled one above the other with extraordinary regularity like the leaves of a book, and dipping to the West at an angle of 35 degrees. The sandy strata contain indistinct fossil plants and particles of coal; the marls are streaked with white veins of calcareous spar, and interspersed with iron pyrites. It was here, that I discovered the first *Belemnites* upon New Zealand. Those Belemnites are found of various sizes; the largest being finger's length and finger's thickness. They are all distinguished by a deep furrow, and are excellent representatives of the family of *Canaliculati* (d'Orbigny).[1] Besides Belemnites there are found in the same strata *Aucella plicata*, Zitt., *Placunopsis striatula*, Zitt., and other small Bivalves.

[1] The same species, only in a smaller variety, I found afterwards on Kawhia Harbour. Nearest related to the New Zealand species is the *Belemnites semicanaliculatus* of the lower chalk.

I consider that complex of strata to correspond to the Lower Green-sand or to the Neocomian of the French. Unconformably above the inclined Belemnite-strata are deposited horizontal banks of a whitish-yellow, very finely granulated sandstone of the tertiary age, containing a great number of fossils such as: *Brissus eximus*, Zitt., *Schizaster rotundatus*, Zitt., *Cidaris sp.*, *Waldheimia lenticularis*, Desh., *Terebratulina sp.*, *Pecten polymorphoides*, Zitt., and sharks-teeth. [1] I recommend this locality, so abounding in fossils, to future collectors, who will find here yet a good many curiosities, which have escaped my notice in consequence of my short allowance of time.

Belemnites Aucklandicus from the Mouth of the Waikato.

January 25. Already in Auckland I had heard of the existence of coal on the West-coast, about seven miles South of the mouth of the Waikato, and I therefore set out to-day for the purpose of inspecting the designated spot. I followed the "Queen's Road," a much trodden foot-path leading from the Waikato along the West-coast to Taranaki, along which a native carries every fortnight the mail from Auckland to Taranaki. The way leads from the Mission station over meadows and fern-hills to the heights of the coast-plateau, about 700 feet above the level of the sea. From the height reached first, a beautiful view opens northwards over the coast-region between the Waikato and Manukau. The mail-steamer was just puffing in front of the entrance to the harbour; but, this excepted, there was no other sail in sight upon the blue sea, which in long, double and triple, white foaming lines of breakers is bordered by the rows of sand-hills at the coast. The road continues over hills, at times through ferns, at times through tracts of bush which extend from the valleys up to the heights. Upon a point, called Mahunga, an unobstructed prospect opens suddenly

[1] All the fossils I collected, are described and delineated in the volume of the scientific publications of the Novara Expedition, comprising the Palæontology of New Zealand.

towards South. I stood as if spellbound. The sky was cloudless, the atmosphere so transparent and clear, that whatever rose above the horizon, was plainly visible. Before me rose, emerging from the blue ocean, the old volcanic cone Karioi, 2370 feet high, the huge corner-pillar at the South-side of the entrance to Whaingaroa Harbour, from its foot to its top densely covered with forest; farther to the interior, in the blue distance, the trachytic Mountain Pirongia, situated on the Waipa River; farther to the South, Albatross Point on the South-side of Kawhia Harbour; then at a still greater distance faint outlines of land extending far to the West; and farther yet to the West upon the blue sea I descried a gigantic cone with snow-clad summit. This was Mount Egmont, the famous Taranaki Mountain. I experienced an indescribable joy, at beholding, for the first time, that magnificent volcanic cone; although at a distance of not less than 116 sea-miles, yet so distinct, that I could plainly discern the regular ribs, running down its slope; and even now, in the middle of summer, its summit was still to a great extent covered with snow.

Farther ahead we met by the road-side some few huts, surrounded by luxuriant bushes of the flax-plant. The native, who accompanied us, whose name was Kuki, i. e. Cook, explained to me, that this was a Maori-camp, and that the huts had been erected for the purpose of providing a shelter for travelling parties. One of the Phormium bushes was tied together. This excited the attention of our Kuki; he untied the bush, scanned the leaves singly, and broke out into a loud laugh of joy. On going up to him, I saw that the leaves had all been written on. Kuki explained to me, that they bore the names of his friends and acquaintances in Whaingaroa, who had but lately passed that way, and that upon a second leaf a young Maori had traced a fond greeting to his beloved. Kuki now likewise cut his name upon it with a muscle-shell:

Na Kuki

91 + 75 + 73 [1]

[1] The Maoris like employing numerical ciphers for writing. The vowels *a, c,*

and very carefully tied the bush up again. Having once had my attention directed to this peculiar Maori custom, I found similarly inscribed Phormium bushes in the most different parts of the country. We might be inclined to laugh at this harmless folly of the Maoris; but in fact, in Europe such folly is carried out to a still greater extent, by cutting names or initials into the bark of trees, and by scribbling on ruins, rocks, caves and the like, with the idea of immortalizing oneself.

Soon afterwards we met the mail-carrier on his way from Taranaki to Auckland. He was a robust Maori, who carried the mail-bag, bearing the royal seal, upon his back; a glowing slow-match in one hand, and a powerful stick (turupou) in the other. Upon my asking for what purpose he carried the heavy stick, he replied defiantly, "to defend himself against Pakehas," and passed on. After a walk of three hours we descended to the beach; and proceeding along the strand in a southerly direction two or three miles farther, we soon arrived at the locality previously pointed out to us.

While the strand consists of fine magnetic iron-sand, the higher rocky coast-banks present instructive sections. The deepest layers visible, consist of greenish-gray conglomerate and sandstone. In it are embedded a great many short, but thick, logs entirely silicified. In other places chunks are found with their bark turned to coal, thin layers of bituminous shale, and small coal-seams. These coal-seams lie just at the high-water line with an inclination of 10 to 15 degrees westward to the sea; and at several points the heads of the strata are seen to rise above the sand of

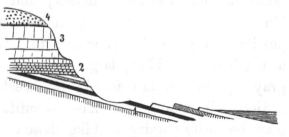

Section on the West coast.

1. Strata with coal and fossil ferns. 2. Tertiary lime-stone.
3. Tertiary sand-stone. 4. Sand-hills.

i, o, u are designated by 1, 2, 3, 4, 5. Then follow the consonants of the Maori Alphabet, h = 6, k = 7, m = 8, n = 9 etc. b, c, d, f, g are not found in the Maori Alphabet, which consists of only 14 letters.

the beach. The coal is a fine glossy coal with conchoidal fracture; different from the brown coal at Drury, and already resembling rather a black coal. The piece which I knocked off of one of the strata emerging from the sand, is lined in two directions, perpendicular to one another, with leaves of calcareous spar, as thin as paper, and is thereby divided into a number of small dice one or two lines in thickness. The layers, however, measuring only a few inches, are not thick enough to be of any practical importance.

Above the coal bearing strata there lie gray marl-banks containing many very remarkable fossil plants. Unfortunately, the banks at the surface are so much broken and crumbled, that, for want of the necessary instruments for digging deeper, I only succeeded in collecting a few distinct pieces with very neat fossil ·ferns, which have been named *Asplenium palæopteris* by Prof. Unger. The topmost beds consist of tabular limestone, replete with foramniferæ and bryozoes, which towards the top change to banks of a finely granulated, yellowish-white sandstone, which breaks in large squares and reminds one of the Quader-sandstone of Saxony or Bohemia; but which, from the fossils enclosed, we infer belongs to the tertiary formation.

A similar section may be observed at a point lying but a few paces north of the spot, where we had reached the beach. Passing northwards along the beach, one has to climb over large sandstone and limestone blocks, and then arrives at a vertical bluff, against which the surge beats, and beyond which it is impossible to proceed. Amongst the rocky masses scattered here at the foot of the bluff, large blocks of calcareous marl of greenish-gray appearance are found; by breaking them with a hammer, a rich collection of the most beautiful specimens of fossil ferns may be easily obtained. Like those at the previous locality, however, these also belong to only one species, but differing from the former and not corresponding to any of the species now living upon New Zealand. Prof. Unger has named it *Polypodium Hochstetteri*. This locality, as well as that of the Belemnites on the Waikato, was formerly not known at all. Our Kuki was utterly lost in astonishment, when he saw, that ferns, "trakau," were growing not only in the woods, but also in the rocks of New Zealand. Besides the load I had given him to carry, he put another extra-piece into his pocket, which he showed to every native we met, every time exciting by it no small degree of surprise.

Heavily laden with the treasures found, and quite satisfied with the results of the day, but at the same time also very tired and hungry, we arrived at sunset again at our quarters on the Waikato. But here, what a disappointment? The little pig, which we had bought in the morning, to have it roasted for supper, had escaped from the old hag of a Maori-woman, — who had offered to attend our cooking, — into the woods; and for the day there was nothing left us, but the scanty remnant of a ham, which we had brought with us from Waiuku.

January 26. — Our provisions had gone out, so we resolved to return to Waiuku, choosing the water-route on our way back.

Polypodium Hochstetteri, a fossil fern.

For a compensation of 15 shillings the natives agreed to convey us in a canoe to Awaroa Creek. Thus we passed up the Waikato, along the sand-hills of the right bank. Two miles above Maraetai the river grows very broad, and contains many marshy islands covered with raupo (*Typha angustifolia*). The left river-bank with its fields of luxuriant growth presents a lovely aspect, and the white church-steeple of the missionary station Kohanga glistened from afar a friendly welcome.

After a trip of one and a half hours, partly upon the open river, partly between the river-islands, we turned off into the Awaroa Creek, the military road of yore and now the modern commercial road of the natives. We had the tide with us, and the passage up the narrow bush-creek, which is just wide enough for two canoes to pass each other, would have been one of the most agreeable, but for the millions of blood-thirsty mosquitoes, which almost drove us to despair. We were therefore glad, when, on our emerging from the swampy bush into the more open grass-land, that plague abated in some measure. But the channel be-came now so shallow, that the native had to get out, in order to lighten the canoe. A rope having then been made from phor-mium-leaves tied together, the canoe was tugged along by this means, until we reached the landing-place at Waiuku.

January 27. — From Waiuku we set out to Mauku, [1] and found hospitable quarters at Major Speedy's, a late officer of the Bengal-army, who now has settled down as farmer in New Zealand.

"By the banks of the Mauku we've fixed our abode,
Where its serpentine current runs down to the sea,
Through the bush and the fern we've opened a road,
And made up our minds to live happy and free.
The world's cares and pleasures are easily seen
To be fitful and vain as the foam of the sea;
We care not for either, — our minds are serene —
By the banks of the Mauku we're happy and free."

Thus sing the merry settlers of Mauku, the happy neighbours of Waiuku. Let him who intends writing novels about the farmer life of the colonists of New Zealand, take up his quarters here; let him make himself at home in the farmhouses of the Mauku-district, so abundantly blessed with rosy daughters, and he will never want matter to suit his purpose. There live the Speedy's, the Vickers, the Crispes, and whatever the names are of all the amiable families in that neighbourhood. Quite romantic is the situation of their snug and comfortable country-seats at the edge of the bush. Bush alternates with meadows, gardens and fields, waving their rich, luxuriant growth upon fertile basaltic ground.

[1] Mauku means "clear of uku", where there is no white clay found.

Upon the heights charming views open to the Manukau Gulf and as far as the volcanic cones on the Isthmus of Auckland. In the valley there are fresh brooks, which, as they plunge headlong over columnar rocks of basalt, form cascades bordered most beautifully with the richest vegetation of New Zealand ferns. And whoever happens to be as lucky as we, will be accompanied thro' woods and fields by pretty young maidens, the choicest flowers of Mauku, mounted on their fleet steeds; and when he takes leave, a blue-gumtree will be planted for him in remembrance of the days and hours spent in idyllic bliss and merriment.

From Mauku a speedy, but rather fatiguing ride across uniform fern-land brought us on January 28. to Drury, and thence we returned by the Great South Road to Auckland.

CHAPTER XIV.

On the lower Waikato; from Auckland to Taupiri.

Bush travelling in New Zealand. — Supply of provisions. — Other articles of equipment. — Fern and flax. — Departure from Auckland. — Drury Hotel. — Mangatawhiri. — Sueking pigs. — Thus far and no farther. — Our embarkation. — The Waikato, the main artery of the country. — Maori-politics. — Boat-songs. — Tiutiu. — Pukatea. — Eels. — Lake Wangape. — The Pah Rangiriri. — Lake Waikare. — The river-island Taipouri. — Brown coal at Kupakupa. — The Taupiri range. — The Mission station on the Taupiri.

The journey through the interior of the North Island, which I will describe in the following chapters according to the contents of my diary, is but a small one, considering the distance travelled over, about 700 English miles. In European countries, where rail-roads and steam-boats are at disposal, one might travel over the same distance in a few days; but compared with the rapidity and the luxurious ease with which such a journey is performed in Europe, travelling in New Zealand is slow and laborious, and a lengthened expedition into the interior cannot be undertaken without some preparations. Roads passable for vehicles, lead to a distance of only a few miles from the towns; and passages practicable for horsemen, at least in the interior of the North Island, were not many. The horse, which to the traveller upon the extensive, open plains in the interior of Australia is totally indispensable, is by no means of the same service in New Zealand. In many districts, it would not only want the necessary feed, but the difficulties arising from the nature of the ground are also such, that the horse must soon prove to the traveller a burthen rather than a help. Almost daily he has to pass over mountain-torrents and steep river-banks; or through swamps and morasses. The slender paths of the natives lead over hills and mountains in steep ascent and descent, rarely in the valley, nearly always along the ridge of mountain-heights. Where they cross the bush, the clearing is just broad enough for one man to wind himself through. An eye used to European paths, will scarcely recognise those Maori-trails, and man and beast would be in continual danger upon them — the horse, in danger of sinking into the deep holes between the roots of trees, and of breaking its legs; the rider, of being caught among the branches, or strangled among the loops of the "supple jack." Hence there is no other choice left but to travel on foot; and it requires full, unimpaired bodily strength, and sound health, to pass uninjured through the inevitable hardships of a longer pedestrian journey through the New Zealand bush, over fern-clad hills, over steep and broken headlands, through the swampy plains and cold mountain-streams of the country. Whatever the traveller needs for his individual wants, he must carry with him, and therefore must be limited to the most necessary articles. Now and then, a solitary European squatter may be met with; and more frequently still, a Mission station. On all these occasions the traveller will meet with a cordial welcome, and hospitable treatment, and transiently he will enjoy even the comforts of civilized life; but, taken as a general thing, he must resign them all; he must learn to find pleasure in living in the open air with the skies for a canopy and the earth for his table and bed. Following the example of the Maoris, he must "go back to first principles" and to the simple demands of the children of nature; and it is to this truly primitive simplicity that a journey in New Zealand owes its indesirable charms.

The woods and fields of our antipodes are but scanty hunting-grounds, which at best yield small birds and wood-pigeons. Along rivers and lakes various kinds of wild-ducks are found, and nearly all the rivers are abounding in eel and crawfish. But this is all, upon which one can calculate during the trip as occasional contributions to his frugal repasts. Any one, therefore, that would depend, for subsistence, merely upon hunting, or upon what little is furnished him from the vegetable world, would be exposed in the interior of New Zealand to the same danger, as those much lamented men of dauntless spirit, who lately, upon the Burke Expedition through the continent of Australia, after they had success-fully attained their object, were, on their way back, doomed to die of hunger. And in fact, such cases have really happened, especially in the interior of the totally uninhabited South Island; and several expeditions endeavouring to penetrate from the East coast across the chain of the Southern Alps to the West coast have found the greatest difficulty in barely supporting life even. A sufficient supply of provisions is therefore always one of the first and most important questions for a longer journey. Upon the North Island, where there are natives living in small villages and settlements all through the interior, this is a matter of no trouble. From station to station the necessary supplies are carried along, such as are obtained from the Maori settlements, which but rarely are more than a few days' journeys apart. Pigs, especially, can be had everywhere, and upon our three months' trip we killed not less than thirty or forty heads, and were always enjoying good health while feasting on fat, juicey roast-pig. But if the season following the gathering of crops, and beginning with February, is preferred for the journey — which is best suited also on account of the pleasanter weather, and less inconvenience suffered from in-sects, as the mosquitoes generally vanish more and more by the setting in of March — there is also no want of fruit and po-tatoes. By that time, even flour is to be had in some parts, and the traveller has ample chance of occasionally changing his uniform every-day-fare of pork and potatoes, and regaling himself with "dampers."[1] Eggs and milk, however, are very scarce in the Maori settlements. Money and tobacco are the current means of

---

[1] "Damper" is a dough made from wheat-flour and water without yeast, which is simply pressed flat, and baked in the ashes; according to civilized notions, rather hard of digestion, but quite agreeable to hungry woodsmen's stomachs.

exchange, for which provisions can be bought from the natives. Whatever is required besides the provisions already mentioned, must be provided from the very first start for the whole journey. Of chief importance are salt, sugar and tea. The latter supplies must be carefully kept from becoming wet, owing to the prevailing dampness of the climate, and the frequency of heavy rains. For this purpose we had a special tin-chest made of such a size, that, when filled, it was about as much as one man could well carry (about forty pounds); we had moreover sent extra-supplies ahead to several stations on the coast which we were to pass on our journey. Tea was usually made three times a-day for the whole company, consisting sometimes of thirty persons; and I know of no other beverage, which during fatiguing foot-excursions produces so refreshing and invigorating an effect, as good, strong tea; or which at the same time is as easily prepared. Even the natives have become so much used to tea, that they generally carry a supply of it with them on their journeys. Tea, pork and potatoes, therefore, were our chief articles of food; or rather, with rare exceptions our daily bread.

For camping out we were equipped in the best manner with tents and woollen blankets. For tent we found cotton-stuffs to be the most suitable, being denser and less heavy than linen. It was cut so that it could be spread like a roof over a $\sqcap$ shaped scaffold constructed of three poles. We carried three such tents with us. A fourth large one was intended for the natives who accompanied me; it was, however, but rarely used, the natives generally preferring to sleep under the open sky, gathered around a large fire, which they kept up during the whole night. The woollen blanket representing my bedding I had sewed up in triple folds at the feet and sides into a kind of sack, so that on one side the blanket was double, on the other single, — an excellent invention of experienced "bush-men." By getting into this sack one is not only sheltered from troublesome mosquitoes and other insects, but has, moreover, according to the weather, the convenient choice left, to turn the double or single side upward, and thus to suit oneself to a cover more or less warm. In this shape, one large, woollen blanket yields the same amount of comfort, that at other times is obtained from two or three blankets; and, together with an air-cushion of caouchouc, composes an excellent travelling bed in a very compact form. A caouchouc cover, which at night was spread upon the bare ground as an underlayer for bed, and in daytime served as water-proof covering for my baggage, also rendered excellent services.

Natives are decidedly the best travelling companions. I had engaged for the whole time of the expedition twelve stout, young Maoris, who were bound by contract to remain with us through the whole of the tour. As wages, each of them received, besides board and a pair of shoes, half a crown a-day. The baggage was distributed so that each man carried about thirty to forty pounds. Owing to the difficulties of the ground we had to pass over, a man could not easily carry more upon a longer journey. Each carried his proper bundle; and likewise in camp, each one had once for all his proper work assigned to him. One assisted in putting up tents, another fetched firewood, carried water, etc. Thus each of them knew after the very first few days of travel, without a special order, precisely what he had to do; and none could be idle at the expense of his fellow. With pleasure I bear testimony to these Maoris, that they always proved themselves willing and of untiring energy, preserving under all circumstances their excellent good-humour; and that their faithful services contributed in a large measure to the final success of the expedition. Their names were Awaroa, Ngakapa, Dominiko, Poroa, Te Kura, Te Kahukoti, Mehana, Paurini, Te Tawera, Timoti, Te Kanihi, and Pateriki. During the journey our company was increased by three Maoris and one European servant, so that we usually numbered twenty-two heads; which number, however, increased sometimes to twenty-five and thirty, whenever special guides and extra-carriers were necessary.

The necessary guides and bearers having thus been properly engaged, and every thing being suitably provided for the expedition, the journey itself presents no further difficulties, although its progress is, according to European notions, rather slow. Fifteen miles with a numerous travelling-company, may be considered a fair day's work considering the miserable roads. As to safety, I really know no uncivilized country upon the earth, where one can travel so safe and secure, as in New Zealand. Robbers and thieves are as little known there, as wild beasts and venomous serpents; and as Nature herself, having produced here no poisonous plant or venomous beast, is harmless in all her creations, so also the native is harmless in his whole conduct and all his actions, unless war or revenge rouse his wild passions. One can, therefore, travel with perfect safety, and tranquilly lay down one's head to rest in the mountains or in the valleys, in woods or field, wherever one may be, when evening or night sets in.

The only plague are mosquitoes and sandflies. The former, by the natives called Waeroa or Ngairoa, are no other than our gnats *(Culex)*, which in swarms of thousands of millions live in the damp bush, and, at the edge of the bush, along the creeks or upon clearings, quite obscuring the air by their dense swarms; shunning, however, the sea-coast and dry fern-heaths. In summertime, from December to February, they cannot be kept off day or night; but in March they already commence being more scarce, and in winter they disappear entirely. The sandflies, on the other hand, the Ngamu of the natives, small midges *(Simulium)*, are most frequent on the sea-coast; but they are met also all through the interior of the land, on sandy river-banks, and upon dry heaths. The very districts, which are clear of mosquitoes, are infested by the sandflies. Their sting is keener than that of the mosquitoes, but is not attended by any swelling of the part stung; and with the last ray of the sun the sandflies disappear entirely, so that at night at last one is rid of that plague. But, sometimes, certain other still more unwelcome guests intrude at night — rats. They are found even in quite uninhabited countries, and gather after the very first night around the camp. To their running at night leisurely over his head and body, the traveller will easily become used; but eatables must be carefully kept out of their reach by hanging them upon poles.

Although nature, as I have already said, offers but little in the shape of food; yet it furnishes two things for the comfort of the traveller, which he does not learn to appreciate to its full extent, until, after a trip through New Zealand, he is travelling in other uncivilized countries, where he has to do without them. In the first place I mention the common fern *(Pteris esculenta)*, which growing all over the country, can serve as an underlayer for the couch. A bed made of such ferns, provided one understands the arranging of it,[1] feels as elastic and soft, as the best spring-matress. The second is the flax-plant *(Phormium tenax)*. It can be employed wherever leather thongs or straps would be otherwise used; it can likewise be made into wicker-baskets, girths, etc. When more than a dozen bundles have to be strapped daily, such an article of almost universal application, which is everywhere and at any moment at hand, is indeed of invaluable advantage. The excellent climate, also, and the abundance of water and wood in every part of the country, greatly facilitate travelling. One suffers neither from heat nor cold; nor are there any fever-countries to be avoided. Swamp-fevers are totally unknown, and scarcity of water, the terror of those travelling in the interior of Australia, is out of the question upon New Zealand. But rarely will the choice of a camping-place for the night cause embarrassment; the proper place can always be easily found, where wood and water are close at hand, and where the weary traveller can enjoy his repose, secure from the blood-thirsty mosquitoes. Not even tent-poles did we need to carry with us, having almost always the opportunity, of procuring such upon the places where we encamped.

Everlasting will the recollection of those scenes be to me, when, after the troubles and trials of the day we encamped at the edge of the woods by a roaring mountain-stream; when the fire

---

[1] Dry ferns must be selected for this purpose, and the plants must be arranged so, that the roots and stalks are turned downwards at an angle of about 45 degrees. Thus the woody stems act exactly like so many elastic springs.

blazed up brightly, and the natives were singing their songs; then, every thing lay hushed in repose, till, with the dawn of another day, the birds of the woods, the Kokorimoko, and the Tui merrily warbled their orison-lays. I love to look back to such scenes, to our river-excursions in the well-manned canoes of the natives, to our stay in their Pahs, and to our peregrinating through the bush in the shade of trees, which are strangers to every other part of the globe, — I look back to them all with a pleasure, which makes me feel most sensibly, how far superior the enjoyments of nature are to all the pleasures of refined life.

On the 7th of March we were ready for travel. We set out upon the Great South Road. This road, — now the great highway to the interior of the country, — was at that time finished only as far as Drury, twenty-three miles from Auckland. Drury itself consisted of an inn and a church. At a greater distance there were some scattered farms. The Drury Hotel or "Young's Inn," so called after the name of the proprietor, was, so to say, the last out-post of civilization towards South. During the presence of the Novara in the harbour of Auckland, this hotel was the head-quarters of the Novara Expedition, whence excursions were made to the coalfields near Drury and to the Waikato; and a huge Austrian flag was floating from its gable. This time also it was our place of rendez-vous, and our starting point.

March 8. — Started for the Maori settlement Mangatawhiri on the Waikato, twelve miles from Drury; thence we were to proceed in canoes up the river. The road leads from Drury in a southerly direction over the Drury flats and then ascends to a wood-clad plateau, which forms the watershed between the Manukau Harbour and the Waikato River. On the sides of small dales the last farmer settlements are met, and thence the traveller penetrates deeper and deeper into the forest. The road was just being made. The recent cuts displayed basaltic con-glomerates totally decomposed into ferruginous clay. One of the labourers presented me with some interesting beetles,[1] which he had caught among the felled trees. Upon the last height before reaching Mangatawhiri, there is a charming prospect into the Waikato country; but the road itself from this spot was a mere clearing through the bush; the felled trees still lying promiscuously across the road; and we could not help laughing heartily at the amusing notion of a wood-cutter, who had written upon a gigantic trunk, which blocked up the whole passage, in large charcoal characters: "XXII miles from London."[2]

Mangatawhiri we found almost entirely deserted. The greater part of the male population was absent on a trip up the Waikato, and our first negotiations for hiring a canoe convinced me, that the hitherto hospitable and obliging savages had become perfectly civilized individuals, calculating and bartering. A continuation of our journey was for to-day quite out of the question, and it was not until late in the evening, that Captain Hay succeeded in closing a bargain with the natives on these terms, that for a compensation of £6 they promised to furnish us by next morning a large canoe, which was to convey us up the river as far as Taupiri.

The village numbers about twenty huts with about 100 inhabitants, who are enjoying considerable wealth. They very recently had a neat flour-mill built by an Englishman, on a small stream running by the village, which cost them not less than £400. The volcanic soil of the neighbourhood is extremely fertile, and there is no scarcity of horses, cattle and pigs in these parts. Less edifying was the abominable filthiness we had to notice in the Maori-huts. A number of them were vacant; we wished to select one of these for our night-lodgings; but they all were teeming with all sorts of vermin. We determined at last to occupy one of them, after it had been well scrubbed and scoured.

[1] Among them especially the large *Prionoplus reticularis*, next *Brenthus* (*Nemocephalus*) *barbaricornis* with a long proboscis; beautiful goat-chafers such as *Coptoma variegatum*, *Coptoma lineatum*, and the trunk-chafers *Rhyndodes Ursus* and *Rhyndodes Saundersii*.

[2] The road was completely finished and macadamised by soldiers as far as Mangatawhiri, in 1862, in consequence of the Maori-war.

Yet, what we had to suffer all night despite the previous cleansing of that Augian stable, I will rather pass over in silence. To me it was from the very start an impressive lesson, never again to prefer a Maori-hut to my tent.

A Maori-girl at Mangatawhiri.

The women and maidens of the village had, in honour of our distinguished presence, dressed themselves to the best of their ability, and donned their choicest attire; there were some really pretty forms and faces among them. But there prevails a strange custom among those women. The sucking-pigs, vulgo farrows, are in great favour with them. They nurse and fondle them with as much tenderness, as our ladies their lap-dogs; they even grant those favourite pet-pigs the same privileges on their breasts, that are generally extended only to babies. In like manner the Indian women are said to nurse young apes.

In the vicinity of the Maori-village the township of Havelock has been laid out, which for the present consisted of only a few houses. This settlement is considered by the natives as the farthest southern boundary, to which the Pakehas are entitled to extend their possessions. "Thus far and no farther," they say. With consistent obstinacy they manage to prevent every attempt towards the continuing of the Great South Road; and during the Taranaki insurrection of 1861, William Thompson, the leader of the Waikato tribes, declared it even to be a case of war, which would result in the breaking out of hostilities also along the Waikato, should the Government advance its troups into the interior farther than Mangatawhiri, or continue the construction of the road under the protection of armed forces. This circumstance alone would suffice to prove the importance of this place, which can only be fully appreciated hereafter, when the Waikato river shall have been opened to European commerce.

March 9. — Rose early; the Mangatawhiri creek offered a refreshing bath; but as customary among Maoris we had much talking before we were ready to decamp. The natives endeavoured, to extort from us a few additional gold-pieces, before they showed any inclination to convey us further. It was not until 9 o'clock, that we commenced to move on. We then proceeded along a small current of water wading a considerable distance through mud and swamp, until we came at last to the promised canoe. The canoe was an entirely new one, wrought from the trunk of a Kahikatea pine *(Podocarpus dacrydioides)*, sixty-one feet long, four wide, three deep, and large enough to hold our whole party together with the entire bulk of baggage. First it was cleansed; then the bottom was covered with fresh ferns, and the baggage distributed with all due cautiousness as equally as possible fore and aft, to the right and left, for the benefit of securing the necessary balance to the rather unstable craft. At last every thing was in readiness, and we Pakehas were directed to take our place in the middle. But the creek was here still far too narrow and too shallow, and the load too heavy, to render the propelling of our Waka,[1] by means of the paddles, possible. We had first to be pulled through the mud. The scene was a highly amusing one,

---

[1] The Maori-name for canoe is waka. Generally those wakas are wrought from the much more durable red Totara-wood (*Podocarpus totara*). Such canoes are said to last three generations, while Kahikatea-canoes last at the most ten years, since they are gnawed by bore-worms.

thanks to the good humour exhibited by the Maoris, even when they sank sometimes waist-deep into mud, and could work their way through the swamp only with the greatest difficulty. Amid singing and laughing, and amid the wildest joking, the canoe was pushed ahead. Finally after the lapse of two long hours, as the creek grew wider and deeper, we were afloat. Now, after all belonging to the expedition had been seated in the canoe, I counted not less than twenty-four of us, whom besides a heavy load of baggage and provisions, the "dug-out"[1] had to carry. In the fore-part sat the Maoris, twelve in number, each provided with a paddle, in the middle the five Pakehas. Behind us four Maori-women with two children had crowded in, who wished to meet their husbands, expecting to fall in with them upon the Waikato; and the helm was managed by Captain Drummond Hay, whom the Maoris jestingly styled a "Maori Pakeha," because he had acquired certain Maori-accomplishments to perfection, and understood especially the management of the paddle as well as any of them. Thus we paddled ahead towards the Waikato, all of good cheer, and with all those feelings of sanguine hope, which the successful start of an interesting journey is wont to call forth.

The narrow creek soon turns into a river about 100 feet wide, the Mangatawhiri river, which about four miles below empties into the Waikato. Round about nothing but moor and swamp; the water is dark-brown, and only low hills, partly covered with bush, the spurs of remoter mountain chains, interrupt the far-extending low-lands. It presents the picture of a New Zealand swamp-land scenery. The strokes of the paddle are scaring up wild ducks and water-hens;[2] they fly up or endeavour to hide themselves by diving; but the keen eye of the native espies them even among the densest reed-thicket, the sharp-pointed paddle serves him as javelin; he hits his mark with certainty and never misses; and thus one

---

[1] "Dug-out" is the name of these canoes among the backwoodsmen of North America.

[2] The Sultan-hen of brilliant plumage (*Porphyris melanotus*), or the Pukeko of the natives; the New Zealand bittern or Matuku (*Botaurus melanotus*), and the wild-duck Parena (*Spatula rhynchotis*).

after the ofter is flung into the boat as welcome booty. A luxuriant vegetation of water and swamp-plants (especially Raupo, *Typha angustifolia*) borders the channel, and where the river-banks arise, there Ti-trees *(Cordyline)*, flax-plants *(Phormium)* and the magnificent Toetoe-grass *(Arundo)* with its silken flags, mingled with violet-blooming Koromiko-bushes *(Veronica)*, form a gayly checkered copse-wood, in the rear of which, at the foot of low hills, dusky Kahikatea-woods are spreading, — genuine swamp-woods, for the Kahikatea pine prefers swampy soil. Where the river makes a bend towards South, the scene suddenly changes. The Maoris ply their paddles more vigorously; swift as an arrow the canoe darts over the waters, and with loud shouts of joyful welcome we greet the Waikato.

The impression made by the sight of the majestic stream is truly grand. It is only with the Danube or the Rhine, that I can compare the mighty river, which we had just entered. The Waikato[1] is the principal river of the North Island. Both as to length of its course, and quantity of water it surpasses all the others. The pieces of pumice-stone, which its waters are continually carrying along, piling them up on the banks and at its confluence with the sea, point to its origin in the vicinity of the extensive volcanic hearth in the centre of the island. Its sources spring from the very core of the land; its waters roll through the most fertile and most beautiful fields, populated by numerous and most powerful tribes of the natives, who have taken their name from it; and no second river of New Zealand has such

---

[1] *Waikato* means literally: running water, or streaming water, in distinction to the *Waipa*, tranquil water, the chief tributary of the Waikato. The average velocity of the Waikato in its lower course amounts still to four or five miles an hour. The Danube, which is known to be a very swift stream, averages at Vienna likewise about four or five miles an hour. The Waikato itself is not abounding in fish; but several of his more placid tributaries are. Some sea-fishes are roving far up the river; moreover there are eels found, and small species of *Eleotris*, Inanga of the natives. Very abundant is the river in fresh-water shells. *Unio Aucklandicus* is fished up from the bottom in great quantities by the natives, who are very fond of that food. Also species of Hydrobia and Latia (*Hydrobia Cumingiana Fisch., Latia neritoides Gray*) are living in great numbers among the river-grass.

an importance, as the grand thoroughfare for the interior of the country. The Waikato is in truth the main artery of the North Island, and this grand stream is wanting but one thing, i. e. an open, unobstructed entrance from the sea. While a great many other large rivers of New Zealand, as, for example, the nearest neighbours of the Waikato, the Piako and Waiho, or the Wairoa in the North, are emptying into protected bays of the sea, widening near their mouths into broad estuaries, by which the sea penetrates far into the interior of the land, and where the regular change of ebb and flow enables larger and smaller vessels to pass from the sea into the river, and from the river into the sea: there are huge sandbanks piled up in front of the mouth of the Waikato, upon which the sea breaks in foaming surges. This is a matter of great importance; for those sandbanks, which prevent the passing in and out of larger vessels, are a natural bulwark for the natives. They look upon the Waikato more than upon any other river of New Zealand, as being the river exclusively their own. Never, up to the time of my journey, had a boat of European construction been known to float upon the proud Native-stream,[1] the Mississippi of the Maoris. Two Mission stations, the one near its mouth, the other at the Taupiri, were at that time the only European settlements on the banks of the river, where the Maori-king had taken up his abode. From his residence at Ngaruawahia, where the Waipa mingles its waters with those of the Waikato, the national flag of Nuitireni was proudly floating in the breeze, and from among the bushes of the flax-plant, the toetoe-grass and the ti-trees, the Maori-huts were everywhere peeping forth, now single, now in clusters of miniature villages and surrounded by thriving plantations. Flats are alternating along the course of the river with fern-hills, or with dusky wood-clad mountain-ridges, and picturesque landscape-sceneries are developing themselves there where the river in a narrow gorge of rocks is breaking through the mountain chains. The Waikato, at the junction of the Mangatawhiri, has a breadth of about half a mile; it

[1] Now steamers are playing on the Waikato.

encompasses several wood-clad islands, and after having passed in an almost precise S. N. direction through extensive low-lands, — the *lower Waikato Basin*, — it makes here a sudden bend to the West. It breaks through a low coast-range, and empties twenty miles farther below on the Westcoast into the sea.

To-day it was the second time that I visited this spot. The first time on December 31. 1858, while the Novara was yet at anchor in the harbour of Auckland. On a short excursion from Drury I then passed down the river together with my travelling companions to the Maori settlement Tuakau on the right bank of the river, a distance of a few miles. There, in a Maori hut, we had celebrated the fleeting hours of the parting year in a manner, which no doubt has left an indelible impression upon the memory of each of us. At that time there resounded by turns, amid the merriment of social glee, national songs, German student songs and popular songs, English and Irish airs, and the melancholy love-ditties of the Maoris. We thought of our loved ones at home; no cherished friend was forgotten, when our glasses were ringing to repeated toasts and to the sincerest congratulations and well-wishes for the New Year. I did not dream then, that I should spend many a night yet, without those friends, in the huts of the Maoris, in the bush or upon the fern-clad hills of New Zealand; or that I should be allowed in this new year, to trace the course of the beautiful river up-stream into the very heart of the country.

After we had turned from the brownish peat-water of the Mangatawhiri into the green waves of the broad, open Waikato, we proceeded up the river. Owing to the swift current in the middle of the river, we kept close to the right bank. The water showed a temperature of 70° Fahr.; but its surface presented an uncommon appearance, large masses of pumice-stone drifting upon the river, which were collecting behind transverse trunks of trees. They were scattered fragments, sometimes the size of a man's head, of a white, coarse-grained pumice-stone,[1] Pungapunga of the natives.

[1] Despite the enormous masses of pumice-stone, which are found in the inte-

It was some time, before our Maoris could be induced, to proceed up the river. The sight of the Waikato awakened too many recollections within their minds. They had a thousand different things to relate to each other. Every canoe, that hove in sight upon the river, was hailed, or hailed us. Of course, the Maoris also are curious to know, what news? — "Whence?" and "whither?" and "who are you?" were their queries. — One canoe came close up to us; it was full of natives, dogs and pigs; and dogs, pigs and natives, all seemed struck with amazement and awe on seeing Pakehas upon the Waikato. The news of our travelling up the river had, as I found out afterwards, run ahead of us with amazing velocity, even without mail or telegraphic communication. Enough of the busy chatting to and fro having been done at last, the paddles were again dipped in the water. Poroa, assuming the part of a Kaituki,[1] commenced to sing a boat-song strophe after strophe;

---

rior of the North Island piled up to several hundred feet, in some places even 1000 feet and more, it is nevertheless shipped from the Liparian Islands even as far as Auckland, because the native pumice-stone is too coarsegrained for practical use.

[1] Kaituki signifies the leader in a canoe, who by singing and various gesticulations incites the crew to ply their paddles, and denotes by the rhythm of the song he chooses, the greater or lesser rapidity of stroke desired. Such a song is called Tukiwaka. In large war canoes, manned sometimes by 60 or 70 men, there are generally two Kaitukis acting as leaders, one placed near the bow and the other the stern. In addition to their voices, they have in the hand some native weapon which they brandish in time. They either sing by turns, one responding to the other; or they sing together, extemporizing at the same time various jokes and witticisms, by introducing into the traditional songs new verses having reference to the momentary situation. It is remarkable to see, how the pullers are in this manner guided in keeping time. With as regular strokes, as if managed by one hand, the paddles are moving on both sides, and with the same regularity the bodies of all the pullers are moving now forward now backward; and as the time increases in velocity, those motions also become faster and more energetic, until at last with an almost convulsive tossing forward and backward of the head and the whole upper part of the body, their hair streaming in the air, the whole crew in wild chorus is repeating the last syllables or words of each verse chanted by the leaders. The sight of such a war canoe fully manned and decked with festal drapery, while, propelled by the simultaneous strokes of 60 or more paddles, it darts along almost with the velocity of a steam-boat, produces an imposing, but also an uncomfortably savage impression. It has the appearance of one body with a hundred arms and as many feet, every part of which is alive and in motion, — like a gigantic centipede upon the water.

slowly at first; then faster and faster; and the paddles kept time
to the tune of the song. After a trip of two hours we landed on
the left river-bank near a small settlement Tiutiu or Tutu. Our
repast consisted of potatoes, bread and tea. The inhabitants of the
place, quite in contrast with the usual curiosity of the Maoris,
seemed to take but little notice of the arrival of so large a travelling company; and upon entering one of the huts, I met two
elderly tattooed men so absorbed in their game, that they did not
look until I accosted them. The game which the two old fellows
were playing, was no other than our game at draughts translated
into Maori style. The "men" of one party were represented by
small potatoes cut in two, called Riwai; those of the other, by peach-
kernels, called Pititi. Instead of the draught-board they had a
piece of board, upon which not even squares were marked,
and the game itself they called Teraku. They, however, most
readily gave all the information required, and conducted me through
the plats surrounding the huts, upon which turnips (Tonapi), melons (Hue), cucumbers (Kumokuma), maize (Kaanga), potatoes
(Rapana) and peach-trees (Pititi) were thriving exceedingly well.
But with special care had the Kumara-fields[1] been attended.

After three o'clock we continued our journey. The country
presented little change of scenery, and nothing of special interest.
Flat alluvial-tracts, partly covered with Kahikatea-forests, alternate
with low rows of hills, at the most 300 to 400 feet high, presenting, where they extend to the river-banks, denudations of soft
horizontal layers of sandstone and clay. At 6 o'clock in the evening we arrived at the Maori settlement Pukatea on the left bank,
and here for the first time we pitched our tents for the night. The
natives regaled us with dried Waikato-eel (Tuna maroke), which
with them are considered quite a delicacy. Upon journeys great
quantities of such eel are always carried with, as part of the provisions, in bundles of twenty to thirty. Those halfsmoked Tuna
marokes, however, although I greatly relished, the fresh eel, in

---

[1] Kumara = *Convolvulus chrysorhizus*, the so-called batata or sweet potato; three
different varieties of that plant were designated to me as Pehu, Monenehu and Orangi.

which all the rivers of New Zealand are abounding, did not suit my taste.

March 10. — During the night we had heavy thunder-showers, which, however, were after all not sufficient to penetrate our tents; whereas innumerable mosquitoes found their way into them and cruelly robbed us of sleep. Next morning the sky looked gloomy. Tattered clouds came lowering upon the landscape, and were gathering themselves into new clouds of rain, until later in the day a keen western breeze swept the murky sky, and prepared a very pleasant evening for us.

With a fat pig on board, which had been purchased at Pukatea, we continued our expedition. We kept along the left bank, passing the mouth of the Opuatia, and further on the small river island Tarahanga with the ruins of a formerly well fortified Pah. A solitary falcon had perched itself upon the tall palisades, which have remained standing at the Northend of the island, — quite the right emblem for a fort in ruins. Thence we drew nearer and nearer to the middle portion of the lower Waikato Basin, where upon extensive plains on both sides of the river numerous smaller and larger lakes are scattered about. First of all, a few miles from the left or western bank of the river, lies Lake Wangape.[1] At the outlet of the lake into the Waikato, on the Wangape creek, there is situated a Maori settlement of the same name. We proceeded a short distance up the creek hoping to find upon a neighbouring hill a point, from which we might enjoy a prospect over the lake; but we had to return without having accomplished our object. The water of the narrow, almost stagnating creek showed a temperature of 72° Fahr., the Waikato having only 70° Fahr. The river here makes a considerable bend towards East, and on a second similar bend, which restores the river again to its south-northerly direction, on its right bank, lies the Pah Rangiriri, at the mouth of the outlet of Lake Waikare.

Rangiriri is the chief point in the lower Waikato Basin, being situated almost in the centre of the basin. Here we halted. But we

[1] The name signifies a surface or an extension of water.

found the Pah which is encompassed by a row of palisades, entirely deserted, nor was a single soul to be seen in the huts outside of the enclosure. All the inhabitants had moved into the country during the summer; they were scattered about in the smaller settlements in the vicinity, where they have their lands and fields, and are said to congregate here only on Sundays to attend church. It is not until after all the crops have been gathered, that they assemble again in the Pah for the winter. The church we found few paces to the rear of the Pah, und I was quite surprised at the sight of the neatly constructed and clean house, in which every Sunday a congregation of Maori Christians assemble together for worship, a native preaching the sermon. A few hundred paces farther, rises the Rangiriri hill, an elevation only about 100 feet above the Waikato, from which a magnificent view all around is opened of a large portion of the lower Waikato Basin.

This was to me quite a welcome point for planting my azimuth-compass and commencing magnetic bearings which, continued on my onward journey, yielded me a triangulation, forming the basis for the construction of the topographical map of the southern portion of the Province Auckland, such as is found annexed to this work. To a great distance the fertile river-valley is seen with its changes of plains and hills, of woods and fern-heaths, encompassed round-about by nearer and remoter heights, with a view, in front, of the broad surface of Lake Waikare, on the East-shore of which steep hills with numerous patches of bare ground arise, while above them, at a still greater distance, dark wood-clad ridges are seen. In the middle of the Lake Waikare a saline mineral spring is said to rise, bubbling up sometimes to the height of three or four feet.[1] To the South, the Taupiri range with the conical Pukemore closes the horizon; and towards the South-west above the spurs of the Hakarimata range the trychyte-stock of Pirongia with its many peaks is looming up in the distance. The broad belt of the Waikato can be traced southward as far as the point,

---

[1] Perhaps the lake has received its name from it, Waikare meaning bubbling water.

Waikare Creek near Rangiriri, on the Waikato.

where the river, after breaking through the Taupiri range, enters the low-lands of its lower basin. On my return from the Rangiriri hill, I found some well-executed photographs ready of the land-scapes along the banks of the river, which gave me great pleasure, because after these first successful attempts I had reasons to hope, that I should not have to regret having brought with me an artist with his apparatus, the transportation of which was necessarily connected with great difficulty during the overland journey. The Maoris, on the other hand, had killed the pig and promised us a juicey roast for the evening.

At about 2 o'clock in the afternoon we started again. We met quite a number of canoes on the river, and in one of them the Maori women, who had come with us from Mangatawhiri, at last found their long-looked-for husbands again. The shipping of them was the work of a few minutes. In exchange for the women with their children we received from the other canoe four stout young men on board, and now we proceeded up the river at full speed. At sunset we put in at the West-side of the river-island Taipuri, the largest island in the Waikato, by some extremely scanty look-ing huts. The inhabitants very kindly brought us melons and apples as a token of welcome, helped us pitch our tents, kindle a fire, and before night had set in, we lay snug and comfortable in our camp, and the cook brought us the promised roast-pig.

March 11. — Pouring rain delayed our departure. It was not until 11 o'clock, when the sky seemed to be clearing off, that we could proceed on our course. We were drawing nearer and nearer to the range, which closes the lower Waikato Basin towards the South. Gray fogs and rain-clouds were hanging over the moun-tains, and some dark blots in the gloomy picture were all that designated the narrow mountain pass, by which the river breaks through the mountains. When, after a bend of the river two of the first advance-heights have been passed, — then the valley opens, and the wood-clad top of the Taupiri, from which the range has received its name, becomes visible.

In Auckland, already, I had heard of beds of coal appearing

along the margain of this range, which seemed to deserve a closer examination. I, therefore, gave orders to land at Kupakupa, a small Maori settlement on the left bank, just where the plain strikes the mountains, and soon found a guide, who conducted us to the place. The name of the locality, where the coal seam crops out, is called Papahorahora, it is situated about one mile South of Kupakupa on the slope of a ridge of hills rising in the rear of the kainga, [1] in a height of 180 feet above the river. The natural opening was formed by a rupture at the upper end of a brook-defile leading to a pond on the west side of the village. Immediately below the yellow clay which covers the declivity of the hill, a horizontal bed of brown coal has been laid bare to the depth of 15 feet. The whole seam, however, is probably still deeper by

NW   *Taupiri 953'*   SE.

Brown coal Formation.   Clay-slate.

Section across the Taupiri range.

several feet, the sole of it being hidden from view. The locality is as favourable for mining as could be wished. The quality of the coal [2] is precisely the same as that of the Drury and Hunna coal near Auckland. Future explorations will show, that the same coal-bed continues also through the hills opposite on the right bank of the Waikato. [3] At any rate, there lies a considerable store of fuel, which will be raised as soon as European settlements have commenced to extend over the beautiful lands on the lower Wai-

[1] *Kainga* is the Maori name for a settlement.

[2] The Rev. Mr. Ashwell at the Taupiri used this Kupakupa coal already years ago for domestic purposes, and found the same fossil gum in it, which is so frequently met with in the Drury coal.

[3] I must mention here, that it has been related to me, that between Lake Wangape and the West-coast there is a point, at which smoke is continually issuing from the ground. Perhaps it is nothing else than a coal-bed, which has spontaneously ignited and has been burning for years.

kato, and steamers to navigate the river. It is a rich treasure reserved for generations to come; lying at the very threshold of the portal which leads into the interior of the North Island.

The forced passage of the Waikato through the Taupiri range forms this portal, the scenery being remarkably picturesque and grand. The mountains rise from the lower Waikato Basin to a height of about 1000 feet above the level of the sea, and, south-wards, shelve off abruptly towards the middle Waikato Basin. Rug-ged ridges, steep declivities and deeply cut ravines characterize the landscape on both sides of the river. The rocks, protruding from the river-bank, present a sharp-edged, variously fissured mass of silicious slate of a great geological age. The course of the river, however, is not perceptibly hemmed in by the mountains. On the right and left of the river there are still broad alluvial banks.

Kaitotehe, Mission Station of the Rev. Mr. Ashwell at the Taupiri.

After a short passage through the mountains we landed oppo-site the Taupiri peak close by the Mission station. The missionary's dwelling is situated on the left bank at the foot of the mountains upon fertile alluvial soil quite hidden behind trees. How cheering it was here, to see once more a European house for the first time since we had left Mangatawhiri; and how invigorating and charm-ing was the view of the beautiful site and landscape, which here suddenly seemed to assume larger dimensions! Like a new country there lay spread out before us the flourishing lands of the *Middle Waikato Basin*. We had arrived at the first principal station of our journey, distant from Auckland about 80 miles, and pitched our tents close by the river-bank. The Rev. Mr. Ashwell happened to be absent from home; but Mrs. Ashwell very kindly offered us hospitable quarters in her house, — which offer we accepted most gratefully, passing some days there most agreeably.

CHAPTER XV.

The Waipa and the West Coast.

The Taupiri. — The Middle Waikato-Basin. — Keeping the Sabbath. — Waikato and Waipa. — Residence of the Maori King. — The chief Takerei. — Terraces. — Whatawhata. — The Wesleyan Mission station Kopua. — A Maori-wedding. — Kakepuku. — To Whaingaroa. — Whaingaroa Harbour. — The township of Raglan. — Mount Karioi. — The Aotea Harbour. — The Kawhia Harbour. — Ammonites and Belemnites. — The New Zealand Helvetia. — A Northumberlander. — Roads to the Waipa. — The bush. — Back to the Waipa.

March 12. — The Taupiri, as viewed from the Mission station, is a steep conical hill with a sloping-flat top, which rises immediately on the opposite, the right or East-bank of the Waikato, forming the most prominent point upon the boundary between the lower and the middle Waikato Basin. We crossed the river, and through ferns and bush reached the top in the space of an hour. My sanguine expectations of a brilliant prospect were not disappointed. Only to the North-east the continuation of the range obstructs the prospect into the distance. In all other directions the eye can freely rove far over the land. In clear weather even the Tongariro and the snow-capt peak of the Ruapahu are visible in the South. Although we were denied this pleasure, the sky being somewhat cloudy, yet the view of the adjoining country alone sufficed to impart to me an idea of New Zealand other than I had hitherto entertained.

Here I beheld for the first time an extensive low-land spreading into distant mountain-cones and remote mountain-chains. I was reminded of the fertile plains of Hungary or of the Rhine-valley, where the Neckar and the Main mingle with the Rhine. Like mirrors resplendent with the broken rays of the sun, you see glistening numerous lakes, and the serpentine currents of large rivers winding through the plains. The Waikato is the main-river in this richly watered, fertile basin, which, in its extent from the western to the eastern mountain-ranges, I call the Middle Waikato Basin. The Waikato flows through this basin in the direction from Southeast to Northwest, a distance of 35 miles. Immediately at the foot of the Taupiri the Mangawaro joins it, running trough large swamps in the East, and 5 miles farther up it receives the Waipa, the principal tributary from the Southwest. Up the Waipa towards Southwest and South, along the western coast-range, the low-land extends as far as the slopes of the pictu-resque trachytic mountain Pirongia, and the spurs of the Rangitoto-range, where from the midst of the plain the Kakepuku rises with a regular conical shape. To the South-Southeast, where the Waikato enters from the southern table-land into the basin, there rises majestically the trachytic Maungatautari. Thence springs up a low range of hills, the Maungakawa-range, separating the eastern portion of the basin — the plains of the Piako and Waiho, rivers of considerable size emptying into the Hauraki Gulf, — from the Waikato and Waipa plains. But it is only the steep margin of the eastern coast-range, from the Patetere Plateau to the Aroha moun-tain, that forms the eastern boundary of the basin. This whole basin was previous to the last elevation of North Island, which was probably connected with the volcanic eruptions in the centre of the island, a bay of the sea, extending from the Hauraki Gulf far into the interior. The steep margin of the surrounding ranges has continued to this day displaying the sea-shore of old, and the singular terrace-formation on the declivities of the hills and the river-banks within this basin is the result of a slow and periodical upheaving.

[1] The designation of the Waikato in the middle Waikato Basin as Horotiu is wrong. Horotiu is the name of a river-district, 15 miles above Taupiri.

The Maori school of the Rev. Mr. Ashwell at Taupiri.

How much more charming would the smiling fields of this beautiful country have appeared to us, could we have beheld it studded with European towns and villages! The middle Waikato Basin is the home of the most powerful and most numerous Maori-tribes, whose king had fixed his residence by the confluence of the Waikato and Waipa at Ngaruawahia, and whose metropolis, — if we may use the expression, — may be considered Rangi-awhia, a large settlement, situated between the Waipa and Waikato in the southern part of the basin, which has attained its importance especially by its extensive corn-trade and horse-breeding. In future decades this blessed region will be the granary of the North Is-land, — a real Eden for agriculture and the breeding of cattle, to which, in this respect, no other part of New Zealand might easily compare. The height of the Taupiri above the level of the sea, as resulting from my barometrical measurements, amounts to 983 feet; it consists of a hard blackish-gray slate-rock, polyhedrally fissured. The top of this hill will be one of the principal points for a future triangulation of the country.

March 13. — Sunday. — The Sabbath in New Zealand is kept by both Europeans and natives, with still greater puritanic severity, than in England. Sunday is Ra tapu, a holy day, on which first of all it is not allowed to travel. A violation of the Sabbath would be censured with double severity in a gentleman on account of the bad example set by it to the natives. I for my part complied easily and willingly with this strict Sunday-law, since a day of rest in the week is a real matter of neces-sity to a pedestrian. Only once was I made to feel the severity of the law, when, compelled by a scarcity of provisions, I travel-led on a beautiful Sunday morning a few miles farther to a Maori-village, where I hoped to supply the deficiency, but, after all, was literally compelled to fast there till Monday morning in ex-piation of my trespass. This time the day of rest was not only pleasant to me, but the truly beautiful manner of keeping the day at the Mission station made also an edifying impression upon my mind. The missionary school numbered 94 pupils, 46 girls and

48 boys of different age. The church is a pretty specimen of a Maori building; its walls neatly plaited of various vegetable products of the country,[1] and its door-posts and gable-beams gayly painted. At 11 o'clock divine service commenced. Two and two, the school-children, all neatly and cleanly dressed, came march-

Church of the missionary station at the Taupiri.

ing to church in a long procession. A large number of men and women from the adjoining villages and settlements followed them. The service consisted of a hymn, prayers, and a sermon delivered by a native with much spirit and lively action. Church-service over, I witnessed the Sabbath-school exercises, and was not little astonished at the geographical acquirements of those Maori-children. They very readily traced to me, upon a map of Europe without names, the course of the Danube pointing out the situation of Vienna; and quite correctly answered my questions as to the active volcanoes of Europe. At 2 o'clock dinner was had in the common dining-room. The Sunday-fare consisted of potatoes and

[1] The favorite material of the Maoris for building purposes is Raupo (*Typha*), a kind of flag or bulrush, which grows in great abundance in swampy places. The roof is generally made of toetoe (*Arundo*) a very coarse cutting grass, that grows on the edges of swamps and creeks, whilst the necessary sticks are provided from Kiekie (*Freycinetia*), a liana, the stems of which can be employed like Bengal cane. The whole edifice is merely tied together with flax.

pork. At 4 o'clock, there was English worship in the school-room, and, this also finished, I was shown various pieces of work from the hands of the Maori-girls, who are here instructed in different branches of domestic work, while the boys are trained for agriculture and all sorts of useful trades. A very pretty article of industry of the Taupiri Mission station are mats and carpets wrought of New Zealand flax in different colours. The children generally remain in the institute, until they are grown. Although the results of their education do not always realize just expectations, yet, it is only with the greatest respect and admiration that we can behold the sacrificing activity of such a missionary family upon New Zealand, whose members have all an equal share in the education and civilization of a people, that but a few decades ago was steeped in barbarity.

March 14. — My collections, increased by contributions from the amiable ladies of the missionary's family, and by numerous birds, which a native by the name of Hone Papahewa had procured for me, I had an opportunity to dispatch to Auckland by the regular Maori mail-carrier. Then we decamped. Through the kind intercession of Mrs. Ashwell we had obtained an excellent canoe and the necessary number of paddles, and were thus enabled to continue our course up the river. Upon the fertile banks of the Waikato, above the Taupiri, there appears settlement after settlement with beautiful farm-lands, — Hopuhopu, Kaumatuku, Pepepe, and what other names there are of those places. Owing to the strong current we proceeded but slowly up the river; and it was not before 12 o'clock, that we arrived at the confluence of the Waikato and Waipa, about 5 miles above Taupiri. Upon the point of land between the two rivers lies the Maori residence Ngaruawahia.[1] From political reasons, however, it was not the intention of Captain Hay, to pay a visit to His Maori Majesty, King Potatau te Wherowhero, at the very beginning of our journey; and consequently we passed by, without seeing more of the remarkable place

[1] Ngaruawahia signifies literally, "a region abounding in fire-wood."

than the flag-staff rising from among the bushes, on which on occasions of special solemnity the national flag is hoisted.

Leaving the Waikato we turned into the Waipa. The names of both rivers are indeed characteristic: Waikato, the streaming water; Waipa[1] the quiet or placid water; for while the Waikato runs at the rate of four to five miles per hour, the Waipa has at its mouth at the most a velocity of a half to one mile. The Waipa is in regard to water-supply greatly inferior to the Waikato; nevertheless it had, despite the low water mark, quite a distance up the river, a depth of from 8 to 12 feet. The temperature and the colour of the water in both rivers were likewise remarkably different. The Waikato showed 68⁰ Fahr., and its water was light-green and clear; while that of the Waipa showed the dark-brown colour of peat-water, and a temperature of 70⁰ Fahr.[2]

The passage on the Waipa between high banks on the right

[1] The term *pa* is used by the natives, when somebody is holding his breath, or fails to answer a question; it moreover signifies a "fortified village." Waiho, the third of the principal rivers of the middle Waikato Basin, signifies the new river; the Piako has its name from a kind of Kaikatea-tree, which used to grow in the river-bottoms, and was known to the natives by its extremely pleasant fruit. Of this fruit a chief once remarked: they are like Piako's eyes. But if the fruit is like Piako's eyes, then, according to the conceptions of the natives, the tract, upon which the tree stands, must be Piako's body, and thus the whole region and the river received the name of Piako.

[2] To Professor J. Smith at the University of Sydney I am indebted for information as to the results of an analysis of the water from both rivers. The analized sample of Waikato-water was taken on January 24. 1856 near Arowhena, about 40 miles above the mouth of the Waipa. It had a temperature of 63⁰ F., a pleasant taste, and was quite clear. The sample of Waipa-water was gathered on January 29. 1856 near Whatawhata, 15 miles above its mouth, and had a temperature of 64⁰ F. The water while fresh, had neither smell nor taste; but when on March 26. the bottle was opened, a decided smell and taste of sulphuretted hydrogen was observed. The analysis of both samples yielded the following result:

	Waikato.	Waipa.	
fixed ingredients	4.8	7.04	grains per
volatile, principally organic substances . .	1.6	3.20	gallon.
	6.4	10.27.	

The fixed ingredients were carbonate and sulphate of lime, chloride of sodium, and traces of alumina, magnesia and iron. The Waipa-water contained more than the Waikato-water. Hardness of the water according to Clark's scale: Waikato $= 1^0 5$; Waipa $= 1^0 3$.

and left, presented little that was attractive or worthy of notice. At 2 o'clock we landed near the old Pah Tekohai, for the purpose of paying a visit to the influential chief Takerei, who, as Captain Hay thought, deserved proofs of attention and respect on account of his loyal sentiments. The reception was very ceremonious. We found Takerei together with his friend Tara Hawaiki, and seated ourselves next to the two chiefs upon the unrolled mats. A long conversation on politics was commenced between Captain Hay and the chiefs, of which I understood nothing. I made my own silent observations. Never had I seen a handsomer Maori-head, one of nobler shape, than the proud head of Takerei, nor colder or more austere lineaments than those of his countenance, which was tattooed all over. No trace of a smile or even of friendly sympathy flitted over the man's marble face during our stay of several hours. There he sat cowering, a dirty woollen blanket abound him, smoking his pipe, and casting wild, and gloomy glances about him. At the same time he gave the natives, who were passing to and fro, short, quick orders. There was something extremely imposing in the proud, austere mien of the man, who appeared to me as though wrought of steel; but at the same time also something exceedingly savage. And yet, Takerei is said to be a man very kindly disposed towards the Europeans; one, who, I was told, would have nothing to do with the Maori-king- movement, who donated a large piece of land for the establishment of a missionary school, and who was using all his influence to prevent the introduction of spirituous liquors (Wai pirau, or stinking water) into the country. It has been related of him, that he refused to let canoes pass, which had come up the river loaded with spirits. Takerei had ordered a repast for us, and there was no possibility — unless we chose to be guilty of the most unpardonable breach of Maori-etiquete — of taking our leave, before having eaten with the chiefs. At length, potatoes, eels and milk were placed before us, and with dusk fast settling into night we were able to continue our travel. The moon shining brightly, we proceeded as far as Karakariki. There we found the inhabitants of the village in a large hut gathered around two brightly blazing fires; it was a real gipsy-scene; but the news that Pakehas had arrived, roused them all to their feet; and till late in the night our tents were beset by crowds of curious spectators.

March 15. — At the very first dawn of day, the Maoris paid their morning devotion; their chants resounding as far as our tents. Then they went to work, — they were engaged in building a new church, — and not until after a few hours labour did they sit down to breakfast. Such is the custom in all Maori-villages. The breakfast consisting of potatoes and fish, was prepared by the women for all at once; the breakfast-scene itself, however, was highly amusing, since the dogs, pigs and cats of the village were also all socially admitted to the meal.

After breakfast we continued our course upon the Waipa. The farther up the river, the more distinctly appears the terrace-formation on both banks, which commences already below Karakariki.

Terraces on the lower Waipa.
a. Clay and sandstone strata with lignite. b. Pumice-stone gravel and sand. c. Recent river alluvium.
d. Bed of the Waipa.

The Waipa-valley, it must be remarked, has so to say, two stories. The river-bed is bordered by steep banks, which at low water-mark are from 20 to 30 feet high. After these have been climbed, the first story or first terrace has been scaled of a fertile alluvial plain diligently cultivated by the natives, which is crossed by the river in various bends and curves, and represents, in some measure, a second, broader river-bed running in plainer lines, more elevated, and sometimes flooded by the river at heavy freshes. The natives call this first terrace Te Kotai. From this first story a steep margin, 30 to 40 feet high, leads to the second terrace or the second story, a large plain extending on both sides of the river, and gradually blending with the flat-rolling hills of the Waikato basin. The plain of the second terrace has also been cultivated nearly the whole length of the river; the huts and villages of the natives are seen upon it, and here and there a small remnant of woods is still standing. Characteristic of this second terrace is — since the river-bed does not rise towards South in the same proportion as the plains of the Waikato basin — that up the river it rises higher and higher, until finally a third terrace is inserted between the first and second stories.

The geological features are extremely simple, and seem to remain the same through the entire middle Waikato basin. As lowest bed there appear layers of clay and sand with bituminous shale, which in some places enclose trunks of trees changed to lignite. Few miles above Karakariki the shale passes over into thin beds of argillaceous shale containing numerous fossil plants. They, however, would not admit of being gathered, the mass crumbling two readily. Worthy of notice are also some sporadic pumice-stone fragments, which are found embedded in it. These and similar strata seem to point to the fact that the whole middle Waikato basin was but recently a shallow bay of the sea, or a far extending estuary, at the bottom and on the margins of which those layers were formed. The topmost layer consists of gravel and sand-banks, for which pumice-stone, obsidian and various volcanic rocks have furnished the principal material. The sand contains moreover large quantities of magnetic iron, and small quartz-crystals as clear as water, originating partly from pumice-stone, partly from trachytic rocks (rhyolite). These quaternary layers of gravel and sand compose the broad plains of the second terrace, while upon the first river-terrace recent river-alluvium is deposited. A similar terrace-formation recurs with all the rivers of the middle Waikato basin. On the Waikato itself the terraces commence already below Kirikiriroa, on the Waiho in the vicinity of Mount Aroha.

About noon we halted by the Maori-village Whatawhata. Here I met a native Austrian, C. L. Strauss, who, he told me, had formerly been a Government official in Triest. By what adverse fates he was cast upon the shores of New Zealand, I do not know. He has been living among the natives for the last twenty years, and is married to a Maori-princess, a near relative to King Potatau,

a very corpulent and goodnatured woman. Of this influential
connection he seems to have made good use, and to have acquired
a considerable fortune by dealing in corn; he spoke Maori and
English with perfect fluency; but had quite forgotten the German.
He cherished the hope of being able to return in a few years to
his native home. The teacher of the place, Hamona (Samuel) Nga-
ropi also presented himself to us. Having been educated in a Wesleyan missionary school, he was numbered among the more refined
Maoris; he carried a watch, for the alleged purpose of setting an
example to his fellow-natives by a proper attention to time; and lived
in a neat little house of entirely European style.

The trachyte-mountain Pirongia on the Waipa.

From Whatawhata we proceeded twelve miles farther up the
river; and then pitched our tents on the right bank upon the first
terrace of the river, at the mouth of the Mangaotama. We had
closely approached the foot of Pirongia; and resplendent in the
mellow radiance of the evening-sun, the mountain-mass with its
many peaks and ravines, — an ancient, dilapidated volcano,
2380 feet high, — presented a magnificent sight. This trachyte-
mountain imparts to the lower Waipa-country the characteristic

charms of that landscape. The eye never tires of gazing at it, as it always assumes new forms from each new point of view.

March 16. — The peaks of the Pirongia had during the night put on a cloud-cap, — to the inhabitants of the plain a sure sign of approaching rains, which in fact were not long in coming and then lasted throughout the entire day. [1] We proceeded a few more miles up the river to a place called Rore, and there met another European, J. Cowell, whose wife was the sister of Toetoe, the chief of Rangiawhia, who was one of the party on board the Novara on her voyage to Europe. Farther up lies the old Pah Taurakohia or Matakitaki, famous in Maori history for a terrible scene of blood-shed during the wars with Hongi about the year 1825, in which more than 1000 persons are said to have perished. Further up, where the Waipa-river cuts through the spurs of the Pirongia mountain, rapids commence; the first or lowest are at the mouth of the Mangauika, a tributary from the left; the uppermost, immediately below the mouth of the Manga-rewarewa and Moakurarua, two tributaries from the Pirongia-side. The latter bear the peculiar name of Whare o moti, house of Moses.

Partly owing to the rapids, over which it would not have been easy to move up the heavily loaded canoe, partly to the pouring rain, I preferred to go from Cowell's on foot, letting the natives paddle up the river by themselves as far as our next station, the Wesleyan Mission station Kopua, situated about ten miles farther up the river. About half-way, we came to the settlement of another European, named Turner, an old seaman, who 30 years ago had been shipwrecked with a small coasting-vessel at the mouth of the Waikato. He was married to a Maori-woman, who had

[1] The natives say, that Karioi, 2372 feet high, near Whaingaroa Harbour on the West-coast, is preparing rain for Pirongia, and that Pirongia is pouring it over the Waipa and Waikato countries. This, however, holds good only with rain coming with a Northwest wind. When, on the other hand, the Maungatautari puts on the cloud-cap, then the natives say, that the mountain is inquiring of its neighbour Aroha, whether it also is ready for rain. When Mount Aroha also is wrapped in clouds, then rain is certain; but in this case rain accompanied with N. E. wind.

borne eleven children, seven robust boys, and four pretty girls, as I had ample chance to convince myself by the specimens present in the sitting room.

At 4 o'clock I arrived at the Mission station, and together with my companions, was received by the venerable Rev. Mr. Alexander Read in his house with a most cordial welcome. The evening was passed with conversations about the manners and the relative capacities for civilization of the natives.

March 17. — The Mission station is situated 30 miles from the junction of the Waikato and the Waipa, on the left bank of the Waipa at a considerable curvature of the river. Few miles to the Northwest rises the picturesque Pirongia mountain; and to the East at a distance of two or three miles upon the plain by the right bank of the Waipa, the beautiful, regular, conical Mount Kakepuku. The district is not only extremely fertile, but, as to scenery, no doubt the most beautiful portion of the Waipa country. The weather having turned pleasant again, we agreed upon an excursion to the top of the Kakepuku. The name Kake or Kakipuku is said to signify swelled neck. The top of the mountain the natives have moreover designated by the name of Hikurangi, arch of heaven. According to their views Kakepuku is the wife of Pirongia, and the smaller truncated cone Kawa rising beside the former, is in turn the wife of Kakepuku. A quarrel having arisen between the couples, they are now standing single and isolated.

Rev. Mr. Read was so kind as to accompany us on the excursion. The top can be easily reached from the missionary's house in the space of three hours. By the bank of the Waipa we met a large company of Maoris, numbering about 200 persons, assembled together for a nuptial feast. There were tents erected, and long rows of tables placed, densely crowded with guests and heavily laden with bread, potatoes, maize, pork, tea and all sorts of fruit. At such festivals, which are generally fixed upon for the time immediately following the gathering of crops — it being a time of plenty, — they feast for three whole days, and drink "*in dulci jubilo.*" But often bitter want is felt afterwards, just before the new harvest. We could only with considerable difficulty manage to withdraw from the urgent invitations to partake of the feast. A canoe conveyed us over the river. On the opposite side the way leads over fertile plains to the foot of the mountain.

Kakepuku.

Past the southern foot flows the Mangawero, a river famous all over the country for its abundance of eel. Along the river we noticed numerous eel-traps, in which the natives are said to catch sometimes in one night more than a thousand eels. The water of the river is a clear mountain-water. Upon a ledge of rocks consisting of hard trachytic tuff, which extends across the bed, the river can be crossed dry-shod. Then begins a gentle slope, which encompasses like a wall the conical mountain, and is separated from the latter by swamps. This outer wall may be justly considered as a tuff-cone or tuff-crater, like those occuring with the Auckland volcanoes. The cone itself rises at an angle of 20 degrees. Blocks of a trachydolerite containing numerous pyroxene-crystals are scattered along its declivity. A convenient foot-path leads upward through the fern-bushes past a spring of 64° Fahr. temperature. The top is said to have formerly been fortified and cultivated; only on the Southwest-side there is a small tract of forest remaining, which the chief, who is the owner of that ground, had ordered to be spared. This sylvan grove welcomed us to its cooling shade, and was moreover found to be rich in small, but also rare land-shells. Besides numerous small species of Helix, *Realia (cyclostoma) turriculata* Pfr. and *Daudebardia (Helicophanta) Novoseelandica* Pfr. are found here frequently. [1]

The prospect from the top is grand; but owing to the tall bushes we were obliged to shift our position several times, to enjoy a view all around. In clear weather, the Ruapahu, Tongariro, and the Taranaki mountain, those three gigantic cones of the North Island, are said to be visible. We regretted very much, that such was not the case at this time; yet, also the nearer environs presented points enough to keep me busily mapping for several hours. The beautiful, richly cultivated country about Rangiawhia and Otawhao lay spread out before me like a map. I counted about ten small lakes and ponds scattered over the plain, and the church-steeples of these places were seen arising from among orchards and fields. Verily, I could hardly realize that I was here in the interior of New Zealand.

The height of the top I fixed at 1531 feet above the level of the sea, or at 1358 feet above the Waipa-plain. Several deep ravines runn from the top towards the foot, and two deep abysses on the top of the mountain seem to designate the old crater of the extinct volcanic cone, at the Southeast foot of which there lie trachytic cones of smaller size, the Kawa, Tokanui, Ruahine, Puketarata, and others.

On our return to the missionary mansion we had the pleasure of meeting with Mr. Williamson, the Superintendent of the province, who had just arrived here with his travelling companions from an excursion to the eastern districts of the province. The house of the missionary, therefore, was crowded with guests.

March 18. — A pure, cloudless sky smiled upon us to-day; the atmosphere so clear, that from Mr. Read's garden the snow-

---

[1] *Daudebardia Novoseelandica* Pfr. is a new species established from my collection by Professor Dr. Pfeiffer in Cassel.

clad giant-head of the Ruapahu could be distinctly seen looming up above the hills and mountain-terraces in the South. I beheld the mountain for the first time. In viewing it from the missionary's house, it lies to the South 7⁰ 55′ East, at a distance of 80 miles. Its top appeared like a broad, snow-covered roof, and is said to be never clear of snow. Mr. Read, however, assured me, that during the week past there had been several bare patches of rock visible, and that the mountain had now, towards the end of summer, already donned a fresh robe of snow.

As I intended, before continuing my journey from the Waipa further South, to visit also the harbours on the West Coast, and first of all, the Whaingaroa Harbour, I had the pleasure, in company of the Superintendent, of again travelling down the Waipa as far as Whatawhata, from which point the shortest and easiest road lead across the coast-range to Whaingaroa. We did not reach Whatawhata before night, and set out next morning, March 19. on our road to Whaingaroa. There I met with a little accident, as we were just on the point of starting, but fortunately without any serious consequences. As I was about to go through the plain overgrown with ferns to the margin of the river-bank I fell unawares into a pit above 6 feet deep, one of those dug by the natives close to their settlements, for the purpose of storing away potatoes. My left knee pained me considerably in consequence of the fall, and limping as I was, I could proceed but slowly. The Superintendent, therefore, who was obliged to arrive at Whaingaroa at a stated time, was soon ahead of me with his party, while I followed slowly.

The road, one of the best and most frequented New Zealand paths, leads across open fern-land, hill up and hill down to the foot of a steeply ascending range, the southern continuation of the Taupiri and Hakarimata range, and forming the boundary between the Waipa country and the West Coast. At noon we camped at the edge of the woods by a clear mountain-stream Toketoke. The passage across the mountains lay for the most part through bush, which was pregnant with the balmy fragrance of an orchidee growing on forest-trees, and which was in full bloom. The highest point of the pass is according to my measurement 853 feet above the sea. The range consists of the same kind of slate, as the Taupiri peak. On the opposite declivity, upon an open place called Te kapa ama hanga, a beautiful prospect is opened over the environs of Whaingaroa Harbour. At sunset we reached the foot of the mountains, and crossed the Waitetuna River (eel-river). This is the principal river flowing from the coast-range to the harbour. A second, smaller river, emptying farther North into the inlet of the harbour, is, like the latter, called Whaingaroa.[1] The banks of the Waitetuna are formed by clay-marls of the same description as the strata at the mouth of the Waikato containing Belemnites. We had hoped to be quite near to our place of destination; but we had a long distance to go yet in the dark, besides a difficult passage across a deep creek; and it was not until 9 o'clock at night, that we arrived at the house of Captain Johnston. The captain himself was absent from home, but had made ample provisions for a hospitable welcome in case of our arrival. The comfortable quarters were the more acceptable to me, as from violent pains in the knee, which rendered me utterly unable to walk, I was obliged to keep my bed for two days, and not until March 22 had I so far recovered the use of my limb, that I was able to mount a horse, and thus proceed to Raglan, the future port-town close by the entrance to the harbour.

Whaingaroa Harbour is a narrow inlet of the sea with numerous branches, six to seven miles long, the eastern half of which is divided into two parts by a far-jutting peninsula. The Whaingaroa river empties into the northern branch; the Waitetuna into the southern. The harbour is passable only for small crafts of 60 to 80 tons, which generally anchor close to the entrance. By means of boats, however, communication is practicable far up into the

---

[1] Whaingaroa signifies long-pursued. The name has reference either to a war-party, which pursued the enemy through the branches of the sea extending far into the land; or simply to the long branches of the sea, which can be traced far into the interior.

remotest side-branches. At neap-tide, the harbour is almost entirely drained of water; broad surfaces of mud are laid bare, and only narrow channels continue to retain some water. The Maori population in the neighbourhood is estimated at 400 inhabitants; the number of Europeans, on the other hand, who have settled in this country, was stated to me as amounting to 122, among them 20 farmer-families. About a mile inside of the Heads, on the Southside, a township was founded by the name of Raglan. The place, however, numbered as yet only six or eight houses, among which, of course, a tavern and a mercantile store. Not far from Raglan, likewise on the Southside, there is a Mission station, where we were received with a very friendly welcome by the Rev. Mr. Wallis. Opposite, on the Northside, there is a Maori village Horea and an old Pah.

The banks of the Waitetuna in the vicinity of Captain Johnston's house consist of light-gray clay-marl of tertiary age, containing fossils. I succeeded here, in company with my friend Dr. Haast, in gathering species of *Turitella*, *Isocardium*, *Natica*, and *Turbinolia*, together with some beautiful *Foraminiferæ*, in which the clay-beds are greatly abounding. On the Southside of the harbour basaltic hills arise, while Raglan is situated upon soft, ferruginous sand-stone, formed of quicksand. Opposite to Raglan, on the Northside of the harbour a picturesque formation of tabular limestones appears, which, corroded by the sea, present the most singular forms, such as towers of 60 to 70 feet high, walls and the like. These limestones belong to the same tertiary formation as the limestones near Drury, and are full of fossils, which, however, are difficult to knock out. On passing, from the North-head along the West Coast in a northern

Cristellaria Haasti Stache, enlarged, from the banks of the Waitetuna.

direction to a distance of some miles, the traveller arrives, behind a corner of basaltic rocks, at a very rich locality of fossils embedded in a sandy tertiary limestone. The rocks are full of large oysters and of species of *Terebratula, Pecten, Balanus* etc.

Viewed from the Northhead, the Karioi mountain presents itself in its full beauty as a mighty corner-buttress jutting far into the sea on the Southside of the harbour. It is a trachyte-stock like Pirongia, 2372 feet high, wood-grown from foot to top, its declivities furrowed by ravines, and with a broad, serrated top.

Mount Karioi on the entrance of the Whaingaroa Harbour.

Along the eastern foot of this mountain our path led us on the 23[d] of March to the Aotea Harbour, the Superintendent returning to Auckland. The distance between Whaingaroa and Aotea is ten miles. The road is good, and cut so broad through the woods, that it admits the use of horses. When ascending from Opotoro creek along the Mata-brook the first height has been scaled, the traveller arrives at an open place in the woods where an European house stands. Then, after crossing a gently sloping wood-clad indentation of the mountain-ridge, which extends from Mt. Karioi to Pirongia, he reaches the open hills on the Northside of the Aotea Harbour. The bush upon that saddle appeared to me very beautiful and luxuriant, no doubt merely because the road had been cut out so broad, that I really had a full view of the forms of the single trees, which is hardly possible on the ordinary bush-paths. Having been delayed on our way, we were obliged to camp in the bush, and, therefore, did not reach Aotea Bay until the morning of March 24.

The Aotea Harbour is an estuary, expanding into a shallow bay within its narrow entrance from the sea, two to three miles wide and extending about six miles into the land, which at the time of ebb becomes quite dry, save some narrow water-courses. We happened to arrive at neap-tide, and consequently had the pleasure of wading, surrounded by innumerable swamp- and water-fowls, through the mud-fields strewn with mussels and shells. Thus we arrived at the Maori-village Rauraukauera situated on the Northside, and received at the Wesleyan Mission station Beechamdale a hearty welcome from Mr. Skinner. The buildings, especially the church, presented a quite dilapidated appearance; but the more beautiful and smiling was the aspect of the apple and fig-trees and of all the lands round about the village. The school numbered only 16 children.

About 270 native and four European families are said to live by the shores of Aotea Bay. Dieffenbach, in 1841, estimated the number of natives as still amounting to 1200. The two most influential chiefs of the country are Te Kanawa, with his Christian name Kihiringi (Kissling), on the Northside of the harbour, and Te Maratua or Pinarika (Abimelech) on the Southside. The acquaintance of the former was not uninteresting to me. He was almost blind and must have already numbered over eighty years. His descent he derived from the crew of the Tainui canoe, which, as the legend goes, was stranded on Kawhia Harbour; he talked to me a great deal about wars and the numerous tribes, which in days of yore had populated the country. Upon my asking, what had become of all those people, he replied with the greatest calmness and composure possible: "we have eaten them all up," and when I requested him to write his name for me upon a sheet of paper, he wrote with the left hand and in gigantic characters, but reversed, so that the paper had to be held to the light in order to decipher the characters upon the other side. How the old man had acquired that kind of writing, I do not know.

Small schooners of 18 to 20 tons are the means of communication between Manukau and Aotea. Larger vessels are prevented
from passing in by the sand-banks in front of the entrance to the
harbour. The Aotea Harbour also terminates towards the East
in two creeks or rivers, the Pakaka North, and the Makamaka
South.

The geological features are very simple, but instructive, since
the two formations, which on the Whaingaroa lie separate, can
here be observed lying one above the other. The best section
is exhibited at a white cliff on the Southeast side, visible at
a great distance, and called Orotangi by the natives, which
signifies, that here rocks fall down with a rumbling noise. At
the bottom, lies grayish marl about 40 feet thick and of the same
description, as on Whaingaroa Harbour, with few fossils. I found
an *Inoceramus* and some few *Pectens*. Above banks of calcareous
sandstone are piled full of fossils. It is the same formation, to
which also the tabular limestones of Whaingaroa belong; at the
neighbouring Puketoa-cliff those sandstone-banks are contiguous to
the level of the sea. The fossils collected here belong to species
of *Pecten*, *Spondylus*, *Cuculaea*, *Terebratula*, *Pollicipes*, *Scalaria*
and *Schizaster*. This tertiary marl and sandstone formation forms
round about the Aotea Harbour a ridge of hills furrowed by
innumerable little gullies. Close by the Heads the quicksand is
piled up in hills 300 to 400 feet high, and at the water-edge
here and there thin lignite-beds are visible.

I reached on the same day yet the Kawhia Harbour, a distance of only four or five miles. The road to this harbour leads
from the settlement Te Kawa Kawa, on the South shore of the
Aotea Bay, across the fertile, richly cultivated Karikari-plateau
and thence past sand-hills and small fresh-water ponds abounding
in eel. With the ferry-man on the Northside of the harbour we
found comfortable quarters for the night.

March 25. — Rose early; when I stepped from the house
upon the beach, the Kawhia Harbour lay before me in full morn-
ing-splendour, a large surface of water, smooth as a mirror.

Not far from the house, by a sand-hill covered with bushes stands the memorable limestone-block, which according to the Maori-legend is a part of the canoe Tainui, in which the first immigrants from Hawaiki came to Kawhia Harbour. The place was formerly tapu, and the Maoris, when paying a visit to the relic, keep at a respectful distance. Yet among them there seem to be some collectors of curiosities; for even from this stone block con-

Limestone Block Tainui on Kawhia Harbour.

sisting of tabular limestone, and projecting slantly from the ground to a height of about four feet, all the corners had been knocked off. Close by, there are some magnificent specimens of the Pohutukaua-tree growing; and about two miles North of the Northhead, close by the sea-shore, some warm springs are said to rise, which the natives call Puias.

The Kawhia Harbour is the last extensive estuary on the West Coast. It is six to seven miles long, three to four miles wide, and intersected by numerous navigable channels, between which at low-water shallow sand and mud-banks are laid bare. The principal channels are: on the Northside, the Kawhia Channel which again is divided into three branches: the Oparau, Kauri and Awaroa Channels; in the middle, the Rakaunui Channel, and on the Southside, the Waiharakeke Channel. All these channels lead to small rivers of the same names, emptying from the East into the harbour. The entrance to the harbour is hemmed in to within half a sea-mile by a neck of land, Te Maika, protruding far from the Southside. In front of the entrance are sand-banks which admit only the passage of small vessels. The coast-navigation is carried on partly by Europeans, partly by natives. Six European families had settled down at various points of the harbour, and the number of natives has been estimated at 500 to 600. They are said to be zealous adherents of the Maori king, to place annually a certain sum of money (I was told, 136 pounds sterling) at the disposal of the king, and to have made the attempt to exact toll from the European coast-cruisers entering the harbour, and tribute from the Europeans settled among them. The original tribes have been here also dislodged and supplanted by Waikato-tribes. The present inhabitants belong to the tribes Ngatimahutu, Ngatimaniapoto and Ngatihikairo.

I was taken over by the ferry-man to Takatahi on the Southside of the harbour, where I lodged with Messrs. Yates, making excursions thence in various directions for the benefit of geological studies. Close by lived an English custom-house officer, Mr. Schultz; and we were always sure of passing a pleasant evening.

While, upon an excursion from Takatahi along the strand in the direction of the Rangitaiki Bay towards Southhead, I was examining more closely the rocky banks of the shore, consisting of steep-rising banks of shale and sandstone, which gradually slope towards East, I had the pleasure of discovering the first New Zealand Ammonites. But notwithstanding a long and tedious search I had to be content with two specimens, both belonging to the same species. I have called it *Ammonites Novoseelandicus*. It has its nearest relatives in European forms belonging to the Neocomien. A second fossil of very frequent occurrence here, a large ribbed *Inoceramus*, I called after my friend Haast *Inoceramus Haasti*. It was evident, that these shales belong to the same formation as the Belemnite-bearing beds on the Waikato; yet

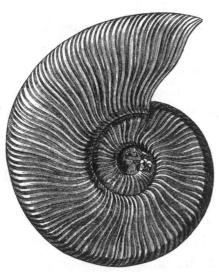

Ammonites Novoseelandicus from Kawhia Harbour.

I searched, in vain, for Belemnites. I was the more surprised, therefore, to discover a rich locality of Belemnites, on a second excursion in an opposite direction, up the Waiharakeke Channel, at the Ahuahu Point on the Southside of the channel, not far from the Wesleyan Mission station. The cliffs are here almost 40 feet high and consist of grayish-brown shale, the steep-rising strata of which alternate with hard calcareous banks. Upon the mud-bank at the foot of this cliff numerous belemnite-fragments are scattered about, and after a short search among the banks complete specimens are also found. It is the same species as on the Waikato, *Belem. Aucklandicus*, only a smaller variety of it. To the Maoris of the neighbourhood those belemnites were well known; they told me, that their children gather the sharp stones to play with them; they called them Rokekanae, excrements of the fish Kanae *(Mugil Forsteri;* mullet), which they thought was in the habit of leaping out of the water and leaving those stones behind on the shore.

At the missionary station close by we found Mr. Schnackenberg, a German countryman, a native of Hanover, who was sincerely delighted at our unexpected visit, but who, we found, had utterly forgotten his mother tongue. In the garden I saw lemon-trees bearing beautiful fruits. While we were amusing ourselves with discussing the past and future of New Zealand, the chief Te Nui-tone (Newton) arrived in a war-canoe, a genuine old warrior, of the number of Toas (heroes). He had just returned from a journey he had made to Taranaki for the noble purpose of bringing about, if possible, a peace

Belemnites Aucklandicus, smaller variety from Kawhia Harbour.

between two tribes of that country who had been waging war against each other for several years past; and was in a very depressed state of mind. He expressed it as his opinion, that affairs in Taranaki were growing worse and worse, and peace was entirely out of question; on the contrary, he apprehended, that the English Government would join one party to make war upon the other. [1]

March 26th I sailed to the Rakaunui River. [2] While the whole Southshore of Kawhia Harbour consists of shales containing ammonites and belemnites; on the northern bank of the Waiharakeke, the same light clay-marl with strata of limestone-banks, that pre-

---

[1] This war broke out in 1860.
[2] Rakaunui, equivalent to big bush.

sents itself on the Aotea and Whaingaroa Harbour, form the river-banks, and extend in nearly horizontal strata over the whole South-east side of the harbour as far as the Awaroa river. On the Rau-kaunui river the limestone-banks reach the level of the water, and present along the shore picturesque rock-sceneries in the shape of towers, walls and ruins, from which circumstance the settlers have given to this portion of the Kawhia Harbour the name of "New

Limestone rocks on the Rakaunui river, Kawhia Harbour.

Zealand Helvetia." Those picturesque columnar rocks, cleft and weather-beaten in the most different shapes and forms — now forming long and high banks, now picturesque islands and pro-montories — always present new and interesting views along the winding creek, while in the valleys and gullies leading to the creek, the villages of the natives lie scattered about, and luxuriant wheat- and maize-fields greet the eye. I reckon this country among the most charming and fertile districts that I have seen upon New

Zealand. The character of the landscape seems to continue the same far up to the steeps of the mountains; for even to heights of at least 1000 feet above the harbour, walls and crowns of white rock are to be seen emerging from among woods and bushes; hence the name Castle Hills for these mountains; the natives call them Whenuapu.

This country is noted also for its numerous caves. One of them, Te ana hohonu, the deep cave, is situated on the peninsula between the Rakaunui and Awaroa rivers close to a small Maori settlement. The entrance to it was entirely closed up by dense bushes of Koromiko *(Veronica speciosa)* and Tutu or Tupakihi *(Coriaria sarmentosa)*. Likewise of the Warengarara fern *(Asplenium lucidum)* with its beautiful glossy leaves, specimens of a most luxuriant growth are found. After we had removed the bushes, we found a kind of natural tunnel, leading in a north-eastern direction about 100 yards, but then getting so low, that a man could proceed only by crawling. The natives assured me, that farther-on the cave grows larger and wider, containing beautiful stalactites, and then separates in three branches. However, it did not seem worth while to continue the exploration of the same. A second cave was pointed out to me as the cemetery of the Ngatitoa-tribe, to which the famous Maori chief Rauparaha belonged. It is said to be crowded with Maori corpses, shriveled and dried up like mummies. This cave, however, is, at yet, strictly tapu, and no admittance granted to it. The tertiary limestones occur also on the Northside of the Kawhia Harbour at the Tawara Bay and on the Pati river. On the latter and on the Awaroa river there are likewise places abounding in fossils of the kind, that we have already met with on Whaingaroa and Aotea. On the Southside of the Kawhia Harbour limestone forms the rugged steeps of the Castle Hills.

March 27. — Pursuant to a cordial invitation, I spent the Sunday at Mr. Charleston's, on the Northside of the harbour. In him I became acquainted with the oldest settler on Kawhia Harbour, a Northumberlander by birth, a hardy, energetic man, who seemed to have been destined by Nature for the rough work of a primitive settlement upon New Zealand. During the 19 years of his sojourn upon New Zealand, he had acquired considerable wealth by trading in corn and cattle, and, radiant with proud self-consciousness, he conducted me through his splendid orchard and vegetable garden, thence through a poultry-yard teeming with chickens, ducks and geese, and finally to his meadows and pastures, where horses, cattle, sheep and pigs bore testimony of a considerable property. He expressed the wish, that my artist, Mr. Koch, would paint his house and garden, with the liberal remark, that he was not particular a to a pound or two, if only the apples in the garden were painted quite nice and red. In the afternoon neighbouring settlers with their families arrived on a visit. Likewise a native, calling himself John Wesley, holding the office of a native-assessor on Kawhia Harbour, and aping in a ludicrous manner the gentleman or dandy, had arrived with his sponse, so that the house was quite crowded. There being no scarcity of pretty young ladies gracing the social circle, and the cellar of the house harbouring excellent English ale, the evening passed off very amusingly.

March 28th I started on my return to the Waipa. Three roads lead from Kawhia Harbour across the ranges along the southern foot of the Pirongia to the Waipa-valley. The traveller will choose one road or the other according to his point of starting on Kawhia Harbour or his place of destination on the Waipa. The southern route is the Awaroa-road. It leads from Awaroa Creek through a romantic limestone, and sandstone country and is connected with the roads to the upper Mokau district. But in order to arrive at the Mission station Kopua on the Waipa, the other two roads are preferable. The Kauri-road, starting from the Kauri Creek, appears to be less woody and montainous, than the Oparau-road, and is therefore specially to be recommended to horsemen, while the Oparau-road is the shortest and most direct route for pedestrians, but offering difficulties to horses, since two very deep and steep ravines are to be passed. Captain Hay and the photographer, who had started from Takatahi on the Southside of the harbour, chose the Awaroa-road; whilst I with Haast and Koch took the Oparau-road.

Mr. Charleston having kindly placed his boat at our disposal, we set out with the beginning of spring-tide. The Oparau-river empties into the Kawhia Harbour between the peninsulas Tiritiri-matangi and Otururu. A close acquaintance with the channels between the extensive sandbanks is necessary in order to reach the river. Our adventurous pilots set us repeatedly upon the sand, before arriving in the narrow river-bed. We could proceed only a short distance farther up, and then landed close by a mill. The mill is called the Mangapapa-mill, after a small brook, on which it stands, and belongs to a Frenchman. The country on the Oparau is extremely fertile and well cultivated. The soil consists of de-composed trachytic tuff and conglomerate, extending from the trachyte-stock of the Pirongia as far as the Eastside of Kawhia Harbour. We proceeded a few more miles up-hill, and, when evening had set in, we pitched our tents upon a potatoe-field at the edge of the bush.

March 29. — The night was star-light and very cold. Early in the morning I saw, for the first time during our journey, frost upon the ground. The road carried us higher and higher up into the mountains, now through bush, now across open heights cov-ered with fern, from which a charming view is had of Kawhia Harbour, and a retrospect upon the country just passed through. After crossing several mountain-streams, the traveller enters into the bush, through which he ascends to the mountain-ridge forming the water-shed between the West Coast and the Waipa. On the right of the road, after having reached the highest point of the pass (1585 feet above the sea), there stands a gigantic Rata tree *(Metrosideros robusta)*, about 100 feet high. The root-stock forms a cone 8 feet high; then the trunk begins, the circumference of which at a height of 4 feet above the root-stock amounts to 46 feet 8 inches. The tree appeared to be perfectly sound, and was en-tirely overgrown with parasitical Bromeliaceæ. In the bush I gathered a great number of beautiful mosses and ferns, especially various species of *Hymenophyllum*, with which the mammoth-trunks were covered all over. The mountain-ridge itself is a southern spur of the 2381 feet high Pirongia range, and consists like the latter of volcanic rocks, especially of trachydolerit with crystals of pyroxene, and of basalt. In the dusky woods and gloomy ravines of this mountain-stock the Kiwi *(Apteryx Mantelli)* and the Weka *(Rallus australis)* are said to be still quite frequent. The natives told me also of a large kiwi resembling a cassiowary, which they said was sometimes found here.

From the height of the pass we were obliged to descend into a deep, wild ravine, called Ngutunia. Hence once more up-hill to a second, broader ridge, and for about four miles through the dark bush. About 5 o'clock p. m. we reached the end of the woods — and there, radiant with the soft light of the evening sun, lay before our wondering eyes the landscape on the Waipa. With pleasure we greeted once more the Kakepuku and the missionary's house, which was blinking up to us so cheerfully. But, although it appeared quite near, we had yet a long distance to go. A second, deep ravine had to be passed, then after a walk over extensive fern-heaths, we came to a tributary of the Waipa, and it was not until 9 o'clock in the evening that we reached again our hospitable quarters at Mr. Read's.

CHAPTER XVI.

From the Waipa through the Mokau and Tuhua districts to Lake Taupo.

On the upper Waipa. — Terrace-formation. — Orahiri. — Hangatiki. — Caves with Moa-bones. — The stalactite-cave Te ana Uriuri. — Sabbath breaking punished. — Manga-whitikau. — Limestone plateau with caves, subterraneous river passages, and funnel-shaped holes. — Puke Aruhe. — Stay at Piopio; rubbing of noses. — Moa-bones. — Wairere rapids. — Pukewhau. — The Mokau river. — Bush and swamp. — A wood-colony. — Maori cooking-stove. — Tapuiwahine mountain. — In the Ohura and Ongaruhe valley. — On the summit of the Ngariha. — View of Tongariro and Ruapahu. — A Tangi in Petania. — Puketapu. — Nothing but bush. — Pumice-stone plateau Moerangi. — Arrival at Lake Taupo.

To reach lake Taupo from the Waipa, I had the choice of two roads. The one, *via* Otawhao, past the foot of the Managatautari, along the Waikato valley reaches lake Taupo at its Northend. This is the road taken by Dieffenbach in 1841, and over which the mail is now carried every fortnight by Maoris from Ahuriri on the East Coast to Auckland. The second road leads up the Waipa through the upper Mokau and Wanganui country to the Southend of the lake. This is longer and much rougher than the former; as it leads, however, through rarely frequented districts, it promised many interesting items. I was the less reluctant to decide upon the second route, because I hoped to succeed in gathering there some moa-bones; and consequently I fixed upon the 1st of April as the day for continuing our journey.

Thus with a new month also a new chapter of our journey opened. Up to the present we had after a few days' journey repeatedly found hospitable quarters in European houses, with settlers or missionaries; but now the European settlements ceased and we had to travel to lake Taupo through a country exclusively, and for long distances but thinly, populated by natives. My guides representing to me that the road was very difficult to travel, that it lay through numerous swamps and woods, up and down steep mountains, rendering the transportation of the heavy photographic chests impracticable, I proposed to Mr. Hamel, to travel direct to Lake Rotorua, thence to Lake Tarawera, and to meet us again at the latter place after having taken views of the most interesting points. Thus we parted: Mr. Hamel, accompanied by four natives, set out in the direction of Otawhao; my route lay in a southerly direction up the Waipa, and after the departure of the photographic company, my party numbered only eighteen. My servant James, whom I had been obliged to leave behind sick at Mr. Read's, I had supplanted by a robust young fellow, named Harry Turner, — a first rate specimen of the Maori-European mixed race, who would have been an honour to any grenadier-regiment in Europe.

We took a hearty leave of Mr. Read and his family, and when upon starting we waved, after German manner, a final, cordial farewell to the amiable ladies of the house, one of the Maoris said: "now look, these gentlemen are real European chiefs; they have manners, indeed; the others are only slaves." My knee still smarting considerably, and the nature of the ground admitting of riding on horse-back, I had made use of the kind offer of Mr. Read, who lent me one of his horses. Not far from the missionary station we had to cross the Waipa; the water in the river was so low, that I could easily ford it, my companions crossing in canoes. Our road lead thence along the right (East) bank up the river. The landscape on both banks of the river attracts attention by the extreme regularity of the terrace formation, presenting itself in three successive levels. In its sinuous course the river intersects a broad alluvial plain, rising from 12 to 15 feet above its bed. A steep bluff, 20 to 30 feet high, leads from this first terrace on to a second extremely fertile, and well cultivated plain;

Terraces in the upper Waipa valley.
a. Trachytic tuff. *b.* Pumice-stone gravel and sand. *c.* River alluvium. *d.* Bed of the Waipa.

and another bluff, 80 to 100 feet high, leads to an extensive table-land consisting of trachytic tuffs, above which wood-clad ridges arise, and at a greater distance a number of extinct volcanic cones, such as the Puketarata, the Taurangakoho and others. From the terraces every vestige of the primitive forests has long disappeared; only isolated Ti-trees, Phormium-bushes, here and there groups of Kahikatea-pines, and the settlements of the natives, interrupt the broad, monotonous fern-covered plains.

The first small settlement we came to, about five miles from Read's, is called Awatoitoi, it consisted of six scanty huts. One mile farther up on the left bank is Tuahu. Another mile farther the Waipa receives the Mangaoronga, a considerable tributary from the right. Upon the peninsula formed by the confluence of the two rivers towers the ancient Maori-castle Tohorewa with steep cliffs, 60 to 80 feet high. The rugged rocks, upon which in olden times a pah stood, consist of a compact trachytic tuff containing pumice-stone. Large blocks of it are scattered around. Having crossed the Mangaoronga, we passed along the foot of the old pah on to the height of the table-land, which the natives here call Kareauwaha. There is a beautiful prospect far over the broad valley of the Waipa with its terraces and native settlements. On the right bank of the river is Kourapirau, meaning a place, where craw-fish cease to be found. The scattered huts farther up belong to Mohoanui and Orahiri. By an isolated conical rock, named Haereuku, our path turned towards the river. We crossed it at a place very shallow in consequence of sand and gravel-banks, and pitched our tents on the left bank near Orahiri.

Orahiri is one of the chief points on the Upper Waipa. Having received two considerable tributaries from the left, the Mangapu[1]

[1] The word Manga, so frequently occurring in river-names, signifies branch of a river, tributary.

and the Mangawhero, the river takes a south-easterly direction towards the Rangitoto-range. The valley here changes to an extensive plain covered with Kahikatea-forests; in the background rises the wood-clad Rangitoto-range forming the water-shed between the Waipa on one hand, and the Mokau and Wanganui rivers on the other. The river-pebbles lead us to suppose, that the Rangitoto-range consists of shale, sandstone and marl, and consequently is not of a volcanic origin.

April 2. — In the morning the mountains were shrouded in dense fogs, which vanished towards noon. The sky, however, remained veiled all day, so that the sun could scarcely peep through. It was not until sun-set, that the misty veil arranged itself in chains or streaks of fleecy cloudlets, all shooting from Northwest, and lit up by the magnificent red and yellow lustre of the parting sun, thus imparting to the sky a quite peculiar appearance, and promising a bright and clear morning.

We set out from Orahiri at 8 o'clock. The little Orahiri creek is dammed up, and forms a mill-pond which abounds in eel. Our road lay across the dam. Here we left the Waipa valley, and turning in a south-westerly direction to the heights on the left bank of the Mangapu, we reached the Mangapu valley itself about six miles from Orahiri, near the Maori village Hangatiki. Here the Mangapu receives the Mangaokewa from the right. The steep ridge between the two rivers, at the foot of which the church of Hangatiki stands, is called Pukeroa. The inhabitants of the settlement belong to the Ngatihuiau-tribe, and have the reputation of having acquired to perfection all the tricks and wiles of European peddling Jews. Yet, they very kindly brought me some grapes, the first I saw and tasted in New Zealand; I found them, however, very sour.

The landscape here assumes a new character. From the broad, open valleys of the upper Waipa, the traveller, in passing up the Mangapu, gradually enters the mountains and hills bordering the plains of the Waipa and Waikato to the South and Southwest. Hangatiki is situated just at the foot of the higher table-land which, though its surface appears very much cut up and broken, nevertheless presents a very simple geological structure. The lower beds, appearing in the valleys, consist of stratified limestones of the same description as those, with which we became acquainted on the West coast (Whaingaroa, Aotea, Kawhia), and above them lie thick beds of trachytic tuff. The limestone abounds in caves; the volcanic tuff on the other hand forms compact banks resembling sandstone, with the white fragments of pumice-stone looking from afar like fossil shells.

At Hangatiki I heard of three caves in that vicinity. Te ana o te moa (moa-cave), Te ana o te atua (ghost-cave), and Te ana uriuri (dark cave). The moa-cave is about one and a half miles Southwest of the kainga Parianiwaniwa, which is about five miles from Hangatiki in the direction of W. 24° S. The ghost-cave lies at the road to Raraoraro, a distance of one mile from the Maori settlement Rotomarama. Both caves abound in stalactite formations. They had been repeatedly visited by English officers: the last time, in October 1852, by Dr. A. Thomson, Major Hume, and Captain Cooper, who went there in search of moa-bones. Those researches have according to Dr. Thomson's account [1] led to very satisfactory results. Numerous bones were dug out. The natives assuring me, that there was nothing more to be found, I resolved to visit only the third cave, Te ana uriuri, it being less out of my way. This cave lies four miles from Hangatiki on the road to Pukemapau and Paripari. Its entrance is 24 feet wide and 20 feet high. It is situated at the northern foot of a hill composed of tabular limestone, the slope of which was overgrown with bushes of the New Zealand coffee-shrub, — Karamu of the natives (Coprosma lucida Forst.), — full of red berries. The limestone formation attains here a thickness of at least 200 feet. Just at the entrance beautiful stalactite formations are to be seen, grown over most picturesquely with ferns as far as daylight peers into the cave, presenting thus the appearance of an artificial decoration. To the right a stalagmite-pillar, 7 feet high, spreading its top like a table, and

[1] Edinburgh New Philosophical Journal Vol. LVI. p. 268—295.

overgrown with the most beautiful ferns looks like a real flower-table; and on the left an equally decorated stalactite represents a flower-vase suspended from the ceiling. Thirty yards from the portal there is a precipice 20 feet deep, down which, however, it was possible to climb. Below, there is a water-pool with a temperature of 55° Fahr. The vault of the apartment, about 40 feet high, is decorated by stalactites. Beyond the water-pool, a narrow passage leads farther into the interior of the cave. Owing to the scantiness of our illuminating apparatus, however, we were unable to proceed very far; and we moreover greatly regretted our not finding a single vestige of moa-bones.

We had to return a distance of about three miles by the same way we had come; then we turned off to the village Pukahuku on the Mangaokewa Creek, and about sun-set reached a small settlement Tahuahu on the right bank of the Mangapu. Having sent my Maoris ahead by the direct road, I found our tents already pitched. An enormously corpulent woman and a Maori teacher, with a whole pack of dogs for their company, were the only inhabitants present. Beside my tent there stood an odd half decomposed figure carved of wood; it was designated to me by the natives as a Tiki, marking the tomb of a chief.

Tiki, a Maori tomb.

April 3. — Sabbath-day. — A beautiful day, indeed; as though all Nature had been created anew. Who could under such circumstances remain idle in his tent? I determined on travelling on despite the Sunday, having moreover another special reason, as our provisions were nearly exhausted, and the Maori schoolmaster at Tahuahu could not be prevailed upon by any means, to sell us some; on the contrary he deemed it his duty to lecture us on the keeping of the Ratapu. I consequently set out on my way to Mangawhitikau, a settlement five miles farther up the valley, where we arrived just at the close of the morning services. I had cherished the hope of finding the natives there less bigoted; but was sadly disappointed in my expectations. At first they refused the use of a canoe for crossing the river near the village; next, they would not allow us to put up our tents. After a long and tedious talking to and fro, we at last carried both points; we had, however, to put up with the disagreeable necessity of fasting for to-day, the teacher of the place forbidding most rigidly the sale of provisions. Those Maoris seemed to esteem more highly the commandments of their missionaries, than the words of their Lord and Master: "The Sabbath was made for man, and not man for the Sabbath." It was, however, the first and last time that I undertook to travel on the Sabbath-day in New Zealand.

April 4. — It was not until to-day, that the inhabitants of Mangawhitikau grew more sociable and friendly. Already before break of day we had bought a fat pig from them, and were thus enabled to indemnify ourselves at breakfast for the involuntary fast day.

The Mangapu divides near the settlement into its three constituent branches. The eastern is called Mangarama, the middle Mangapu, and the western Mangawhitikau. At their confluence, situated 230 feet above the level of the sea, the valley opens to a basin, shut in all around by towering limestone-rocks. The Mangarama flows through a large swamp clothed with Kahikatea woods; the other two little streams, after a long subterraneous passage, suddenly issue from under the limestone rocks close by the village. The place, where the Mangapu issues forth, bears the characteristic name of Tenganui, long throat. The subterraneous course of the river is said to be four miles long. At a low water-level the natives, with torches (kapara), made of the resinous wood of the Rimu pine, crawl far into the underground channel for the purposes of catching craw-fish. The temperature of the water at the outlet of the river is 53 ⁰ Fahr. The river before its disappearing underground is said to run through a romantic, narrow gully in the limestone-range Mahihinui, by the Pah Pehiop. The course of the Mangawhitikau is similar to that described. It runs close by the Pah Nikau through the narrow, rocky ravine Huruhuru,

80 feet deep, bridged over by a Totara-bridge; and then it disappears for a distance of one to two miles, to reappear again at Mangawhitikau.

The river bed is formed by the clayey strata underlying the limestone formation. The limestone contains the same fossils, that are found at Whaingaroa and on the Kawhia Harbour, attaining however in these parts a thickness of 300 to 400 feet. The limestone formation between the West Coast and the Upper Waipa and Mokau districts forms a plateau, rising to a height of 1000 feet above the level of the sea, remarkable for subterraneous water-courses, caves and deep funnel-shaped holes, called by the natives Tomo. Some of them are filled with water, as the Rototapu (holy lake) near Mangawhitikau, which, as the natives say, was made by evil spirits for the purpose of catching men in it. In several of the caves repose the bones of generations past and gone; they are therefore held sacred, and Pakehas are not allowed to enter them. It is really astonishing to observe how minutely the Maoris know their country, and how they have named not only each plant, bird, or insect, but also nearly every place. Every single cave and cleft, every rock and every hole in these parts has a special name with some legend or other attached to it. In order to learn the names and legends, it is necessary to inquire of the older chiefs; the common man knowing but little, and the younger generation caring no longer for them. My source of information was in Mangawhitikau the Chief Reihana te Huatare, a stout man, kind-hearted, and of intelligent countenance, who, on my asking him the geography of the surrounding country, traced me a figurative map upon the ground, representing the rivers by little sticks of wood, and the mountains by small stones.

In winter, at a high water level, the Mangapu river is navigable for canoes establishing during that season an uninter-

---

[1] Tomo signifies crumbling, sinking down, and is therefore very appropriate. This limestone region also abounds in land-shells and lizards. The natives told me of a large green lizard, said to be found here. I, however, could observe only the little Moko-Moko (*Mocoa Zelandica* Gray, or *Lampropholis Moco* Fitz.).

rupted water-route to the Waipa and Waikato, of which the natives avail themselves for commercial expeditions to Auckland. A portion of the villagers were just dragging along a large canoe, which they had hewn out in the woods, to the river, in order to ship 30 fat swine to Auckland after the setting in of the rainy season. In the afternoon I succeeded in getting off.

We had now to cross the dividing ranges between the Waipa and the Mokau districts. The road lay across the heights between the Mangapu and the Mangarama, partly through bush, partly over open fern-land. Two or three miles from Mangahawitikau we passed through the settlement Mania. The huts were deserted, fields and meadows utterly neglected; for the Maori-custom demands, that a place, where an eminent chief has died, shall remain uninhabited until the years of mourning are passed. The deceased nobleman, whom Mania was mourning for, was Huatare, Reihana's father. Stately peach-trees and the luxuriant growth of ferns thriving here in bushes to 14 feet high, indicated an extremely fertile soil. From Mania we ascended a steep slope, through a magnificent forest, full of the loveliest fern-trees, to the summit of the Puke Aruhe. There stood in olden times a celebrated Maori-fort. The only remnants left of it are deep moats and ditches, which, being most deceivingly overgrown with ferns, are very apt to prove fatal to the unsuspecting traveller; there are likewise numerous round stones such as the natives are wont to use in their mode of cooking. The height of the summit is 877 feet above the level of the sea. Thick banks of trachytic tuff forming the rounded tops of the mountains are overlying the tertiary limestone formation. From the open height we enjoyed once more a lovely retrospect as far as the Kakepuku over the parts travelled through during the past few days; and on crossing the range we entered an entirely new country.

Next we traversed a limestone country about one mile wide, and attracting special attention not only by the innumerable funnel-shaped holes (Tomo), amongst which the path winds along, but also by its peculiar vegetation. It was the first grass-plain I met with in New Zealand. The ferns, the usual growth of all open tracts of land, were wanting here as far as the limestone extended; while they covered with their thicket all the tuff-hills rising above the plain. At the southend of the plain, sharp rugged limestone-rocks, called Terore awairoa, form a small defile. Thence we ascended a wood-clad hill, and in the evening reached Takapau, a settlement consisting of a few deserted raupo huts.

April 5. — During the night a heavy gale blew from South-west with frequent gusts of rain; but towards morning it cleared off again. Descending from the height on which Takapau is situated, we came past the ancient Pah Whararipa. The rock, upon which the Maori-stronghold stood, arises with steep ascent from an extensive swampy plain; its lower part consists of tabular lime-stone, the top of trachytic tuff. On the Eastside, there issues from a cave the Waitoatoa, a tributary of the Mokau, whose district we had just reached. The road continuing partly through marshy river-bottom, partly across the rugged crests of the hilly range that intervene between the several valleys, reaches the Mokau valley not far from Piopio, at the mouth of the Mangakohai.

Piopio is a small settlement consisting of only three huts. Nevertheless we found a large company of Maoris congregated there for a festival. On our approach, they came out to meet us bearing baskets full of potatoes and meat. A continuation of our journey was for the present quite out of the question, until the very last meat-basket [1] was emptied. I had to resign myself to my fate and, moreover, was obliged to rub noses with an extremely pleasant old woman, the honourable spouse of the great Mokau chief Ngature, who kindly administered to me a fat piece of pork, potatoes and apples, and invited me to seat myself by her side upon the unrolled mat, amid a group of weather-beaten faces, which without any further disguise might have appeared most cre-ditably in the wellknown scene in Macbeth. The occasion of the

[1] The baskets are called Patua, and are made of the bark of the Totara-tree; they were filled with meat and fat to the brim.

festival was the exchange of presents with a tribe dwelling on the lower Mokau.

In the afternoon I ascended a height, not far from Piopio, on the right bank of the Mokau, and close by the settlement Mairoa, for the purpose of sketching and mapping the surrounding country. Opposite, on the left bank of the river, rises a wood-clad ridge, Kahuwhera; from the summit of which a beautiful view is said to be had of the Tongariro and Ruapahu. In the evening I joined the natives as they sat clustered about the fire. The topic of our conversation were the caves with moa-bones in the upper Waipa and Mokau districts. From their recitals I learned, that in former times those caves abounded in moa-remains, and that in nearly all caves of the limestone-range not only single bones were found in great quantities, but even whole skeletons; they, however, were of the opinion, that long ago every spot and track had been picked clean, and that there was nothing more to be found now. They informed me, that their forefathers already had searched those caves for moa-bones; that they employed the skulls as boxes for preserving their paint-powders; the large bones being either used as clubs or worked into fishing-hooks. Then the Pakehas came offering such enormous prices for the old bones, that the Maoris collected for sale the very last remnants. Nevertheless I do not doubt, that many such curiosities may still be found in those parts; but a great many of those caves, being used as the final resting-places of their departed ancestors, are held sacred and secret by the Maoris; moreover, it would have cost more time, persuasion and money, than I could afford, to prevail upon the natives, to conduct a Pakeha into the tombs of their ancestors, and aid him in searching for moa-bones.

April 6. — With the rising sun the whole Maori-party had started off. Each of them had a share in the gifts, and went home with some present or other. We, too, were ready to set out earlier than usual. Our road lay through the swampy plain toward the heights bordering the valley. Here stands the village Poroporo; it was utterly deserted. At a sharp bend of the Mokau river, we again struck the broad plain of its valley, which displays here the singular existence of numerous, circular water-pools, poetically styled by the natives Karu-o-te-whenua, eyes of the earth. The "eyes of the earth" have a diameter of 12 to 20 feet; they are holes in the turf-moor that covers the bottom of the valley. The water stagnating therein is generally covered with a beautiful reddish liver-wort (*Marshantia macropora*). Mitten Thence the road slopes a second time toward the heights on the right river-bank, across a low range of picturesque limestone-rocks, [1] and strikes the river again at the grand Wairere-falls.

The Mokau, after winding in numberless curvatures, through a broad swampy valley 450 feet above the level of the sea, plunges here, foaming and roaring, over successive banks of slate into a narrow rocky gorge, a depth of about 150 feet. It is worthy of observation, that above these falls only the common eel is found, while below them other kinds of fish are met with. The water was so low, that we could without danger ford the river on a ledge of rocks just above the falls. When the water rises, the river has to be crossed in canoes at a quiet spot some distance farther up. We took a bath in the river, the water showing a temperature of 61° Fahr.; and arrived about noon in the Pah Pukewhau, one mile from the left bank of the river, a charming point perched upon the wood-clad slope of the mountain.

Te Hikaka Ngatirora (*vulgo* Ngature), the chief of the place, received us most solemnly and ceremoniously in black attire. He seemed to be fully aware, what motives and interests had brought me into his country; for first of all he conducted me to a place, where the sight of the steaming Tongariro by the side of the snow-clad Ruapahu struck my wondering eyes. Although only the top-most peaks were visible above the dark wood-clad mountain-ridges towards the Southeast, [2] the view greatly interested me, for here I

---

[1] On these limestone-rocks there grows quite a peculiar variety of *Asplenium bulbiferum*, which shows, how extraordinarily the ferns vary according to their respective locality.

[2] The Tongariro lies from Pukewhau to the East 30° 35 South; the Ruapahu precisely to the Southeast.

beheld for the first time the regular cone of the Tongariro volcanoe rising at an angle of 30⁰, and bearing the special name of Nga-uruhoe; and I plainly saw the curly clouds of white steam whirling up from the crater. The natives assert, that they have never seen black smoke; that, however, the clouds of steam are often much larger, especially in winter-time. The Tongariro was clear of snow, while of the Ruapahu nothing was seen but its snow-clad summit. In perfectly clear weather Mount Taranaki, also, is said to be visible from here. We were, as yet, about 50 miles distant from those volcanic cones, and in the same direction, through sombre wood-grown mountains, lay our onward route.

But before straying off into that region of forests and swamps, we must add a few remarks concerning the Mokau river.

The Mokau is, next to the Waikato, the most important river of the West Coast, emptying into the sea without forming an estuary. Its mouth is half-way between the Kawhia Harbour and New Plymouth. Its size is equal to the Waipa. Both rivers have the same length, their course being about 80 English miles; and both have their source upon the Rangitoto range at an elevation of about 2600 feet. Thence the Waipa flows in a northerly, the Mokau in a south-westerly direction. The upper Mokau valley with its numerous smaller side-valleys spreads between ranges of wood-clad hills into wide plains, partly marshy, partly covered with grass, and very fertile.

Section at the upper Mokau River.   *Wairere falls.*

a. Clay-slate. *b.* Tertiary clay. *c.* Tertiary limestone. *d.* Trachytic tuff with pumicestone.

The bottom of the valley is composed of tertiary clays, impervious to water, above which at the slope of the hills tabular limestones protrude, the heights themselves being formed of trachyte-tuff. At the Wairere falls below the tertiary clays slates of the same description as those at the Taupiri form the masses of rock, over which the river falls. Below the falls the valley becomes very narrow. Several miles further down, the river receives the Mo-kauiti, the little Mokau, its principal tributary, from the left. The latter, the course of which we shall follow on our road from Pukewhau to the Tuhua district, rises on the dividing ranges between the Mokau district and the sources of the Wanganui river. A few miles below the mouth of Mokauiti near Papatea or Whakatumu the river begins to be navigable for canoes, although numerous rapids render the passage difficult and even dangerous. From the mouth to Whakatumu up the river the natives reckon two days journeys; consequently the distance may amount to about fifty miles. The principal settlements on the river between those two points are Motuka-ramu, Mangatama, Mangakawhia, and immediately at the mouth the Pah Te Kauri. The old Missionary Station opposite the Pah is now deserted. Above Mangakawhia, twenty miles from the mouth of the river, along the banks seams of brown coal are said to lie open to view.

Pukewhau is a central point in the upper Mokau district. From here the roads diverge in different directions. Our destination was Lake Taupo, or, for the present, the Tuhua district towards East-South-East. The country we had to traverse presents the character of a table-land cut up by deep valleys into long and steep ranges; its height from about 1500 to 2000 feet above the level of the sea. The road lies through swamps and marshes, over hill and dale, and cannot be found without safe and expert guides. Ngature provided us with a guide to Horitu on the Mokauiti, — a distance of three miles, — in the person of a charming Maori girl, who led us knee-deep into swamp and mud. Here we met Te Wano, Ngature's brother, who himself became our guide thence for the next few days. In the evening we encamped on the left bank of the Mokauiti, about three miles from Horitu, at the edge of the bush. Our camping-ground was, according to my observations, 473 feet above the level of the sea.

April 7. — The night was cold, with a heavy dew-fall; but the day was clear, with the sun shining brightly. And, indeed, we might well congratulate ourselves, for precisely like the country, which we were to traverse to-day, I fancy Germany must have looked in the times of Tacitus *"silvis horrida aut paludibus fœda."* It was a wilderness of swamp and woods in the full sense of the word. A road or path was entirely out of the question; on the contrary, we had to work our way through, as well as we could, through deep swamps and gloomy, marshy woods.

Close to our camping-ground, we passed the Mokauiti and entered a dark, stately forest. We made a hard shift to dig our way through, over the smooth texture of roots in the sombre twilight of the virgin forest, when suddenly close by us a shot was fired, and from behind a gigantic Kahikatea-trunk a human figure stepped forth a double barrelled gun in the hand, and with a startling mien and savage gesture; in short, a brigand, such as only the most horrid brigand-story of a heated imagination could picture; and moreover, — yonder lay the whole band encamped around a brightly blazing fire, all armed with guns. But we remembered, that we were in New Zealand; the band of brigands was after all but a peaceable Maori-party shooting pigeons. We saluted each other with a friendly „tena koutou," exchanged tobacco for some of the finest wood-pigeons, — a capital dish for our dinner, — and passed on. After the lapse of an hour we had to cross the creek again, and, leaving the bush, came to the deserted huts of Poporata. Thence we plodded onward for another hour along the left bank of the Mokauiti in a southerly direction, through an abominable swamp; then, for a third hour through a still worse swampy bush. If one would wish to punish criminals in New Zealand in a very severe manner, it would only be necessary to chase them up and down in such a bush, where they would sink at every step into marshy holes, tearing their legs with the knotty roots to the very bone. This punishment might be termed "running the roots." We passed the Mangawhata Creek, and about 11 o'clock reached the fern-hills, whence a view opened over the wood-clad heights we had to cross. Another hour of travelling brought us through the bush to the heights of a steep sandstone-ridge Tuparae, on the opposite slope of which we reached a clearing and a small settlement, called Puhanga, which I found to be 937 feet above the level of the sea.

I believe, we were not less astonished to find in that lonely wilderness a colony numbering fourteen persons, than the Maoris were, to see Pakehas to visit them. We evidently created quite a sensation; the children seemed for the first time to have seen white men; yet we were received with a cordial welcome, and the women at once proceeded to prepare a meal, cooking it by means of heated stones in holes dug in the ground.[1] It was set before us in newly made baskets — potatoes and pork, as ever and everywhere. Puhanga has an extremely romantic site in a beautiful region of wood-clad mountains; in the rainy season, however, it must be perfectly inaccessible, and the two families, which, as I was told, sought a refuge here, in order to live secluded from the rest of mankind, could hardly have found a place more suited to their purpose.

At 3 o'clock we set out. Our road lay again down-hill through bush, then through a broad, deep swamp; thence over a wood-grown eminence, from which we had an unobstructed view of the higher mountain-range forming the water-shed between the Mokau and Wanganui. Before reaching this range, we had first to pass over a very marshy, grass-grown place, called te Roto, the lake, and then we encamped in a potatoe-field at a Maori station Maro-tawha, four or five miles from Puhanga.

April 8. — We were up early; our guide having announced

[1] The Hangi Maori, Maori cooking-stove, consists of a hole dug in the ground, greater or smaller in proportion to the quantity of meat or potatoes to be cooked in it. At the bottom of this hole, round stones are laid, which have been strongly heated; above them comes a layer of Phormium-leaves or ferns, or cabbage-leaves, if such are at hand. Then follows a layer of meat or potatoes; another layer of foliage, and so on, until the hole is full. Then the whole of it is once more carefully covered with leaves, water poured upon it, which is converted by the heated stones into steam, and finally they hasten to shovel earth upon it to prevent the generating steam from escaping. In this manner the victuals are steamed. For meat, the stove must remain covered one and a half to two hours, while potatoes are done in twenty minutes. Having applied this method very frequently during our journey, I can state from personal experience, that meat and potatoes thus steamed are a savoury dish.

to us beforehand a long, difficult and tedious road. Close by our camping-place we crossed the Mangateka, the last creek on our road, belonging to the sources of the Mokau. For an hour the road was quite level, leading partly through bush, partly over marshy grass-plains and through a thicket of Manowai-bushes; then it commenced to rise. Now we were once more upon a clearly beaten path leading quite steep up-hill through the bush. The mountain-range we had to ascend, is called Tarewatu. The summits of the mountains above us were wrapped in clouds; bad weather was approaching. On the slopes,[1] in consequence of numerous cavings, horizontal beds of shale and finely granulated brown sandstone are laid open to view, containing indistinct and broken fossil shells. The height of the pass I estimated at 1580 feet. The summit, however, was somewhat to the right of our road, and probably reaches a height of 1800 above the level of the sea. From the height we descended a steep slope of about 400 feet into the ravine of the Mangatahua, — the first branch of the Wanganui river, — and on the other side we had to scale another, higher mountain-range, Tapuiwahine (1933 feet). We reached the summit about noon. A hut made of the bark of the Totara-pine afforded us some shelter from the pouring rain. Heavy banks of clouds, piled along the whole horizon, deprived us of the charming prospect, which is said to be otherwise had from this point. A short distance from our road, we passed upon the highest ridge over a perfect tapestry of the tender foliage of *Hymenophyllum* and *Trychomanes*, of a growth more luxuriant than I have ever seen; thence the descent is quite steep. Having crossed the water-shed we encamped in the bush under gigantic Rata-trees at the source of the Waikaka. Here I had ample opportunity to gather a number of beautiful and rare ferns.

Having warmed and invigorated our frames with tea, we continued our journey. It stormed and rained, as though the firmament were ready to burst over our heads. Where the woods opened now and then for a moment, we had a fleeting view of dark ravines and gloomy, wood-clad mountains, along which misty clouds were

---

[1] The natives have two words for cavings: pari-horo and whati-horo.

driving. After a journey of four miles, at a place called Hine-
maori, we came to the open valley of the Ohura, a small creek,
here only a few feet deep, but gradually swelling to a considerable
river, and emptying into the Wanganui about forty miles below.
At 4 p. m. we reached the small settlement Ohura on the right
bank of the river. The country here seems to abound in birds;
for thousands of Tuis (*Prosthemadera Novæ Zelandiæ*), which had
perched themselves upon a group of Kahikatea pines, gave here a
concert, such as we hear in Germany from the starlings, when
they visit the vineyards in autumn.

We consulted, whether we should remain here through the
night, or continue our journey to a larger settlement on the
Ongaruhe river, which, according to the statement of our guide, we
could easily reach that day. I agreed to do the latter, especially because we had to expect, that in consequence of the continued rain the numerous creeks, gushing from the mountains in
all directions to the Ohura, would swell to such a size, that we
should be utterly unable to pass them for a day or two.

The upper Ohura valley is 900 feet above the level of the
sea. We followed the creek in its southeasterly course, choosing,
according to the nature of the ground, now the left bank, now
the right; and a few miles below, where the Ohura shapes its
course more and more South and Southwest, upon the left bank
we turned off over marshy grass-plains into a narrow, deep defile,
which, descending to a depth of about 200 feet took us into the
broad plain of the Ngawaitangirua valley covered with pumice-
stone. The word signifies the two-voiced valley, and is said to
indicate, that from this plain water-courses are running in two
directions, one to the Ohura, the other to Ongaruhe. Dusk having set in, all I could observe was, that at these creeks a terrace
formation begins, similar to that on the Waipa. A grass-covered
side-valley between steep-sloping mountains took us to the Ongaruhe
river. It was dark by the time we reached the river; we only
heard the roaring of the water that here tumbles in whirling
rapids over powerful rocks, and feeling with our sticks the blocks

of rock above the madly gushing river, we were obliged to climb round the steep declivity of the Kawakawa mountain. After we had once more set foot on level ground, where we descried a fire ahead of us, I felt as though we had been marvellously saved from an imminent danger. We had safely reached the Maori settlement Katiaho.

The dogs hailed our arrival with a perfect jackal's howl; the pigs, roused from their repose, were running to and fro; but human voices also became audible, and at length some persons came up to us in the dark, who conducted us to a large house, of which only the roof seemed to protrude from the ground. One after another we slipped in through a low square-hole, and found ourselves in a spacious apartment lit up by two blazing fires and heated to an almost tropical heat (85° F.), in which we were most cordially received by the chief of the place, Taonui, with the surnames Tekohue and Hepahapa, and by the whole people gathered about him, all expressing their unfeigned surprise at being honoured yet so late at night, and in such a weather, with a visit from Pakehas. There might have been twenty or thirty persons in the hut, which number was almost doubled by the addition of our party. The hut in which we found ourselves was a so-called Wharepuni, a conversation- and sleeping-room, such as existed in former times in every Maori village; which, however, have fallen more and more into disuse, owing to the influence of the missionaries, who opposed the sleeping together of old and young, of boys and girls. This Wharepuni was quite new; it had been but recently erected on the occasion of a visit from a neighbouring tribe. It was a real palace in comparison with the miserable raupo huts in other Kaingas. The side-walls were artificially wrought of plaited reeds and rushes; the ground-floor was covered with neat mats, and a row of carved columns supporting the roof divided the large room into two halves. The right side, according to Maori custom was assigned to the guests; and all of us strangers having arrived in a most deplorable plight, wet to the skin, and tired to death, we could well congratulate ourselves on having found so excellent a shelter. We divested ourselves of our dripping garments, and wrapped ourselves after Maori fashion in woollen blankets. Outside, in the cook-house, the meal was prepared, and after supper we chatted together till late in the night. Our Maoris would never tire of relating; nor the inhabitants of Katiaho, of asking questions. On recalling those scenes to my mind now, I can hardly comprehend, that those same men, with whom I sat there, in 1859, so perfectly unconcerned in social conversation, would already in 1860 and 1861 have participated in the bloody wars against the Pakehas.

April 9. — During the night an entire change of weather had taken place. The sun shone cheerfully into the valley, as I stepped out into the open air; the foggy clouds, which still clung to the mountains, vanished beneath the genial rays, and a charming landscape lay there spread out before my wondering eyes. A continuation of our journey, however, was out of question to day, because several of my carriers had sore feet, and all our clothes had to be thoroughly dried.

Two valleys, bordered by picturesque mountains, meet together at Katiaho, the Ongaruhe valley from the North, and the Mangakahu valley from the East. Between the two, opposite the settlement, arises the Ngariha mountain. The Ongaruhe is the main river; the Mangakahu only a small tributary; and Katiaho lies just opposite the junction of the two rivers on the right bank of the Ongaruhe. The latter rises in the Hurakia range, runs through an extensive table-land of pumice-stone, called Tetaraka, where it receives the Waimiha river rising at the Pukeokaku in the Rangitoto range, and reaches near Katiaho very nearly the size of the lower Waipa. It is here 40 to 50 feet wide and at an ordinary stage of water 8 to 10 feet deep. Below Katiaho, at the rapids of Pikopiko and Onehunga, which we had passed in the preceding night, it makes a sudden bend towards West, and 15 to 20 miles farther below, near Ngahuinga (i. e. coming together; junction), it empties into the Wanganui. [1]

[1] Wanganui, from Wanga, opening valley, and nui, large, signifies "big valley."

The terrace formation, so remarkable on the Waipa, is still more marked in these as well as in all the other valleys of the Upper Wanganui district. There are here three terraces in the valley and as many on the declivities of the bordering hills. The former are cut into thick beds of pumicestone gravel, which fill the bottom of the valley; the latter into trachyte-tuff, composing the hills and mountains on the sides of the valley. To the ter-

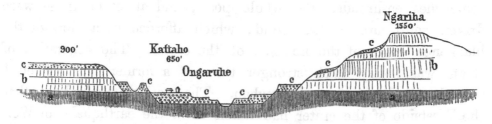

Terrace-formation in the Ongaruhe-Valley.
a. Tertiary sandstone and clay. *b.* Trachyte-tuff. *c.* Pumicestone-gravel.

races on the sides of the valley correspond farther up the valley extensive table-lands covered with pumicestone, and everything indicates to the observer that he is drawing nearer and nearer to a powerful volcanic hearth, from which those huge masses of pumicestone and trachyte-tuff are originating.

Hoping to have a view of the Tongariro and Ruapahu, which we had approached by this time to within a distance of twenty-five miles, and in order to execute another series of observations for the compilation of a map, I ascended the Ngariha on the 10th of April. Although the mountain rises only 900 feet above the bottom of the valley, and is entirely free from woods, the ascent was nevertheless quite difficult, as we had to break our way through ferns of the size of a man. Bathed in perspiration, we at length arrived at the top; but we found ourselves amply rewarded for our toils by the view now presented to our eyes. Even the natives, who had accompanied us from Katiaho, and who had never before scaled this hill, were greatly surprised. The eyes of all turned in one direction. There lay the volcano Tongariro before us, all clear from foot to top. The still active cone, called by the natives Ngauruhoe, with its regular conical form arises majestically from the midst of a circular range shutting it in all around and open only on the Southwest, similar to the Vesuvius encircled by the Somma. The funnel-shaped crater at the summit of the cone could be distinctly seen, indeed almost looked into, the West-side of the crater being much lower, than the Eastside. Consequently the crater presented itself to us in the form of an ellipse, from which continually dense, white steam-clouds arose, which sometimes enshrouded the whole peak, and at other times were driven southward by the wind, which afforded us a view of the blackened edges of the Eastside of the crater. The generation of steam was to-day much stronger than at a former time, when I saw the mountain from Pukewhau. The natives assured me that the Westside of the crater had fallen in at the earthquake of Wellington in 1855, and that at that time also a second crater to the North had been active. Farther North, on the slope of the mountain, a briskly steaming solfatara was visible. The Tongariro was entirely clear of snow. But to the right of the Tongariro arose the towering mass of the Ruapahu; its summit wrapped in dense clouds, and below the cloud-cap, the snowfields of the peak were seen to reach down as far as an absolute height of about 7800 feet. At the base of these mountain-colosses, dark forests extended; but in the foreground, mountains with sharp edges and deeply fissured precipices; and at our feet the valleys with their long stretched terrace-lines. Thus we beheld in one glance the effects of fire and of water on the grandest scale in one and the same landscape-view.

My stay at Katiaho enabled me also to buy from a Maori for the price of one pound Sterling the pelvis of a small Moa which had been found near Teruakuaho a few miles above Katiaho on the Ongaruhe, under a cliff of the Herepu mountain. This was the first Moa relic, that fell into my hands, and I was not little gratified at the lucky circumstance.

April 11 The terminus of our to-day's journey was Petania at the foot of the Tuhua-mountain, a distance of about 16 miles from Katiaho. At the mouth of the Mangakahu we crossed the Ongaruhe by means of canoes, and followed the river up the

EDUARD ADE.

Tongariro and Ruapahu

View from Mount Ngariha towards South-East.

valley in an easterly direction on a miserable road, which, in a manner very fatiguing to the traveller, led continually up and down over the terraces, through swamps, over numerous small creeks, and several times across the river itself. In the valley there was a successive change of grass, luxuriant shrubs and picturesque groups of Ti-trees. At a distance of three miles, the valley turns South-East towards the Tuhua-mountain, the most prominent point in the whole country, about 3400 feet high with a broad platform and a steep descent on its South side. At the bend of the river I counted no less than eight terraces on both sides. The valley here changes its character; it contracts into a rocky gorge shut in by vertical walls of trachytic rock, which present a beautiful columnar array, and through it the river rushes along over powerful blocks obstructing its course. About noon we reached a romantic spot, Koapaiari, where amongst sugarloaf-rocks of trachyte a small creek Komahina forms a picturesque cataract shaded by beautiful groups of trees and bushes. Here we stopped to make tea.

The Mangakahu there taking an eastward course, the road diverges from it, leading in a southerly direction over the terraces and over woodless hills with a steep ascent on to the table-land Pokomotu, 1386 feet above the level of the sea, at the western foot of the Tuhua mountain. This plateau is litterally covered with pumicestone. After having advanced about three miles, we came to the banks of the Piaua, a small stream running in a southerly direction to the Taringamotu river. Here a view was opened over the magnificent landscape at the junction of the Ongaruhe, Taringamotu and Wanganui rivers. The chief ornament of this landscape is the Hikurangi (i. e. ascending towards heaven), at the right between the Piaua and the Ongaruhe rivers, a volcanic cone of a very regular shape, rising from a very gently sloping base steeper and steeper to a height of 800 to 1000 feet, the top appearing as if cut smooth with a knife. Dark woods cover its declivities. The top is said to contain a water basin at the bottom of a funnel-shaped crater, and on the South-Eastside two powerful mineral springs (probably chalybeate) bubble out. In proceeding from the

Piaua valley over low woodless hills, from which the Tongariro and Ruapahu are visible in all their grandeur and majesty, the traveller comes to the Taringamotu valley. We had no difficulty in wading through the river, as it rippled along over its broad shingle-bed, and proceeded up the valley along the left river-bank; crossed the river a second time and were in Petania (Bethany). Such is the Christian name of a Maori village formerly called Te-terenga, and known also by the general designation Tuhua Settlement. It is situated at the southern foot of the Tuhua-mountain, 754 feet above the level of the sea. Our guide had to adjust various articles of his toilet before entering through the enclosure into the Kainga. The Maori-women also, who had joined our party from Katiaho, now began to adorn themselves, putting on an odd looking, towering head-dress made of plumes (Tauwakereru); the women in the village, on the other hand, commenced with doleful cries — which, however, were merely intended to express friendly salutation — to whine their "haere mai, haere mai ki konei", i. e. "come, come to this place", and after we had assembled all together at the place in front of the hut of the chief, they commenced in due form their Tangi, a lamentation, such as I had never heard before. They uttered the most doleful cries of anguish, clasping their hands now over the breast, now over the back, then bending low to the ground; thus expressing, by all sorts of convulsive motions their utter misery and anguish. This scene lasted for a whole hour without interruption; it was enough to move the stones to pity. But we could do nothing else than look on quietly and listen; for during all that doleful wail of lamentation the men stood in silent awe, wrapped in their blankets, without moving a muscle. Such is the Maori-custom, when friends or relatives happen to meet again after a long time of separation; they thus mutually lament their kindred, departed since the last meeting. The doleful scene ended at last in a general rubbing of noses and a cordial invitation to sit down to the meal, that had been prepared in the meantime; it consisted of potatoes, which were served steaming hot, in plaited baskets.

April 12. — Before breaking up I made some few purchases in Petania. From the chief I bought, for five shillings, after much parleying, a Moa leg-bone which he had carried for many years as a kind of club; I also enchanged a red woollen blanket for a piece of New Zealand national costume, a beautiful flax-mantle, Tatara, such as were in former times universally worn, before woollen blankets were introduced. Now-a-days such garments are only to be found in the remoter parts of the interior, where there is as yet but little intercourse with Europeans. They are now but rarely manufactured, the rising generation being unacquainted with that useful art. The distance from Petania to Lake Taupo is estimated a two days' journey. The road, however, is extremely difficult; it leads up and down from valley to valley, from mountain to mountain, across the ridges springing from the Tuhua-mountain in a southerly and southwesterly direction, and through dusky primeval forests. It traverses the sources of the Wanganui, and, ascending higher and higher, it finally reaches the watershed between the Wanganui and Lake Taupo. We were three whole days passing over this route. On the first day, after a most fatiguing passage through deep ravines cut into pumicestone gravel, we crossed the Takaputiraha range (1534 feet high), and encamped on the left bank of the Pungapunga river upon a beautiful grass-plain, called te Patate, 897 feet above the level of the sea.

April 13. — We had now to scale the Puketapu. This mountain is the most remarkable point on the road from Tuhua to Lake Taupo. The ascent is extremely steep. According to my barometrical observations the height of the mountain is to be estimated at 2073 feet. As the summit was covered only with young underwood, I ordered the same to be cut down, and thus gained an interesting view of the sources of the Wanganui, over a sombre mountain-country and wood-landscape, in the back-ground of which the Ruapahu loomed up in all its majesty, its peak wrapt in clouds. Southwest of the Ruapahu another volcanic cone, 3000 feet high, was visible; it was pointed out to me as Hauhanga. To the Northwest and West the Tuhua-mountain and the Hikurangi-cone were the most prominent points. In clear weather, Mt. Egmont also is said to be visible from here. The Puketapu is moreover especially remarkable for the circumstance, that, in the midst of a landscape, in which every thing is covered by volcanic tuffs and pumicestone, it is composed of clay-slate of exactly the same description as at the Taupiri on the Waikato.

After leaving the Puketapu we were continually in the bush; it seemed as though it would never come to an end. Up and down, from ridge to ridge, from dale to dale; we passed the Waipari, then the Waione, cold creeks, the water of which showed a temperature as low as 50° F. Again we had to climb up-hill, over roots and logs in the sombre dusk of the bush, the huge crowns of the tall forests trees shutting out the light of day, and the sky being moreover veiled by dark, dismal clouds of rain. The magnificent fern *Leptopteris hymenophylloides* grows in those damp woods with a extraordinary luxuriance, in the shape of the variety called *superba*. At length we came to a small creek flowing in a direction different from that of all the other creeks we had hitherto passed; it was the source of the Kuratao, running in a N. E. direction towards Lake Taupo; a sign that we had crossed the water-shed, and we hailed with joy the first indication of our having come quite close to our long looked — for destination. It was 5 o'clock p. m. when we emerged once more from the darkness of the bush into open daylight. Involuntarily we all with one accord burst out into a loud shout of joy, as we stepped out upon the open, grassy plain. Of the lake, however, the sight of which we had expected to greet our longing eyes, there was as yet nothing to be seen. But in its place two beautiful mountain-cones, the Kuharua and the Kakaramea, rose before us. We had reached a pumicestone-plateau, called by the natives Moerangi, and I was greatly surprised at finding the result of my barometrical observations to show a height of 2188 feet. We pitched our tents by the banks of the Kuratao river, which, cutting through the pumicestone, forms a ravine about 100 feet deep with triple terraces.

April 14. — The distance from the lake was greater than we had supposed, and the day a dreary one, indicating rain. Travelling by turns through woods and lawns, we struck once more the Kuratao valley at Whakaironui, a potato-plantation at the margin of the plateau. The valley of the river is here already wider by far and deeper than at the place, where we had spent the nigh . We had to climb down over four terraces, a depth of about 400 feet and crossed the river at a point where it forms a picturesque cascade 20 feet high, thence rolling on in a narrow bed between vertical bluffs of trachyte-tuff. A bridge leads below the falls across the defile. On the opposite side we were again obliged to climb over miserable, slippery paths, from terrace to terrace, up to the height of the pumicestone plateau, and arrived at the village Poaru, where we took our dinner. The distance between the latter place and the lake is three miles. After we had plodded along through marshy woods, we came to an open eminence, whence we had the first view of the lake. Like a sea it lay there spread out in the distance, without our being able to discern the opposite shore in consequence of the murky weather. Gently sloping down-hill, and passing along the foot of the Kuharua, — a beautiful wood-clad mountain-cone, — our path led us across a small creek Hauwai on to an elevation, from whence we saw the celebrated pah Pukawa, the residence of the great Maori-Chieftain Te Heuheu, situated beneath our feet at the margin of the lake.

It was a cold and dreary day, and now it moreover commenced to rain. "When strangers come, the mountains weep", is the Maori adage. I should have much preferred to have it read: "When strangers come, the sun doth smile", and that the latter version had proved true in our case.

---

[1] Farther down the river, quite close to Lake Taupo, there is a larger cataract, called Huka, i. e. foam.

CHAPTER XVII.

Lake Taupo, Tongariro and Ruapahu.

Rev. Mr. Grace. — The Maori-Chief Te Heuheu. — Lake Taupo. — Volcanic cones at the South-shore. — Hot springs of Te Rapa and Tokanu. — The Waikato-Delta. — The Volcano Tongariro, its cones and craters. — Mr. Dyson's ascent of the Tongariro. — Legends of the conflict between Tongariro and Taranaki. — Ruapahu the highest mountain of the North-Island. — Pumicestone-plateau. Terraces. — Climate of the Taupo-country. — Legend of Horomatangi. — Population. — Fauna. — The East-shore of the lake. — Rhyolite. — Outlet of the Waikato.

The name Taupo reminds me of one of the grandest natural sceneries, I have ever seen, and at same time of the generous hospitality of the Rev. Mr. Grace and his amiable family. The missionary's house is only a few hundred yards distant from the Maori Pah Pukawa; picturesquely built against a bluff, upon a terrace 200 feet above the lake. Beneath its hospitable roof I passed five days, during which time I was engaged in sketching a detailed map of the lake. Mr. Grace, by virtue of his exact local knowledge, was of great assistance to me in this work; he accompanied me on my excursions, while the arrangements of the excellent lady of the house made us utterly forget, that we were sojourning in the remotest interior of New Zealand. The picture of that happy family circle, blessed with a number of blooming children, was truly calculated to awaken the most grateful emotions. How oddly contrasted with this picture the Maori-character, such as it was represented with all its former pagan splendour in the neighbouring Pah in the person of the celebrated Maori-Chieftain, Te Heuheu!

Long ago I had heard of the great and mighty Te Heuheu, residing in Pukawa at Lake Taupo. His name is known wherever the Maori-language is spoken; for he belongs to one of the oldest and most renowned noble families of the country, and is numbered among the heroes or demigods of his people. He had been pictured to me as a man of considerable talents, as the best and worst fellow at the same time, as proud, shrewd, generous, as a mysterious medley of modern civilization and ancient heathenism. I was curious to make his personal acquaintance, and had arranged with my travelling companions to pay the dreaded potentate of the country a visit of respect and homage in all due form. Having employed, however, the first day of my stay, — on account of the fine weather, — for an excursion to the hot springs near Tokanu, it was not until the second day after my arrival at the lake, that I carried my purpose into execution.

Mr. Grace accompanied me to the Pah. It is situated upon a peninsular projection into the lake, with an enclosure of strong palisades, through which two sliding gates give admittance. Entering through one of the gates we came to an oblong place at the upper end of which a sightly provision-house (I'ataka) first attracted our attention. It was painted red, and based upon four round posts, to protect the provision-supplies stored within from the voracity of the rats. Its front was ornamented with extravagant carvings after the style of art peculiar to the Maoris. Neat flourishes and arabesque-like figures alternated upon the gable-field with grotesque, big-headed and big-eyed human forms. Opposite tho this Pataka stood an unsightly hut without any architectonic ornament whatever, but with a little porch beneath the protruding roof in front, as is customary in all Maori-huts of better quality, — it was Te Heuheu's residence and in the verandah of that Maori-palace sat a man of stern and gloomy mien, wrapt in a dirty woollen blanket, — it was Te Heuheu himself.

My reception at his hands was any thing but a gracious one. It was not until the missionary had spoken with the chieftain, that the latter condescended to extend his hand to me, inviting me, to seat myself beside him upon the unrolled mat. He scanned with his flashing, black eyes the group of my companions, who saluted him most reverentially, and then addressed to me the little amiable question, whether I was aware, that the natives, I had with me as guides and carriers, were not slaves, but the sons of free and independent chiefs, which interrogatory was followed by a long expatiation. Te Heuheu stated that he was always glad to become acquainted with independent Europeans of better rank and station, they always being worthy people, who meet and treat the natives with due respect; that, on the other hand, the common Europeans such as run-away-sailors and other rabble, with which New Zealand was flooded from Europe and Australia, were detested by him as the worst and most contemptible kind of people; that, numbering me among the former class, he had expected my visit already the day before, and made all suitable preparations to receive me with all due form as well as with a cordial welcome; but that, as I did not choose to come at the expected time, it was my own fault, if I found him to-day in his ordinary every-day-clothes.

It required repeated apologies, and various explanations to propitiate the offended pride of the noble chief; yet I must state

to his credit, that he did not bear me a long grudge; for on the same day he ordered a fat pig to be killed for my Maoris, and, moreover, entertained them in his Pah through full five days without accepting any pay whatever in return. He also showed me a magnificent Mere punamu, a battle axe, 15 inches long, and cut out of the most beautiful, transparent nephrite, an heirloom of his illustrious ancestors, which he kept as a sacred relic. He explained to me that this murderous weapon was taken from a hostile chief in bloody combat, that five times already it had been buried with his ancestors, and that the notch on one side of it dated from the last fatal blow struck at a hard skull. A second piece, which he showed me with much pride, was an English saddle, which, with a horse, had been presented to him

Te Heuheu's Mere punamu. Battle-axe of nephrit.

years ago by Sir George Grey, in acknowledgment of his services as guide and travelling companion on a journey to Lake Taupo.

Iwikau Te Heuheu has five wives, and at the time of our visit he expressed himself inclined to take two more. He is the worthy sire of a numerous progeny, his pride and his joy; but, although not quite averse to Christianity, he has always refused to be baptized, fearing to lose by such a step his influence and authority as chief, which is based upon various pagan notions, especially upon his supposed power over the evil spirits of the earth, water and air. He is of middle size, delicately rather than robustly built, wearing his black hair in long locks. His beardless face, but imperfectly tattooed upon the right cheek, with the small sparkling eyes characterized him to me as a man of cunning, calculating shrewdness. He has nothing of the imposing, majestic hero-stature of his deceased brother Tukino Te Heuheu, who is said to have been a giant nearly seven feet high with silvery hair, the great man, to whom the present Te Heuheus owe their fame and authority. Tukino Te Heuheu had met his death by an awful catastrophe in May 1846, in the neighbouring village Te Rapa. He was burried alive with his six wives and fifty-four persons by a land-slip connected with a flood, which occurred during night time. Iwikau had the corpse of his brother exhumed from the entombed village, and accorded him a solemn interment. According to Maori custom in the case of great chiefs, the remains were disinterred after some years, laid out upon a kind of bed-of-state, and presērved in a magnificently carved coffin. The sacred remains were intended to be then conveyed to the summit of the Tongariro; for the deep crater of the volcano was intended to be the final grave of the hero, with the heaven-ascending pyramid of scoriae and ashes for his monument. But the grand idea was but half carried out. As the bearers were approching the top of the ever steaming cone, a subterraneous roaring noise became audible, and awe-struck they deposited their heavy load upon a projecting rock. There the remains still lie. The mountain, however, is most strictly tapu, and nobody is allowed to ascend it.

The present Te Heuheu, moreover, erected in his memory a mausoleum (Wahi-tapu), said to have been a master-piece of Maori-architecture. All we found remaining of it, were at the lower end of the Pah, under picturesque groups of Karaka and Kowai trees, several artificially carved posts with most note-worthy representations, which all seemed to have reference to the inexhaustible manly vigour of the departed hero and to the prolificness of his numerous wives. Within the enclosure of the Pah there live besides Te Heuheu only his nearest kinsmen and friends. In the lower part I noticed moreover, a cucumber-patch, and a piece of a vineyard, started by the Rev. Mr. Taylor, but now utterly neglected. The vines were full of grapes, which, however, were sour and utterly unfit to eat. Outside the Pah various huts are scattered about, the dwellings of Te Heuheu's subjects.

The proud chief returned my visit in an elegant suit of black. Many another hour did I sit together with the interesting man during my stay at Pukawa, listening to his observations and recitals. It is from his lips, that I have the interesting legends, which in the tradition of the Maoris are attached to the Taupo country, and which I shall recite here-after. In his political views, he proved himself a zealous advocate of the national party, avowing most solemnly, that he would never visit again the "Pakeha-City" Auckland, where at his last visit he had been treated like a dog. At our departure he sent me word through the missionary, that he would be glad to receive and keep me again at any future time; that however he warned the Englishman, who accompanied me by order of the Governor, from a second visit to his Pah; stating that he had tolerated the latter only on my account, I being a stranger and not master of the Maori-language.

That was Te Heuheu, one of the few surviving representatives of the old heathen-times, around whose head there is still a faint halo of that romantic heroism, which like a dim, dark legend reminds us of the classical age of a savage cannibal people, hastening, under the influences of European civilization, with rapid strides towards its final extinction.

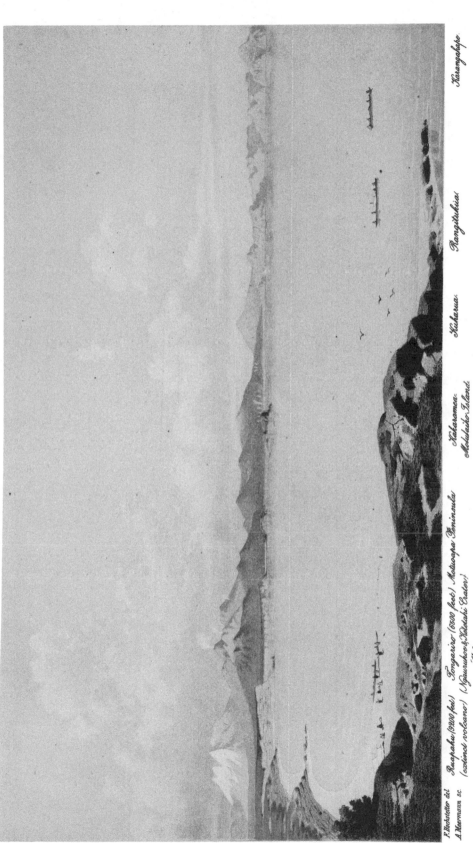

New-Zealand, North-Island.

F. Hochstetter del.
A. Mannen sc.

Ruapahu (9800 feet)
(extinct volcano)

Tongariro (6500 feet)
(Ngauruhoe) Petahi Crater!
Pihanga.

Motuoapa Peninsula.

Hatawamea.
Motutaiko Island.

Taupo Lake.
1250 feet above the sea.
View of the Southern shore.

Kuharua.

Rangitukua.

Karangahape.

Many other interesting personages of the higher Maori-aristo-cracy at Lake Taupo might be characterised, such as Te Heu-heu's rival in Tokanu, the sullen Herekiekie; or the fat postma-ster Puhipi, the David of the Maori Jerusalem (Hiruharama, a village at the Northend of the lake), a famous gormandizer; but let us look now upon the lake and its environs.

Lake Taupo[1] is a real inland-sea, 25 miles long from South-west to Northeast, its greatest breadth about the 20 miles, and of a depth as yet not fathomed. It lies 1250 feet above the level of the sea; this is the mean result of a number of barometrical observations which I made at the shores of the lake. Dieffenbach fixed its height from the temperature of the boiling water at 1337 feet. The lake is everywhere surrounded with volcanic formations. Quartzous trachytic lavas, — which of late have been distingui-shed from the common trachyte by the name of rhyolite, — in the most different modifications of structure and colors (crystalline and vitreous), together with huge masses of pumicestone, are the prevail-ing rocks. They form round about the lake a high table-land from 2000 to 2200 feet above the level of the sea, upon which numerous volcanic cones arise built up of trachyte, phonolite, trachy-dolerite or andesite, and partly also of basalt. The lake itself evidently owes its origin to a break in the plateau, and seems to be of an extraordinary depth especially in its western half.

The West shore of the lake is formed by vertical bluffs of rocks, which near Karangahape, at a promontory projecting far into the lake, attain a height of more than 1000 feet. Upon that side of the lake a landing is practicable only at the few points where little rivers empty into the lake. The long-stretched wooded ridges of the Rangitoto and Tuhua mountains, rising to a height of 3000 feet above the level of the sea, shut out the horizon in a northwesterly direction, and only one point attracts the attention

---

[1] *Taupo* signifies a place, where night and darkness reign; we might here think of ejection of ashes from the Tongariro volcano, which obscured the sky. The natives, however, designate also a scenery of dark, obsidian like rock (rhyolite) at the Northcoast as Taupo; and they say, that the lake had its name from those rocks.

by its rather singular form, — I am speaking of the Titiraupenga mountain, from the summit of which a bare pyramid towers up, resembling a ruined castle. The East shore in its greatest part is flat, and formed by a broad sand-beach, upon which the road leads along the lake. Widely gleaming, white pumicestone cliffs border the strand. Above them extend pumicestone-plains, covered with grass and bushes, which rise in terraces up to the foot of a high wooded range, which under the name of Kaimanawa forms the continuation of the Ruahine-chain in the province of Wellington, and, together with this chain, is to be considered as a continuation of the Southern Alps of South Island. The foot of the range is ten to fifteen miles from the East shore of the lake; behind the wooded ranges, rocky, pyramidal peaks tower to the sky, which attain a height of 6000 feet and more above the level of the sea, and present with their rugged Alpine character a picturesque contrast to the regular conical shape of the volcanic mountains on the South side of the lake. Farther to the N. E. the mountains are growing lower, and bear the name of Te Whaiti. The range in its whole length from Cooks Strait to the East Cape, was and for the greater part is still a *terra incognita*, and if there is anywhere upon the North Island a prospect of finding gold, silver and other metals, it is in those unexplored mountain-chains. I was greatly surprised at the sight of them, because I had not found upon any map of New Zealand even the least intimation of the existence, between Lake Taupo and the East-coast, of such a high chain of mountains. In that range all the numerous and partly considerable rivers rise, which empty into the lake from the East. The detritus, which they carry with them, consists mostly of bluish slate, and of gray sandstone. At the North-end of Lake Taupo the beautiful cone of the Tauhara points out the region, where the Waikato leaves the lake, as a stream of a quite considerable size.

By far the most attractive parts, however, are the southern shores. They are bordered by a successive series of picturesque volcanic cones, behind which the Tongariro and Ruapahu rear their lofty heads. From the South-shore itself those two giants are not visible; but from the East- and North-shores they are every-where seen towering high above those lower mountain-cones, by the natives so well designated their wives and children. Their names are: Pihanga, Kakaramea, Kuharua, Puke Kaikiore and Rangitukua. Pihanga, the eastern one of those cones, is also the highest. I estimate its height at 3500 feet above the level of the sea. Only its topmost peak, cleft by a deep chasm, is woodless, and displays already from a far a crater open towards North. Likewise the Kakaramea, the summit of which is of a red colour, bears probably a crater. Both craters are deemed extinct; but, the volcanic forces below have by no means been as yet lulled to their final repose; for on the northern declivity and at the foot of the Kakaramea it steams and bubbles and boils in more than a hundred places.

I visited those hot springs on the very first day after my arrival at the lake in company of the Rev. Mr. Grace. East of Pukawa, in the rear of a steep promontory, a small bay extends south. The western-shore of this cove is formed by vertical bluffs consisting of alternating horizontal banks of trachyte, trachytic conglomerate and tuff. A small creek, the Waihi, plunges quite close to the South-end of the cove, in a magnificent fall about 150 feet high over this bluff of rocks. At this cascade the moun-tains recede somewhat from the lake; and here already, from the conglomerate-layers forming the beach, hot water, of 125° to 153° F., is seen bubbling forth. By conducting this water into artificial basins, the natives have prepared several bathing-places, the water in which showed a temperature of 93° F. Conferves of a magni-ficent emerald-green cover the places, where the water flows, and silicious, not calcareous, sinter is deposited in them. But strange to say, there is amid those alkaline springs also a chalybeate one of 156·5° F., which deposits large quantities of iron ochre. Above these springs on the side of the mountain, probably 500 feet above the lake, steam issues from innumerable places. The whole North-side of the Kakaramea mountain seems to have been boiled soft by hot steam, and to be on the point of falling in. From every crack and cleft on that side of the mountain hot steam and boiling water are streaming forth with a continual fizzing noise, as though hundreds of steam-engines were in motion. Those steaming fissures in the mountain-side, upon which every stone is decomposed into reddish clay, the natives call Hipaoa, i. e. the chimneys, and it was at the foot of that mountain-side, that in the year 1846 the village Te Rapa was overwhelmed by an avalanche of mud, and the great Te Heuheu perished. The inhabitants of the Pah Koroiti upon the mountain terrace near the Waihi-falls use those steam-holes for cooking their victuals over them. The little cold brook, which empties into the lake at Te Rapa, is called Omohu. The chief range of springs, however, is on the Southside of that cove near the Maori village Tokanu at the river of the same name. From the small mountain-cone Maunganamu to the mouth of the Tokanu river it comprises an area of about two square-miles. It is, impossible, to describe every single point; I will therefore mention only the principal springs.

The powerful column of steam visible far over the lake-shore, which is seen to ascend at Tokanu, belongs to te large fountain Pirori. Pirori signifies fountain, eddy. From a deep hole on the left hand bank of the river Tokanu, a boiling-hot water column of two feet diameter, always accompanied by a rapid development of steam, is whirling up to a height of six to ten feet. The natives, however, told me, that the water was frequently thrown up with a booming noise to a height of more than forty feet. At a few paces from it there is a basin eight feet wide and six deep, covered with a silicious deposit resembling chalcedony; it is called Te Korokoro-otopohinga, the jaws of Topohinga, in which the water is continually boiling. Farther-on we came to a warm creek Te Atakokoreke with a temperature of 113° F., a favorite bathing-place of the natives. On the other side of the creek, there are three basins close together. Te Puia-nui, the large spout, was filled with clear, and but gently bubbling water of 186·8° F.[1] to

---

[1] In reference to a chemical analysis of the water of this spring comp. the Appendix to Chapter XVIII.

the very brim so that it ran over into the second basin. In this basin, eight feet wide, grayish white mud was boiling, which showed a temperature of 188° F. The third basin contained

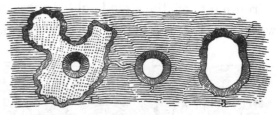

The Puias at Tokanu on Lake Taupo.

again clear boiling water. All three basins are lined with crusts of silicious deposit, and bear to each other a relation of periodical exchange, so that No. 1 is filled, while in Nos. 2 and 3 the water falls, and *vice versa*. The natives likewise assert, that the middle basin which I beheld only as a boiling mud pool, was in March and April of 1846 an immense geyser, throwing out a spout of hot water to a height of nearly 100 feet, so that the village was completely flooded by it. It is evident also from other sources, that in that range of springs continual changes are going on, and that those phenomena are periodical with a great many of the springs, similar to the Geyser and Strokkr in Iceland.

A crust of silicious deposit, three feet thick, under which fine clayey mud is bedded, covers the larger portion of the spring-region. In smaller apertures, from which nothing but hot steam emanates, the thermometer rises to 208° F. Here also the natives employ the steam-fissures for cooking; they have special huts for the winter, erected upon warm ground. They call the hot springs Puia, distinguishing Papa Puia, springs with clear water, yielding silicious deposits, and Uku Puia, the boiling mud-pools and small mud cones. Of sulphuric acid or of sulphuretted hydrogen I could discover in this spring-region only faint traces. I believe, if any one at Tokanu or on the declivity of the Karakamea would endeavour to count the several spots, which give out either hot water, steam, or boiling mud, he would find more than 500 of them. This, however, is only one, nor by far the most important of those numerous hot spring regions of the North Island.[1]

The settlement Tokanu is noted for a magnificent Wharepuni,

[1] See Chapter XVIII.

a remnant of the good old times of the Maoris. The annexed wood-cut is intended to illustrate some carvings on the door-posts.

Wood-carvings of a Wharepuni at Tokanu.

One figure had fallen off the roof, and lay on the ground in dust and dirt. I wished to purchase it from the chief of the place; he, however, was much astonished on hearing my demand, and informed me, that the figure in question represented his grand-father, and that it was utterly impossible for him to sell his grand-father to a Pakeha.

The fertile low-lands about Tokanu are already to be considered as a part of the extensive delta of the Waikato river, which here, at the South-East side, flows into the lake. The river in its delta is divided into four branches; the natives, however, very strangely do not call it Waikato, but Tongariro, like the village situated on its bank not far from the mouth. It is formed by two rivers uniting at the foot of the Pihanga, the one coming from the South bearing the name of Waikato, the sources of which are at the Tongariro and Ruapahu; the other, rising in the Kaimanawa range, and, as it seems, the larger branch, retaining the name Tongariro. Evidently it would have been more correct to name the tributary coming from the Tongariro volcano also Tongariro, and to leave the name Waikato to the main-river. Consequently, according to the Maori nomenclature, the Waikato loses its name for the distance from its junction with the Tongariro river to its influx into Lake Taupo.

The foot of the Tongariro is about twelve miles distant from the lake. Between the volcano and the Pihanga and Kakaramea mountains lying before it, there intervenes a broad valley with the beautiful lake Rotoaira, three miles long. The outlet of this lake, the Poutu, is one of the principal tributaries of the Waikato. Dieffenbach mentions here also another smaller lake Roto Punamu, shut in round about by mountains as in a crater-basin, 2417 feet above the level of the sea. In order to scale the Tongariro, it is necessary to set out from lake Rotoaira. However, the difficulties which the natives oppose to such designs, are the same now, as in 1841, when Dieffenbach was endeavouring in vain to obtain permission to ascend the mountain; or in 1850, when the same thing occurred to Sir George Grey. The mountain is tapu; and even, should Te Heuheu let himself be persuaded, to give his consent to scaling the mountain, Te Herekiekie most certainly would oppose it the more determinately. The only two Europeans, who are said to have been on the top, carried out their purpose without the least knowledge of the natives.[1] I made

[1] The difficulty of ascending Tongariro, Dr. Thomson says, does not entirely arise from its height, or the roughness of the scoriæ, but from the hostility of the natives, who have made the mountain "tapu," or sacred, by calling it the backbone and head of their great ancestor. All travellers who have asked permission of the natives to ascend Tongariro, have met with indirect refusals. The only way to get over this difficulty is, to ascend the mountain unknown to the natives of the place, or even your own natives. Mr. Dyson did this, but his ascent was discovered by a curious accident. During his progress up the mountain he took for a time the little frequented path which leads along the base of Tongariro to Whanganui. A native returning from that place observed his foot-marks, and knew them to be those of a European. As he saw where the footsteps left the path, he, on his arrival at Rotoaire, proclaimed that a European was now wandering about alone on the sacred mountain of Tongariro. The natives immediately suspected it was Mr.

no attempt to extort a permission, or to elude the natives, as the weather was much too bad for such an undertaking, and only a greater expense of time, such as a thorough exploration of the volcano necessarily required, could have led to any satisfactory result.

Nevertheless I deem myself capable of contributing something to a better understanding of the Tongariro system, and will endeavour to describe the several portions of that grand volcanic mountain according to the knowledge which I gained from the West- and North-sides, of course only at a distance of several miles; yet, with objects so grand, the general proportions are much more easily and therefore often much more correctly viewed at a greater distance, than by climbing for whole days along the colossal sides. The Tongariro is not a single mountain like the Ruapahu, but is composed of a whole group of mountain-cones. The beautiful cone, towering high above the other parts of the group, is distinguished by the special name of Ngauruhoe. It bears a funnel-shaped top-crater, the principal active one of the Tongariro. The Ngauruhoe together with the grand circular range, from the centre of which it rears its head, forms the southern portion of the Tongariro system. It is a cinder-cone of the most regular conical shape with a slope of 30 to 35 degrees; the height from the basis to the top I estimate at 1600 feet. It overtops the heighest points of the other parts of the system by about 500 feet and attains probably an absolute height of 6500 feet above the level of the sea.[2] The outer circus, shelving off inside with steep walls, and from all appearances forming a grand mountain-amphitheater with rocky precipices of 1000 feet in height, is opened on the West-side by a broad chasm; and no doubt, it is through this chasm, from the *atrium* between the cone of cinder and its circumvallation the that chief-source of the Wanganui river flows. This seems also

Dyson, and they went to his house, waited his return, and took several things from him. He was now a suspected man, and his conduct was watched.

[1] At the Rotoaira they say *Auruhoe*.

[2] *Dieffenbach* I. p. 355, infers from Mr. Bidwill's observations a height of 6200 feet. No real measurement of the Tongariro has ever been made. Estimates of height exceeding 7000 feet are at any rate over-estimated.

to be the only side, from which the cone is accessible, and those who did ascend the mountain, must necessarily have chosen this direction.

I have never heard of any native having been at the top. The dread of the demoniac powers of the lower regions seems to have kept them from such an undertaking; and the mountain was tapu. As far as I can learn, as yet only two Europeans have succeeded in ascending the Ngauruhoe, Mr. Bidwill in March 1839, and Mr. Dyson in March 1851. Dyson's account of his adventures during the ascent, was communicated to the 'New Zealander' by Dr. A. S. Thomson, as follows:

In the month of March, 1851, a little before sunrise I commenced my ascent alone, from the northwestern side of the Rotoaire lake. I crossed the plain and ascended the space to the northward of the Whanganui river. Here I got into a valley covered with large blocks of scoriæ, which made my progress very difficult. At the bottom of the valley runs the Whanganui river. After crossing the river, which at this place was then not more than a yard broad, I had to ascend the other side of the valley, which, from the unequal nature of the ground, was very tedious, and I kept onwards as straight as I could for the top of the mountain. At last I came to the base of the cone, around which there were large blocks of scoria which had evidently been vomited out of the crater, and had rolled down the cone. The most formidable part of my journey lay yet before me, namely the ascent of the cone, and it appeared to me from the position where I stood that it composed nearly one fourth of the total height of the mountain. I cannot say at what angle the cone lies, but I had to crawl up a considerable portion of it on my hands and feet, and as it is covered with loose cinders and ashes, I often slid down again several feet. There was no snow on the cone or the mountain, unless in some crevices to which the sun's rays did not penetrate. There was not on the cone any vegetation, not even the long wiry grass which grows in scanty patches up to the very base of the cone. [1] The ascent of the cone took me, I should think, four hours at least; but as I had no watch, it is possible from the laborious occupation I was at, that the ascent of the cone looked longer as it was. But whether it was three hours or four that I was clambering

---

[1] *Bidwill* mentions a small grass and a snow-white Veronica, which grow yet on the lower part of the cone.

up the cone I recollect I hailed with delight the mouth of the great chimney up which I had been toiling. The sun had just begun to dip, and I thought it might be about 1 p. m., so that I had ascended the mountain from the Rotoaire lake in about eight hours. I must confess as I had scarcely any food with me that I kept pushing on at a good pace. On the top of Tongariro I expected to behold a magnificent prospect, but the day was now clowdy and I could see no distance. The crater is nearly circular, and from afterwards measuring with the eye a piece of ground about the same size, I should think it was six hundred yards in diameter. [1] The lip of the crater was sharp: outside there was almost nothing but loose cinders and ashes; inside of the crater there were large overhanging rocks of a pale yellow colour, evidently produced by the sublimation of sulphur. The lip of the crater is not of equal height all round, but I think I could have walked round it. The southern side is the highest, and the northern, where I stood, the lowest. There was no possible way of descending the crater. I stretched out my neck and looked down the fearful abyss which lay gaping before me, but my sight was obstructed by large clouds of steam or vapour, and I don't think I saw thirty feet down. I dropped into the crater several large stones, and it made me shudder to hear some of them resounding as I supposed from rock to rock, — of some of the stones thrown in I heard nothing. There was a low murmuring sound during the whole time I was at the top, such as you hear at the boiling springs at Rotomahana and Taupo, and which is not unlike the noise heard in a steam engine room when the engine is at work. There was no eruption of water or ashes during the time I was there, nor was there any appearance that there had been one lately. I saw no lava which had a recent appearance; [2] nothwithstanding all this, I did not feel comfortable where I stood in case of an eruption. The air was not cold — the ascent had made me hot — but I had time to cool, for I remained at the crater nearly an hour. At about 2 p. m., I commenced my descent by the same way that I ascended. A fog or cloud passed over where I was, and caused me to lose my way for a short time. When descending I saw between Tongariro and Ruapahu a lake about a mile in diameter. I could see no stream flowing out of it on its western side. An extinct crater may also be seen near the base of Tongariro. It was almost dark before I reached the Whanganui river, and,

[1] This estimate is at any rate much above the mark; the diameter of the crater can scarcely be more than 500 feet.

[2] *Bidwill* speaks of an entirely new stream of lava at the foot of the cone, about $3/4$ of a mile long, and not yet covered with lichens; he describes the lava as black, hard and compact.

although in strong condition and a good walker, I felt completely done up, and I fell asleep in a dry water-course. The night was cold, but I slept soundly until daylight, when I immediately rose and continued my descent, and at 10, a. m., I reached my residence at Rotoaire, with the shoes almost torn off my feet.

This account agrees in its main-points perfectly with Mr. Bidwill's descriptions. The Ngauruhoe crater, consequently, seems to be at present in the state of a solfatara throwing out continually large masses of steam and other kinds of vapour. The natives know nothing of lava-flows; yet from time to time the crater is said to eject cinders and hot mud, and during such eruptions, now and then, a fiery shine is said to be visible over the mountain.[2] Such is specially said to have been the case in the month of February, in 1857, when the ejection of cinders and ashes lasted from two to three weeks. Such ejections seem to exercise a changing influence upon the topmost craterlip. I always saw the point of the

Peak of the Ngauruhoe, April 1859,
seen from North.   seen from West.

cone such, that it was evident, the western lip of the crater was necessarily much lower than the eastern. A trifling change however, seems to have taken place since, concerning which my friend Dr. Haast writes to me as follows: Mr. Ch. Smith of Whanganui sojourned in the month of December 1859 in Tokanu at Lake Taupo for the purpose of negotiating with Te Herekiekie, the chief of that district, concerning a pasture for sheep. He related

[1] See Dieffenbach I. p. 347—355.
[2] Taylor (p. 225) mentions, that in olden times the natives, whenever they saw fire upon Tongariro, considered it as a command of their Atua (God) to wage war, and that the inhabitants of the coast were then expecting an attack from the Taupo lake.

to me, that in the first days of December, with a cloudless sky, but an oppressively sultry atmosphere, about 11 a. m. suddenly a subterraneous noise like thunder was heard, lasting for the space of 1½ hours; meanwhile, there was not the least concussion of the ground noticeable, such as is generally felt during an earthquake, and Lake Taupo was quiet as before. Only the hot springs of Tokanu were in an unusual commotion, throwing out intermittent spouts of water with great power to the height of 30 feet. The natives at once ascribed the roaring noise to the Tongariro, the top of which, however, is not visible from Tokanu on account of the protruding Pihanga. A week later, on his return *via* Onetapu, Mr. Smith noticed with great astonishment, that the Ngauruhoe, which a fortnight before when seen from the same point, showed an unbroken peak, was now broken in and had two sharp horns. There being nothing seen of an ejection of ashes or other substances, the whole seems to have been an eruption of vapors and hot water, which in connection with an explosion burst the upper crater-lip.

The Ngauruhoe does not reach the limit of perpetual snow; yet, the native assured me, that in winter-time, when the lower parts of the mountain are covered with snow, the latter would not stick to the Ngauruhoe; so that the whole cone seems to be heated from within.

The Ngauruhoe, however, is not the only crater in the Tongariro system. Bidwill mentions, that from the top of the Ngauruhoe towards the north he noticed upon another part of the Tongariro a circular lake. This remark most probably has reference to the truncated peak immediately north of the Ngauruhoe, which the natives designate as Ketetahi, the crater of which acts periodically. In the year 1855, at the time of the earthquake in Wellington, an eruption of ashes seems to have taken place from it, and the mountain is since seen to steam from time to time. From the Northend of Lake Taupo, on the 21th April, I saw large and dense volumes of steam, larger than those from Ngauruhoe, emerging from the Ketetahi crater. To the northwest of the Ketetahi there is a third cone, likewise truncated, and nearly 6000 feet high. Concerning the condition of its top I an unable to give any positive information; I merely suppose, that it also contains a deep crater. On its northern side, at a height of about 4000 feet above the level of the sea a fissure is to be seen, from which, as from the Ngauruhoe crater, dense clouds of steam are continually streaming forth. This seems to be a great solfatara. The hot sulphurous springs of that solfatara are often visited by the natives on account of the relief they experience in cutaneous diseases. The outlet of these hot springs flows into the Rotoaire. A fourth cone, north of the Ketetahi, or northeast of the last mentioned cone, shows on its northwestern slope, at a height of about 3500 feet above the level of the sea, a crater, apparently entirely extinct. From the East-shore of Lake Taupo, to the right of the Pihanga, the dark black hole can be plainly seen.

Although this grand volcano with its various craters, has within the last centuries, as far as it is known, not had any eruption of lava, yet I would not venture to assert, that such might not suddenly recur again. At present it is in the state of a solfatara. Earthquakes of such vehemence, as they have occurred at a distance from this central point of volcanic activity upon the North and South Islands at various coast-points (Wellington, Wanganui, Cloudy Bay), are not known in the Taupo district; light shocks, on the other hand, accompanied by a subterraneous noise, are no rare occurrence.

South of the Tongariro rises the Ruapahu. The feet of the two mountains gently slope together forming a plateau about ten miles wide and 2200 feet above the level of the sea. Upon this plateau four lakes are said to lie, two of them about three miles long, the other two smaller. One of these lakes is called Taranaki; its outlet flows into the Wanganui. The natives have a peculiar legend to the effect, that a third giant, named Taranaki, formerly stood by the side of Tongariro and Ruapahu. They were friends until Taranaki attempted to carry off Pihanga, the wife of Tongariro. This brought about a conflict between Tongariro and

Taranaki. Taranaki was worsted in combat and compelled to flee; he hastened down the Wanganui, drawing the deep furrow of that river. He fled as far as the sea, where he now stands in solitary grandeur, the magnificent snow-capped beacon of Mount Egmont (8270 feet). In his flight, however, two pieces were torn off; and the natives show to this very day two blocks of rock as the fragments torn off the Taranaki, masses, which, differing in their nature from the formations contiguous to the Wanganui, lie at a distance of 18 miles from the source of the river at Waitotara.

The Ruapahu[1] has the shape of a truncated cone, towering up into the regions of perpetual snow. No one has ever ascended or explored it. Nevertheless there can be no doubt as to its volcanic nature, but it seems to be perfectly extinct; there is no trace of a solfatara to be discerned in the distance either at its sides or at the top; and it is totally unknown whether the broad summit forms a plateau, or whether it contains a crater.[2] The mountain is but rarely free from clouds; and if once the weather happens to be clear, large snowfields are seen covering the summit, and running down along the fissures, by which the slope of the mountain is channelled, as though they terminated in glaciers. The limit of perpetual snow in the latitude of the Ruapahu (39° 20′), is at a height of about 7800 feet, and to judge by the colossal extent of the snow-fields even in midsummer, the mountain appears to reach a height of 9000 to 10,000 feet above the level of the sea.[3] At any rate the Ruapahu is by far the highest mountain of the North Island. A portion of the mountain bears the name of Paratetaitonga. At the eastern declivity of the Ruapahu rises the southernmost source of the Waikato. It forms a waterfall according to

[1] The natives about the Rotoaira say Ruapehu. The word seems to have a characteristic meaning. Ru, Rua, also Ruu means concussion, earthquake; pahu means noise; the natives understand by rupahu a man who makes "much ado about nothing." The name Ruapahu therefore probably originates from the fact, that occasions with a subterraneous noise emanate from the mountain, but without volcanic eruptions.

[2] Sir George Grey — I am recently informed — states that he has been to the summit of the Ruapahu, and there found a crater with hot springs.

[3] The English sea-maps state 9195 feet. *Arrowsmith's* map 9000 feet, Taylor's map 10,236 feet.

the statement the natives; and 50 yards from the source of the Waikato, the source of the Wangaehu is said to lie, which flows South, and empties into Cooks strait East of the mouth of the Wanganui river. Its water, the natives say, has a milky color, and a bitter, astringent taste.

The pumicestone plateau, upon which the Tongariro and Ruapahu rear their colossal heads, assumes on the South-East side of the Ruapahu, where it forms the watershed between the Waikato and Wangaehu, the character of a sandy desert. The natives call the plateau Rangipo, and the sandy desert Onetapu. Superstitious notions are connected with this sandy plain. The road from Lake Taupo to Wanganui, the so-called Rangipo-Road leads over it, and the natives have driven pegs into the ground in order to point out the direction of the road.

From the southern foot of the Ruapahu the country slopes gradually towards Cooks strait in the same manner as from the North-end of Lake Taupo towards the Bay of Plenty. It consists on both sides principally of pumicestone, pumicestone-tuffs and rhyolite lavas, and it can be justly said, that the foot of the two volcanic colosses reaches from sea to sea.

Section through the North Island from SW. to NE.

Consequently the Taupo volcanoes arise upon a huge flat cone, which was formed by the first submarine eruptions, and rose only gradually by the upheaving of the land above the sea. In close connection with this rising is the terrace-formation in all the river-valleys of that cone, a phenomenon which is very characteristically marked on the shores of Lake Taupo. The first terrace is at Pukawa about 100 feet above the present level of the lake. It is covered with the sand- and boulder-alluvium of the lake, and so very characteristic that even the natives could not help noticing it.

They say, that in former times before the breaking through of the Waikato to the North, the lake had stood at that height. The second terrace is 300 to 400 feet above the lake, and forms extensive plains round about the lake. Yet, it is only the third stage that leads on to the pumicestone table-land 700 to 800 feet above the lake. The formation of the terraces is most perfect in the Kuratao and Waikato valley upward from the lake, and along the eastern side of the lake.

The climate of the Taupo-district is not so mild, as the climate of the coast districts; especially the winter-season is cold and chilly. The rough blasts so frequent here are partly to be ascribed to the high mountain-ranges of the country.[1] We had sufficient proofs of it. The weather, unusually fine and pleasant in the first half of April, had undergone a total change during our stay at Pukawa. The bright aftersummer was followed by a rough and blustering autumn. On the 15th April, after a long calm, suddenly a N. W. gale broke forth, which became really dangerous during our passage across the small southern cove of the lake near Te Rapa, our canoe being very rudely tossed to and fro by the squally waves of the lake. The Northwester was followed on the 16th April by chilly blasts from southwest, which lasted three whole days, accompanied by heavy showers of rain and pelting hailstorms. In the mountains it snowed. The temperature fell during those storms at night as low as 38° 7 F., nor did it rise even during the day above 54° 5 F., so that we found the chimney-fire at Mr. Grace's quite comfortable. The lake during those days presented the appearance of a wild sea. White-foaming surges were rolling towards the shore, producing a roaring surf as at the open sea-coast; terrible blasts came breaking forth from the ravines and valleys on the South-side of the lake. In mad uproar they swept

---

[1] The inhabitants of the Taupo country distinguish four principal directions of the wind:

marangai = stormy Northeaster with rain,
tuariki = Northwester with rain,
hauauru = West wind,
longo = South and Southwest wind.

over the watery plain, and where they met together from opposite directions, eddies were formed from which the spray was dashed high into the air. Whoever happens to be overtaken by such weather in a canoe upon the lake, is irrevocably lost. The lake, therefore, for the imperfect vessels of the natives, — here called tiwai, — is much more dangerous, than even the open sea, because the fresh water is much more easily stirred up by the wind than the salt water, forming short, tumbling waves; moreover the shores present but few landing-places. The natives therefore, are extremely cautious, venturing on longer trips, only, when they can safely rely upon serene weather. Nevertheless fatal accidents are of frequent occurrence, and every dweller near the treacherous water can tell of cases, when he had a hair-breadth escape from the grasp of the hungry sea-sprite (Taniwha) Horomatangi, haunting that spot, according to tradition, and stirring up bad weather.

Horomatangi is said to be an old man and as red as fire. Thus the natives assert to have seen him. He lives in a cave on the island Motutaiko in the lake. There he watches the passing canoes, dashing forth from his lurking-place as soon as he espies one. He churns up the water into mad surges bubbling up like the big spout Pirori near Tokanu; together with the water he throws up large stones, which falling upon the passing canoes upset them. He devours whatever comes within his reach; carrying on his work of treachery and destruction both in fine and bad weather. The natives point out a place, situated almost in the centre of the lake between the island Motutaiko and Te Karaka Point, as chiefly dangerous, avoiding even in the finest weather to venture here too close to the haunt of the evil spirit. Even when the general surface of the lake appears smooth, the water on this spot is in boiling commotion; in stormy weather it appears as one large patch of foam. The canoes passing over it are said to be turned from their course. These phenomena being real matters of fact, the observer might be tempted to suppose the existence of a spouting submarine spring at that place, or even of submarine volcanic eruptions; for such an explanation, however, further indications are wanting, and it is perhaps nothing but a current disturbed by the transverse influx of several extensive streams of water, which current passes through from the influx of the Waikato to where the river leaves the lake, and makes itself specially felt at that place. By supposing the existence of such a current, we likewise have an explanation for the assertion of the natives, that the canoes at that place are always turned in a direction from southwest to northeast.

Horomatangi has moreover special relatives, the Kaukapapas, distinguished by peculiar attributes, and on that account held in great esteem. Te Toko of Ornanui, a village north of Lake Taupo, is said to be such a Kaukapapa, often disappearing suddenly, reappearing at Lake Rotorua, and returning with equal suddenness. In like manner Te Ihu at Tapuaiharuru is reported as being able to live with Horomatangi under water in the cave on the Island Motutaiko. Incredulous minds, of course, assert, that this cave, the entrance to which is under water, rises higher at some distance from its mouth, being quite dry there, so that a man can very easily live in it. Such and a great many similar stories are in vogue about the lake. — Yet, sooner or later, I thought, the time will come, when well-constructed vessels of European style, when even steam-boats will navigate this magnificent lake. Then the reign of those dreaded Taniwhas will be at an end for ever.

The temperature of the lake I always found to be several degrees higher than the temperature of the brooks and rivers running into it; on the 15th April, at Pukawa, the lake-water had a temperature of 59° F.; during the cold days from the 16th to the 18th of April the temperature fell; and on the 19th and 20th April I observed at several points of the East-shore 58° to 60° F., while none of the tributary rivers showed more than 56° F.[1]

The population about the lake was estimated by Dieffenbach in 1841 at 3200 inhabitants; in 1859 there were only 2000, living

[1] On the 19th April the Waikato had 54° F., the Waimarino 56° F., the Tauranga 55° F.; on the 20th April the Waipehi had 52° C., the Hinemau 55 1/2° F., the Waitanui 54 1/2° F.

scattered in numerous villages about the lake. The settlements are nearly all situated at the mouths of the rivers running into the lake, the natives preferring to cultivate the fertile river-alluvion. The Waikato-delta especially is a perfect granary, while the plantations upon the pumicestone plains are yielding only scanty crops. It is quite characteristic that the Taupo-Maoris, who have hitherto kept comparatively most aloof from European influence, are decidedly to be numbered among the best and stoutest representatives of their race, and that here, as I was told, the families are yet blessed with numerous offsprings.

The Fauna of the lake is very scanty. Ducks, sea-gulls, and other water-birds are found on the shores; and in the water there are only three small kinds of fresh water fishes, Inanga, Koaro, and Kokopu of the natives, species of *Elæotris*, very much like the English whitebait. The latter are caught in large quantities by the Maoris and are considered quite a dainty food. But strange to say, the Taupo district is entirely destitute of eels. The numerous rapids of the Waikato after its leaving Lake Taupo, it seems, are the impediment preventing the migrations of the eels to the lake. Besides those kinds of fishes also a cray fish, Koura, is found; likewise some fresh water shells, *Unio, Cyclas*, and small species of *Hydrobia* only 2 or 3 lines in size. [1]

On the 19th April we took leave of Mr. Grace and his amiable family; and set out along the East-shore of the lake. After passing the Waikato-delta we came to the strand. It consists of loose pumice-stone-sand, and pumice-stone gravel interspersed with small fragments of obsidian, upon which the walking is both difficult and disagreeable. Sometimes pumice stone blocks are found here having a diameter of three or four feet. Those large blocks are gathered by the inhabitants of the country, cut in square-stones or in the shape of bricks, and used for building. Mr. Grace had an ad-

---

[1] Together with these shells, there is also very frequently found upon sandy patches along the borders of the lake, a brown, slightly curved, tubulous shell, belonging to the chrysalis of the *Phryganeæ* (dragon-flies). Taylor (Pl. III. 9) erroneously placed an illustration of it by the side of the shells, designating it by the name of *Corniforma*.

dition to his house built entirely of pumicestone, and highly praised the dryness of the material. The first river to be passed is the Waimarino (placid water). On the northern shore there is a settlement of the same name, where we stopped to dinner. From Waimarino we passed the rocky peninsula Motuoapa, — it is connected with the main-land by a low neck, and was no doubt, in former times an island like Motutaiko, — then the Tauranga river, next to the Waikato the principal tributary of Lake Taupo, and at sunset we reached the Pah Motutere. The Pah is situated upon a low neck of land jutting far out into the lake, and surrounded by a row of palisades of stout posts bearing carved figures. We found, however, not a single inhabitant, because the whole population had gone to a large Maori-meeting at Ahuriri. We pitched our camp for the night in the spacious church (Ware karakia), situated outside of the pah.

April 20. — Motutere is halfway between the South- and Northends of the lake. About three miles from the shore lies Motutaiko, the only island in the lake. In times of war it was always a safe place of refuge for the inhabitants of the neighbouring shores, and is said to be an extremely romantic place of sojourn. The northern side of the island is formed by a vertical precipice. Motutaiko, moreover is said to be the only place in the Taupo-district, where the beautiful Pohutukaua-tree *(Metrosideros tomentosa)* is still found. Just before our departure I had the pleasure, to see the Ruapahu gleaming brightly in the morning sun up to its highest points. From Motutere the shores grow steep and rocky. Twice we had to climb over high projecting rocks, called Poroporo; the path is artifically cut in the brittle sandy mass of stone. It is a kind of trachytic (or rhyolitic) tuff full of quartz-crystals, clear as water, and of the size of a pea. Thence the road leads again down upon the sandy beach, and the traveller arrives at the village Totara (likewise called Hamaria, i. e. Samaria). In the rear of the village are vertical bluffs of a very regular columnar rock. The natives call the rocks Taupo, and the lake is said to have its name from them. It is singular, that these very rocks, by the natives called Taupo, consist of an extremely remarkable kind of rock which has attracted the attention of every stranger travelling along the lake. It is a volcanic rock of a very striking, lamellar structure. Like the leaves of a book, sometimes of a microscopic fineness, the thin lamellar sheets of stone lie one above the other. Grayish black layers resembling silicious schist, of various lighter and darker shades, alternate with pearl-gray, violet flesh-coloured, sometimes even with brick-coloured layers, so that the streaky mass reminds the observer of agate. From the numerous, white, transparent quartz-grains, and small, yellowish-white felspar-crystals (sanidine) enclosed, it moreover receives a porphyritic structure, while in smaller or larger vesicular spaces light-brown mica appears. There can be no doubt as to the genuine lava-character of the rock. As by the stretching and pulling of a mass composed of mixed fusions, artificially streaked glass is produced, so this rock is likely to have originated from a volcanic magma composed of various stone-fusions. My friend, Baron v. Richthofen has, in 1860, described quite a similar kind of rock from the vicinity of Telkibánya, Mad, Tokay, Sarospatak etc. in Hungary, under the name of Lithoidit, or lithoidic Rhyolith, while Dr. J. Roth has named a similar lava upon the Liparian Islands Liparit. [1]

North of the Totara the Hinemau river (al. Hinemaiai), empties into the lake. Its valley is distinguished by numerous, extremely regular terraces. From that point off, the shores of the lake are formed by towering cliffs, the snow-white colour of which had long ago attracted our attention from afar. We now found that those bluffs rising at some places to the height of 300 feet, consisted of pumicestone. Although I had long ago become accustomed to the huge masses of pumicestone, scattered everywhere throughout the North Island, yet, I could not help gazing with astonishment, on beholding here, in the mother-country as it were, where all the pumicestones originate, pumicestone deposited in small fragments

---

[1] What Dieffenbach states in various passages of his work concerning the occurrence of leucit in the lavas and in the sand on Lake Taupo, is a mistake. Those little crystals are either quartz or felspar.

and in colossal blocks to a height of 300 feet. It is undermined by the surf of the lake during northwesterly gales. Then sometimes whole bluffs fall in, covering the lake with their ruins, which are carried by the Waikato to the western coast. Between the pumicestone, now and then also a coarse conglomerate is seen to intervene, consisting of hyaline varieties of the rhyolite group, such as obsidian, perlit, etc. The passage along the foot of the cliffs, in the loose masses of sand and gravel, is extremely difficult, until Tekohaiataku Point is reached, where the strand spreads itself into a broad bay, in which the lagoon Roto Ngaio is situated, with a settlement of the same name upon the narrow strip of land separating the lagoon from the lake. Here, although we had made but a very short day's journey, I had our tents pitched, and was occupied until even-tide with observations for a map of the lake.

Tauhara, extinct volcano at the lake Taupo.

April 21. — After leaving Roto Ngaio, the picturesque Tauhara-mountain with the small cone Maunganamu alongside of it is the main object, which the traveller has continually before his eyes, and to which he is gradually approaching. The path winds along the shore of the bay which, in a beautiful curve extends to the North-West. We had to ford the Waitanui (or Waitahanui), a considerable, rapid river, running for a great distance parallel with the strand; and, two miles from the lake we turned into the interior over steep pumicestone terraces, channelled by numerous dry water-courses. Large blocks of dark-black obsidian-porphyry, interspersed with white quartz and sanidine crystals, are scattered over those plains; and it took us some time until we succeeded in knocking off suitable pieces from the brittle, crumbling mass. We reached the shore of the lake again at the northern-most bay, where the Waikato leaves the lake, and followed the shore as far as the outlet of the Waikato. The strand and the shores are here likewise formed of all sorts of rhyolitic gravel furnishing to the scientific collector various and interesting specimens.

Like the South-end, so also the northern-most end of the lake is remarkable for its hot springs. The lake steams for a whole mile along the shore as though it were a lake of hot water, and when I endeavoured to ascertain the temperature of the water in the surf lashed by a strong West-wind, the thermometer rose to 100° F. The hot water issues forth at various places cementing by its silicious deposits the sand and gravel of the strand, into a solid sand-stone, which covers the shore in large slabs, occassionally 3 to 6 feet thick, and resembling floating cakes of ice. A slight smell of sulphuretted hydrogen is noticeable near the of hot springs. Likewise a warm brook, Waipahihi, which probably rises at the Tauhara mountain, flows here into the lake; forming a steaming cascade, at the spot where it plunges over the last low terrace. The water showed a temperature of 87° to 88° F. A second smaller brook not far from the Waipahihi had a temperature of 81° F.

About noon we reached the point, where the Waikato flows out of the lake. Large as the stream is, which here in rapid course rushes out of the lake, I could not refrain from thinking, remembering the many tributaries of the lake, that the Waikato cannot be the only outlet, but that the elevated Taupo-Lake is a reservoir, from which parts of the water are drained by subterraneous channels, reappearing in the form of the innumerable hot springs, which are found between Lake Taupo and the East-coast.

As the mailroute from Auckland to Ahuriri leads here across the river, there were cables stretched from bank to bank for the benefit of facilitating the crossing of the river, thus establishing a kind of ferry. We crossed over, and arrived in Tapuaiharuru, the residence of the chief Puhipi, the post-master of the lake country, to whom we had forwarded from Auckland a supply of provisions, which we had been eagerly expecting for some time past. Puhipi himself was absent on a journey; but we found our provision depot in best order, so that we had ample chance to have one jolly day yet before leaving the lake.

CHAPTER XVIII.

Ngawhas, and Puias; boiling springs, solfataras and fumaroles.

The country between lake Taupo and the East-Coast. — Analogy between the hot springs in New Zealand and Iceland. — Departure from Taupo. — The Karapiti fumarole. — Orakeikorako on the Waikato, and its geysers. — The Pairoa range and its fumaroles. — The boiling springs on the Waikiti. — Tropical ferns. — Rotokakahi. — Arrival at lake Tarawera. — Mr. and Mrs. Spencer. — The Tarawera mountain. — The warm lake Roto-mahana and its boiling springs. — Rotomakariri, the cold lake. — Rotorua. — The obliging chief Pini te korekore. — Ohinemutu and its warm baths. — The geysers of Whaka-rewarewa. — The solfataras Tikitere and Ruahine. — Lake Rotoiti. — Origin of the hot springs.

Appendix. Chemical analysis of the water and sediments of some hot springs.

From the table-land, upon which, on the North-shore of Lake Taupo the picturesque Tauhara volcano arises, there extends in a northeasterly direction with a gentle slope towards the Bay of Plenty the Kaingaroa[1] plain, an extensive plain, fifteen miles wide and channelled by numerous valleys. Vast quantities of pumicestone cover the almost tree-less plain, the scanty soil of which pro-duces only a meagre growth of grass and low shrubs. It appears, as though in olden times a powerful stream had taken its course over the plain to the sea. On the East-side the plain is bordered by the Te Waiti range striking in the direction of the East Cape; on the West-side by a volcanic table-land cut up and broken by faults and dislocations into a thousand hills and mountains, which

---

[1] Kai-ngaroa signifies a protracted meal. The name refers to the legend about a female relative of the chief Ngatiroirangi, who arranged here a protracted meal, and changed her companions into ti-trees, which are still standing there, but are said to be continually receding at the approach of the traveller.

separate the sterile pumicestone plain from the wood-clad Patetere plateau.

The Waikato, after leaving Lake Taupo, shapes its course North-East for a distance of fifteen to twenty miles, flowing through a broad, terraced valley on the boundary of the Kaingaroa plain. It is not until after the junction of the Pueto river, below the Pah Tetakapo, that the river turns in a keen bend toward. Northwest, and enters the mountainous region at Mount Whakapapataringa. In a deep gorge with numerous rapids the river breaks through the mountains, emerging again near Maungatautari into the broad plain of the middle Waikato-basin.

Along the whole distance from Lake Taupo to Maungatautari the river is innavigable on account of its numerous rapids. The land on both sides consists of trachytic tuff, of pumicestone and of partly vitreous, partly crystalline rhyolithic lavas, the flow and extent of which is to be traced to the volcanic centre of the Tongariro and Ruapahu mountains. While the deep terraced valleys are the result of a long continued erosion by water, we see, on the other hand, the effects of the volcanic fire displayed in an immense number of hot springs, in which the country abounds. If we suppose two parallel lines to be drawn from Lake Taupo, touching its East- and West-shores and extending in a N. E. direction as far as the Bay of Plenty, then these two lines, including the range of hills and mountains situated between the Kaingaroa plain and the wooded Patetere plateau, border likewise the space, upon which from more than thousand places hot vapours arise, calling forth all those phenomena of boiling springs, fumaroles, mud-volcanoes and solfataras, for which the North Island of New Zealand, and especially the "Lake District," are so remarkable. The southernmost point of this wonderful zone of hot springs, which by far exceeds all others in the world in variety and extent, is the Tongariro volcano with its solfataras and the northern end is marked by the ever steaming Island of Whakari in the Bay of Plenty, a distance of 120 seamiles. [1]

[1] Even the natives have very correctly brought the hot springs directly in

The phenomena are similar to those upon Iceland, and as the Icelanders distinguish their hot springs as Hverjar, Námur, and Laugar, so also the Maoris make a similar distinction, although so not quite marked, between Puia, Ngawha and Waiariki. The Hverjar upon Iceland are either permanent fountains, whose boiling water is continually in a state of ebullition; or intermittent ones, whose water shows a vehement ebullition only at certain periods, when it reaches the boiling point, while during the intervals it is in a state of calm repose, its temperature often falling

connection with the still active volcanoes, though they have clothed their conceptions in the garb of a legend. I here recite the legend such as I heard it from Te Heuheu.

Among the first immigrants, who came from Hawaiki to New Zealand, was also the chief Ngatiroirangi (heaven's runner; or the traveller in the heavens). He landed at Maketu, on the Eastcoast of North Island. Thence he set off with his slave Ngauruhoe, for the purpose of exploring the new country. He travels through the country; stamps springs of water from the ground to moisten scorched valleys; scales hills and mountains, and beholds towards South a big mountain, the Tongariro (literally "towards South"). He determines on ascending that mountain, in order to obtain a better view of the country. He comes into the inland-plains to Lake Taupo. Here he had a large cloth of kiekie-leaves tattered and torn by bushes. The shreds take root, and grow up into kowai-trees (*Edwardsia microphylla*, a beautiful locust-tree with yellow blossoms, quite frequent in the Taupo district). Then he ascends the snow-clad Tongariro; there they suffered severely from cold, and the chief shouted to his sisters, who had remained upon Whakari, to send him some fire. The sisters heard his call and sent him the sacred fire, they had brought from Hawaiki. They sent it to him through the two Taniwhas (mountain and water-spirits living underground), Pupu and Te Haeata, by a subterranean passage to the top of Tongariro. The fire arrived just in time to save the life of the chief, but poor Ngauruhoe was dead when the chief turned to give him the fire. On this account the hole, through which the fire made its appearance, the active crater of Tongariro is called to this day after the slave Ngauruhoe; and the sacred fire still burns to this very day within the whole underground passage between Whaikari and the Tongariro; it burns at Matou-Hora, Oka-karu, Roto-ehu, Roto-iti, Roto-rua, Roto-mahana, Paeroa, Orakeikorako, Taupo, where it blazed forth, when the Taniwhas brought it. Hence the innumerable hot springs at all the places mentioned. This legend affords a remarkable instance of the accurate observation of the natives, who have thus indicated the true line of the chief volcanic action upon the North Island. Another legend says: — when Maui stepped upon the island fished out of the sea (Chapt. X.), he took through ignorance some of its fire into his hand, and horrified, flung it into the sea, where subsequently the volcano Whakari arose. The ashes of the volcano, Maui scattered about with his feet, and thus the fire vomiting mountains of the island and the numerous hot springs originated.

considerably. To the Hverjar belong, for example, the celebrated springs of Haukadal, the great Geyser and the Strokkur, and with these the Puias of New Zealand correspond. The word Puia is especially used in the Taupo country, to designate the intermittent, geyser-like fountains of Tokanu, of Orakeikorako on the Waikato and of Whakarewarewa[1] on Lake Rotorua. Puia has moreover the more general meaning of crater or volcano, and is applied to active as well as extinct volcanoes. Namur upon Iceland are the non-intermittent springs, such as the solfataras of Krisuvik and Reykjahlid, having no periodical eruptions; and the same are in New Zealand the Ngawhas, a term especially used for non-intermittent springs, for the solfataras and sulphurous hot springs on the Rotomahana, Rotorua, and Rotoiti. Finally the springs suited to bathing-purposes, the water of which never reaches the boiling-point, and all naturally warm baths are called Waiariki, corresponding to the Laugar of Iceland.[2]

A visit to the Puias, Ngawhas, and Waiarikis was the main object of my journey from Lake Taupo to the East Coast. I had the choice of two routes; either the direct route on the right bank of the Waikato across the Kaingaroa plain to lake Rotomahana (two or three days journey); or the road along the left bank via Oruanui to Orakeikorako, and after here passing the Waikato, along the Pairoa range to lake Tarawera, three or four days journey. Dieffenbach in 1841 had taken the former route;[3] I therefore, chose

[1] One of the intermittent fountains of Whakarewarewa is specially named Te Puia.

[2] W. Preyer and Dr. F. Zickel, Reise nach Island 1862. p. 69.

[3] The most note-worthy points on this route are the fumaroles and solfataras on the western foot of Mt. Tauhara, at the brinks of which sulphur and alum are deposited. One of them is called Waikore, another, whose steam-column is visible at a great distance, Parakiri (the skinner, peeler). Farther on, at the N. W. foot of the Tauhara, lies the Rotokawa, bitter-lake, one mile in length from North to South, and a-half mile wide. The water has a strong taste of alum. At the North-end of the lake dense clouds of steam ascend continually from a number of solfataras lying there. A third group of springs is on the right bank of the Waikato about five miles North of the junction of the Pueto, among them the solfatara Ipukaihimarama and the fountain Te Kohaki. Farther North the road passes by the Maunga Kakaramea, which is steaming up to its very top, to the Rotomahana.

the latter, and the more so, because according to the statements of the natives, a great many interesting points could be visited on this route.

April 22. — The mountains had wept tears on the day of our arrival on Lake Taupo, and to-day they wept again at our departure. Lest we should feel too sorry at parting, Tongariro and Ruapahu were wrapped in clouds, so that we were unable to see them. In the dreariest weather we wandered over the pumicestone terraces on the left bank of the Waikato, through a very broken and treeless country, looking waste and dreary in its scanty garb of grass and fern and with its dry water-courses. Soon, however, our attention was again engrossed. About two miles from Tapuai-haruru we entered the Otumaheke valley. A rivulet of warm water (71° F.) flows through it, and on its left bank, a little off the road, a colossal column of steam whirls high up in the air. It was only with great caution that we could approach the place, from which the steam issued, because around there the bottom of the valley is literally perforated and furrowed by fissures and crevices, from which hot steam streams forth; whilst in the numerous pot-shaped holes about, a gray clay-coloured mud, or turbid milky water can be seen boiling. Besides, the whole ground is heated to a considerable extent, and boiled perfectly soft into an ferruginous clayey mass, from which small mud-cones protrude. The steam and mud-holes seem here continually to shift their positions. The *Lycopodium cernuum* found everywhere in hot climates and around hot springs has settled down also here in luxuriant growth. We safely reached the place, where with immense force, and amid loud hissing and booming the steam is streaming out of a circular hole in the loose masses of pumicestone at the foot of the hill. It is high-pressure steam, without a trace of any other gas, and bursts out through a small aperture in the depth of the circular hole in a somewhat slanting direction, with a sound like letting-off the steam from a huge boiler, and with such force, that branches of trees and fernbushes, which we flung into the jet of steam over the hole, were tossed into the air, twenty or thirty feet high. The natives call this steam-spring, the column of which is seen already from the East-shore of Lake Taupo, at a distance of 12 to 15 miles, Karapiti (encompassed, circular).

The Karapiti fumarole, Taupo country.

After another mile we came to a second valley with its direction towards the Waikato. A little distance off the road, on the right side of the valley, there was likewise steam ascending from numerous places. The ground about these places appeared of a reddish color. No doubt this locality and the one previously referred to are those that Dieffenbach visited in Mai 1841, on his journey from Otawhao to Lake Taupo, and of which he has given us a description in his "*Travels* etc. Vol. I. p. 327—329." Yet, various changes must have taken place since that time, as Dieffenbach at the first named locality does not at all mention the large Karapiti-fumarole, but speaks of a powerful fountain, the water of which was thrown out to a height of 8 to 10 feet and was heated above the boiling-point. Perhaps the fountain has changed to a steam-spring, in a similar way as the "roaring geyser" in Iceland, — not to be confounded with the great geyser — which in former times had periodical eruptions of water, now emits only steam.

At two p. m. we reached the native village Oruanui, at the foot of a wood-clad hill of the same name, and as we found here commodious shelter in a house specially built by a native for travelling Europeans, we determined upon remaining to wait for better weather.

April 23. — Heavy storms and showers of rain had continued all night; but when in the morning the rain ceased and even the sun was peeping through the cloud-banks, we set out again. It was for the first time, since we had left Pukawa, that we again entered the bush. The rain, however, commenced anew, and, by marching onward with all possible speed and without the least delay, we strove to reach as soon as possible the Pah Orakeikorako on the Waikato, a distance of ten miles from Oruanui. We passed the Orakanui valley, climbed a wooded range of hills, and on arriving upon the open height of Tehapua, we had an interesting glance, through smaller side-valleys, at the Waikato valley far below, from which heavy clouds of steam were ascending, the harbingers of a new region of hot springs, the Puias of Orakeikorako. We reached the Pah about noon. It is situated upon an elevation on the left bank of the Waikato, about 200 feet above the river, and had been lately fortified anew on account of impending hostilities between the tribes of the Taupo and some others dwelling more North. The terraces on the descent facing the Waikato had been dug anew; but instead of the palisades of stanch beams, as in olden times, there was only a scanty miserable hedge-fence put up, which, although double and triple, could be almost overthrown with the mere hand. A ridiculous fort, indeed, about which, however, the Maoris of the vicinity made much ado. In the middle of the Pah I had my tent pitched; but a heavy shower of rain inundating the whole ground, I was obliged to seek shelter in the hut of the chief Hori from the terrible tempest, which now burst forth in all its fury. It was the first violent storm with electric discharges since our departure from Auckland, and I was compelled to await in quiet resignation the coming morn, in order to visit the Puias at the foot of the Pah. Hori informed me, that a fortnight ago three faint shocks of an earthquake were distinctly felt here.

April 24. — The tempest had abated during the night. In the morning a dense fog lay upon the Waikato; but it soon vanished; the sun shone brightly into the valley, and now — what a sight! In swift course, forming rapids after rapids, the Waikato was plunging through a deep valley between steep-rising mountains; its floods, whirling and foaming round two small rocky islands in the middle of the river, were dashing with a loud uproar through the defile of the valley. Along its banks white clouds of steam were ascending from hot cascades falling into the river, and from basins full of boiling water, shut in by a white mass of stone. Yonder a steaming fountain was rising and falling; now, there sprung from another place a second fountain; this also ceased in its turn; then two commenced playing simultaneously, one quite low at the river-bank, the other opposite upon a terrace; and thus the play continued with endless changes, as though experiments were being made with grand waterworks, to see whether the fountains were all in perfect order, and whether the water-falls had a sufficient supply. I began to count the places, where a boiling water-basin was visible, or where a cloud of steam indicated the existence of such. I counted 76 points, without however being able to survey the whole region; and among them there were numerous, intermittent, geyser-like fountains with periodical water eruptions. The sketch which I drew on the very spot, can be but a faint illustration of the grandeur and peculiarity of the natural scenery at Orakeikorako.

The region of springs extends along the Waikato a distance of about a mile on both banks, from the foot of the steep Whaka-papataringa mountain South, as far as the foot of the wood-clad Mount Tutukau North. The larger portion of the springs is situated upon the right bank; it is, however, difficult of access, since the rapid river cannot be passed close by the springs themselves, but only far above or below; and then it would be necessary to climb along the steep-sloping river-banks covered with dense undergrowth, where the traveller upon a soil totally softened by hot vapours would be every moment in danger of sinking into boiling mud. I was obliged to content myself with a closer inspection of the springs situated on the left river-bank just below the village.

The most remarkable springs of the whole region are comprised within a large whitish mass of silicious deposit, 120 yards long, and of equal breadth, by the natives called Papa Kohatu, the flat

The hot springs of Orakeikorako, on the Waikato river.

stone, which with a gentle inclination from the foot of the Tutukau-mountain extends into the Waikato. Close by the river bank lies, first of all, the Puia Te-mimi-a-Homaiterangi. The manner, in which we practically experienced the intermittent properties of this fountain, proves sufficiently, how much caution is necessary in approaching such springs for the first time and without expert guides. My travelling companions wished to enjoy the luxury of a river bath early in the morning, and had just deposited their clothes, when suddenly they heard violent detonations, and saw the water madly boiling up in a basin close by. They started back afright, but only just in time to escape a shower-bath of boiling water; for now amid hissing and roaring a steaming water-column was being ejected from the basin in a slanting direction, and to the height of about 20 feet. By the time I had heard of the treacherous geyser, and arrived at the place, every thing had long subsided again into quiet repose, and in the pot-like basin four to five feet wide I only saw water as clear as crystal gently bubbling up. It showed a temperature of 202^{0} F , proved a perfectly neutral reagent, and tasted like weak broth The first water-eruption, which I observed with my own eyes, took place at 11^{h} 20' a. m. The basin, a short time previous to the eruption, was full to the brim. Amid a distinctly audible, murmuring noise in the depth of the basin the water came more and more into violent ebullition, and then was suddenly expelled with great force at an angle of 70 degrees in the direction of S.S.E. to a height of 20 to 30 feet. Together with the water huge volumes of steam burst forth from the basin with a loud hissing sound, partly veiling the spout of water. This lasted one and a-half minutes, then the expelling force decreased, the water rose only two or three feet high, and after two minutes the play of water ceased entirely amid a low, gurgling noise. On now stepping up to the basin I found it empty, and was able to look down to a depth of eight feet, into a funnel-shaped, gradually contracting aperture, from which steam escaped with a hissing noise. Gradually the water rose again; after the lapse of ten minutes the

[1] Te mimi a Homaiterangi literally means the urine of (the chief) Homaiterangi.

basin was full again, and at 1ʰ 36 p. m. the second eruption took place; at 3ʰ 40 p. m. the third, which I had occasion to observe. Consequently the eruptions seem to occur about every two hours. The sediment of this, like of all surrounding springs, is silicious; the recent sediment is soft as gelatine, gradually hardening into a triturable mass, sandy to the touch, and finally forming by the layers deposited one above the other a solid mass of rock of a very variable description at different places both as to colour and structure. Here it is a radiated fibrous or stalky mass of light-brown color; there a chalcedony hard as steel, or a gray flint; at other places the deposit is white with glossy, conchoidal fracture like milk-opal, or with earthy fracture like magnesite. A second Puia, about 30 yards distant from the geyser, is called Orakeikorako. The name is said to have reference to the transparent, shining water. It is an elliptical basin of eight feet length and depth, by six breadth, half-filled with gently bubbling water clear as crystal.

The main spring, however, to which the Papa Kohatu chiefly owes is origin, is close by the foot of the sloping hills. It is a powerful fountain continually bubbling up to a height of two to three feet, the water of which showed a temperature of 209° F., and had a distinctly noticeable smell of sulphuretted hydrogen. The chief accompanying me told me, that for two years after the earthquake of Wellington, in 1848, this fountain was a geyser spouting to a height of 100 feet (no doubt somewhat exaggerated), and throwing up with powerful force even large stones as fast as they were flung in. Three smaller basins close by, which in former times were probably likewise independent springs, are now filled by the discharge of the fountain, and form excellent natural bathing tubs. The water flows from one basin into the other, so that there is a triple choice of temperature. In the first basin I found 116° F., in the second 110° F., and in the third 96° F. The latter, at a depth of 3 to 5 feet has exactly the dimensions of a large bathing-tub; its basin is formed of snow-white silicious deposit, resembling the purest marble, and its crystalline water looked so inviting, that I could not resist taking a bath in it. To these springs, moreover, considerable medicinal virtues are ascribed. At Orakeikorako we met an Irishman, belonging to Port Napier, who told us, that he had been conveyed thither perfectly lame with the gout, and that after a short use of those baths he was entirely restored.

On both sides of the Papa Kohatu, up and down the river, there are, concealed in the copse of the river-banks numerous boiling mud-pools, which can be approached only with the utmost caution, because the softened soil, unprotected by a sheet of silicious deposit, gives way most readily. The largest of these mud-basins I saw a few hundred yards down the river. It has an elliptical form, is 14 feet in length, 6 to 8 feet wide and equally deep. There was boiling in it a mass of mud dyed intensively red by oxide of iron; clammy bubbles of mud rose, burst, exhaling a sulphurous stench, and relapsed — a truly infernal sight. Woe unto him, who here misses a single step! The very thought of it made me shudder; and yet, such fearful accidents have frequently occurred here both to children and adults.

On the opposite river-bank lies the Puia Tuhi-tarata. The discharge from a basin full of sky-blue shining water forms a steaming cascade over strata of silicious deposits shelving off in terraces towards the river and varying in the gaudiest colors, white, red, and yellow. The same scene recurs five ot six times up the river; and intermediate there are points exhibiting periodical eruptions, on some places every five minutes, at others every ten minutes.[1] But wherever bare, reddish patches occur along the river-terrace, there steam can be seen ascending; the same is the case at countless places in a side-valley intersecting the river-terrace. But impossible as it is to see every thing here, it is yet more impossible to describe every thing. Orakeikorako with its hot springs would prove an inexhaustible field for years of observation.

April 25. — Accompanied by Captain Hay I ascended the summit of the Tutukau mountain, rising North of Orakeikorako to

---

[1] The natives have for the most of these springs special names, such as Te Wai-whokata, Rakau-takuma, Whangairorohea, Ohaki, Te Wai-angahue, Te Poho, Wai-mahana.

the height of 2100 feet. The summit is mostly wooded; nevertheless
we found some points, from which we had a good view North and
South. On the height above and in the midst of the woods we
found potatoe-fields and native huts. It is an old Maori custom
dating from the times of war, to establish at remote and less accessible points, usually in large forests, stations and plantings, upon
which the people might fall back in case of need. So there are
also travelling-stations, lying off the road upon hidden paths and
known only to the tribe which has established them.

About noon we set out from Orakeikorako. Te Hori conducted us across the river in quiet water at a passage below the
rapids, giving us a proof of his admirable dexterity, as he had first
to steer the canoe from a place farther up the river through the
eddying and foaming rapids. We stood on the bank and shouted
a loud enthusiastic bravo to the bold chief as he dashed in the
little craft through the frightfully eddying floods.

Through a small side-valley, called Rotoparu, we ascended
the right bank of the Waikato, crossed a fern hill and came into
the Rotoreka valley, a dreary and swampy plain with here and
there an isolated Ti tree. Towards the West the valley is bordered by low, wood-less hills; towards the East, a high rocky bluff
ascends almost vertical, extending in the direction of N. 24° E.,
in a straight line. Above the steep precipice numerous rugged
cliffs tower up, and in the middle of the rocky wall a high, wooded
peak Pairoa (or Paeroa) projects towards the West. After this
prominent peak I have called the whole extent of the bluff the
Pairoa range; and it is easily to be seen, that along this range an
immense dislocation took place, that the almost perpendicular western
side of the range is caused by a "fault" corresponding to a deep
fissure in the earth-crust, and that the low-lands between the Pairoa
range East and the Patetere-plateau West were produced by a breaking or sinking of a large part of the volcanic table-land. In a most
remarkable manner the fissures, and the lines of dislocation are also
indicated by the numerous hot springs issuing along the Pairoa
range, at the foot of the precipice, on its slopes, and even above

on the heights. In warm, dry weather the steam-clouds are less visible; yet the red patches on the sides, devoid of every trace of vegetation, point out from afar the places, from which sulphuric acid, sulphuretted hydrogen, sulphur and steam are continually escaping, producing fumaroles, hot springs, boiling mud-pools and solfataras. On the one end of this remarkable break in the table-land, which can be traced to a distance of 18 miles, Orakeikorako is situated, and on the other end the famous Rotomahana, the marvels of which far surpass every thing else occurring upon New Zealand in the shape of hot springs. Upon the western fissure, parallel to the Pairoa fissure, are the hot springs and solfataras of Waimahana, Rotorua and Rotoiti; and at the North-East side of the break are the lakes Tarawera, Roto Kakahi, Okataina, Okareka etc.

At three p. m. we encamped close at the foot of the Pairoa peak amid a dense growth of Manuka. It was a dismal neighbourhood; for close by our camping-ground lay a terrific basin about 30 feet in diameter, in which a bluish-gray clay-pap was boiling. By the side of this mud-basin, concealed among the bushes, arose a flat mud-cone about 10 feet high, with a regular crater on the top. A heavy cloud of steam, which, accompanied by a slight detonation, suddenly escaped from the crater, attracted our attention. Carefully sounding the ground with our sticks, we approached the mud-crater, and saw a deep, funnel-shaped hole, in which a thick boiling mass of mud rose higher and higher, heaving up in large bursting bubbles. As the mud rose quite close to the brim, we receded a short distance, and then observed a second eruption, during which again steam escaped with a hissing sound, while the mud discharged itself over the margin of the basin. Quite a number of such mud-volcanoes extend on the steaming slopes of the Pairoa peak, playing in the gayest colours of red, white and yellow, while at the top of the mountain a powerful column of steam is seen to ascend, which, as the natives told me, belongs to the great fountain Te Kopiha. I am of the opinion, that this whole portion of the mountain up to the Te Kopiha fountain, — being, as it seems, thoroughly decomposed by hot vapours, — will some day cause a sudden catastrophe by falling in, and covering the Rotoreka-plain with a flood of hot mud.

Our route continued across the marshy plain at the foot of the Pairoa-range. Anxious to reach a convenient place for camping, we marched on till after night fall; but we had after all to be content with a place, where no firewood was to be had. We had arrived at the Waikite creek, where numerous deep pools with boiling water are scattered right and left, close by the road-side, and the natives deemed it a dangerous risk to pass there in the dark. The Maoris with timely foresight had provided themselves with tent-poles from the Manuka-shrubs on the Pairoa, and thus we camped in the immediate vicinity of the hot springs, the seething noise of which I had all night in my ears.

April 26. — The Waikite springs are real boiling wells. In well-shaped, circular holes, six, eight or ten feet wide and equally deep, partly clear, partly turbid water of a milky colour is boiling; in some of them also mere mud. None of them are full to the brim; nor are there any silicious incrustations to be seen. It is in consequence of this peculiarity, that plants can settle and spread upon the inside of the holes, and that the vegetation in them sometimes reaches a depth of four feet. Whatever grows there, grows in a uniformly warm steam-atmosphere. The plants were ferns of luxuriant growth; but forms such as we had as yet observed nowhere else. We were therefore very desirous of gathering them, although the job was not entirely without danger. The most successful method for accomplishing our object was that one of us laid himself flat upon the ground, and, while the others were holding him fast by the legs, gradually pushed the upper part of his body so far over the margin of the hole, that he could reach far enough down with one arm. Our delight on first seeing the beautiful ferns was fully justified; for the result proved they were species,[1] which are usually found only in tropical countries, and singularly occur here, in the interior of the island, isolated at a place, where by means of hot springs the proportions of moisture and temperature

---

[1] *Nephrolepis tuberosa* and *Nephrodium molle.*

peculiar to the torrid zone are produced. The first seeds of those ferns, however, must have been wafted thither by atmospheric currents from the tropical regions of Australia or America, or from the tropical islands of the South-sea. On both sides of the Waikite there are no doubt upward of 20 boiling wells to be counted; their discharges flow into the creek, the temperature of which is consequently considerably raised.

At the Waikite we left the Pairoa line, and following a small valley we ascended in a northerly direction the Whaihorapa-Plateau. We passed two grassy plains, which like drained lakes form basins in the plateau. The first plain bearing the name of Waihorapa, was especially remarkable for a large, at some places widely gaping rent of the ground, which could be traced for a whole mile, in the direction of N. 24° E., consequently parallel with the Pairoa-fissure; moreover for numerous funnel-shaped holes, 20 to 30 feet deep, reminding us of the funnel-shaped holes in the soil of Calabria. Those phenomena are no doubt produced by earthquakes, by which the country is said to be very frequently visited.

The Horohoro Mountain.

Upon an open height we had a clear view of the Horohoro Mountain six miles to the West, which acts a prominent part in the Maori-traditions. It is one of the most striking table-mountains of the country, rising isolated from the plain with vertical sides, a remaining fragment of the formerly unbroken table-land. Thence our road lay through bush for a short distance, and descending through a dry valley we came to the charming shores of the Roto-Kakahi (muscle-lake).

Thus we had reached the first lake of the Lake District, so famous for its numerous lakes and the beauty of its landscape. Like a miniature-picture of the magnificent Alpine lakes of Upper Italy, the little lake with the picturesque island Motutawa in the middle, lies hidden among the mountains. At Teriria, a catholic village on the South-side of the lake, we halted to rest for an hour. From the natives, who received us with a most cordial welcome, I inquired the names of the most note-worthy points on the lake. Their zeal to serve me was so great, that, as a whole crowd were speaking at the same time, there was no possible chance to understand any thing at all, until one of them hit upon the excellent plan of tracing with his knife, after his own fashion, the outlines of the lake upon the sand, and thus to fix the various points of it. Although these outlines did hardly correspond with the real shape of the lake, such as it resulted from my own sub-sequent observations; yet the primitive sketch at the hands of a man, who had perhaps never in all his life seen a map, appeared to me noteworthy enough to copy and present it here.

Maori sketch of the Rotokakahi.

My travelling companions crossed in a canoe over to the N. E. end of the lake, I myself climbing along the eastern shore, — where a path is cut on the steep mountain-side, — for the purpose of making surveys of the various points of the lake. At the other end of the lake, I was not a little surprised at striking a broad carriage-road. It was the first indication of the proximity of the Tarawera Mission station. The road led to a mill; the Wairoa river, an outlet of the Rotokakahi, which drives the mill, was well bridged over; then I came to a settlement, Hereaupaki, laid out in European style, and not far from it, on the left, above at the edge of the bush, lay the hospitable dwelling of the Rev. Mr. Spencer, the central point, from which emanate all the civilizing influences, the cheering proofs of which are met everywhere in the neighbourhood. We had been expected for a long time, and Mrs. Spencer, in the absence of her worthy husband, welcomed us most heartily. We met our photographer, Mr. Hamel, here again in the best state of health.

April 27. — Mr. Spencer's residence, Temu, situated between the lakes Rotokakahi and Tarawera, at the edge of the bush upon an eminence about 500 feet above the level of the Tarawera,[1] is certainly, to every visitor of that country, a place not easily to be forgotten on account of both the amiable hospitality found there, and the exquisite beauty of its site. I too, spent here several days, which I number among the most pleasant during my travels.

It was on a delightful morning, when from the garden adjoining the house my eyes roved for the first time over the smooth deep-blue mirror of the Tarawera lake; resplendent in the radiant sun-light, the landscape appeared to me perhaps doubly beautiful. The scenery of Tarawera lake surpasses in wilderness and grandeur that of any of the lakes in the lake district. The word signifies burnt cliffs. Its general form, exclusive of its deep side-coves, is that of a rhombus, with its main diagonal running from West to East. In this direction it is seven miles long, having a breadth of about 5 miles. The lake is probably very deep; for its shores are mostly rugged, rocky bluffs, shaded by Pohutukaua trees. The chief ornament of the adjoining landscape is the Tarawera-mountain[2] with its crown of rocks, divided into three parts by deep ravines; it rises on the south-eastern side of the lake to a height of at least 2000 feet above the level off the sea. It is an imposing table-mountain, consisting of obsidian and other rhyolitic rocks; and it is not to be wondered at, that its dark ravines and vertical sides have given rise to many an odd story in vogue among the Maoris. Among others, a huge monster, 24 feet long, resembling a crocodile, is said to haunt the clefts between the rocks, devouring every one who dares to scale the mountain. The Rev. Mr. Chapman, a well-known missionary of Maketu, once, despite all

---

[1] Temu according to my measurement is 1502 feet above the level of the sea. The country is said to be frequently visited by earthquakes. Scarcely a year passes without at last some light shocks, recurring various times during the space of 2 or 3 days, and sufficiently alarming to induce Mr. Spencer, to leave the house and dwell out of doors in tents.

[2] The northern portion is called Te Wahanga, the middle Ruawahie, and the southern Tarawera.

remonstrations of the Maoris, ascended the mysterious mountain. After searching for a long time he found a small lizard scarcely span-long, which he took with him to show to the Maoris, and to convince them, what kind of crocodiles really live there. An aged chief on seeing the lizard made the funny remark, that, if that huge monster was no longer seen, it must have been eaten up by cats.

The outlet of lake Tarawera is on the Eastside; it is the Tarawera river or Awa o te Atua, running along the northern foot of the Putauaki mountain (Mount Edgumbe) and emptying into the sea near Matata on the Bay of Plenty. Besides numerous smaller tributaries it receives the discharges of five small lakes; from South-East the joint discharge of the Rotomahana and Roto-makariri (the warm and cold lakes); from North-West the waters of the Okataina and Okareka lakes,[1] and from the West the Wairoa river, which, flowing from the Rotokakahi, at a short distance from the missionary station forms a picturesque waterfall 80 feet high, and empties into the lake through a narrow gorge of rocks. At the shore of the lake there are various Maori settlements, the original names of which, similar to the Taupo settlements, were changed into biblical appellations, such as Ruakeria into Kariri (Galilee), Te Ariki into Piripai (Philippi). A large portion of the land about the lake is still densely wooded; the cultivated portions, on the other hand are said to be very fertile. The principal rock found in the Tarawera country is a finely granulated rhyolite containing black mica, which makes an excellent building-stone and looks very much like finely granulated granite. With the banks of crystalline

---

[1] The Okataina as well as the Okareka lake, both distant from the Tarawera lake about one mile, seem to be connected with the Tarawera lake only by subterraneous channels. The Okataina lake has a very irregular form, with numerous far-jutting promontories. The Okareka lake is the larger; it is about six miles in circumference, and about 60 feet above the Tarawera. It is shut in by wooded heights, has an oblong shape with a valley at each end, and a peninsula, upon which the Pa Taumaihi is situated, projecting into the middle of the lake. Its outlet flows underground for half a mile, and forms, where it comes to light again, the charming waterfall Waitangi.

structure, however, alternate banks of a hyaline texture, in which the mass assumes more or less the character of obsidian.

On the 28th April I started to the Rotomahana, — by land, because the gale blowing from South-West was too heavy, to cross the boisterous Tarawera in the small canoes. The distance is ten miles. The road leads over the heights on the South-shore of lake Tarawera. It is a much frequented foot-path, but very tiresome, because the traveller has continually to climb up and down over the broken ground. On the road we had a view of the regular volcanic cone Patauaki (Mount Edgumbe, 2575 feet high), a distance of about twenty miles to the North-East. The saddle-shaped excavation on the top indicates distinctly the existence of a crater, which, however, is extinct. At four p. m. we reached the northern shores of the far-famed Rotomahana (warm lake).

I do not think, that the impression the traveller receives at first sight of the small, dirty-green lake, — with its marshy shores, and the desolate and dreary-looking, treeless hills about it, covered only with a dwarfish copse of fern — corresponds in any degree with his previous expectations conceived from hearing so much about the marvels of this lake. So it was at least with us. The lake lacks all and every beauty of landscape scenery; that which makes it the most remarkable of all New Zealand lakes, — indeed we may well say, one of the most wonderful points of the world — must be observed quite closely, it being mostly hidden from the eyes of the traveller on his first approach. It is only by the steam clouds ascending everywhere, that he is led to suspect something worth seeing.

We found natives there with a canoe, as Mr. Spencer with friendly care had sent already a week ago provisions to the lake, supposing that we would travel directly from Taupo to the Rotomahana. In the canoe we crossed over to a small island in the lake, called Puai, and recommended to us by the natives as the best dwelling place during the time of our stay about the lake.

Puai is a rocky cliff in the lake, 12 feet high, 250 feet long, and about 100 feet wide. Manuka, grass and fern grow upon it, and for occasional visitors of the lake small raupo-huts have been erected, in which we made ourselves at home as well as circumstances would admit. I believe, however, that if we had not known that others before us had lived for weeks at that place, we should hardly have been induced to spend a single night there after a close examination of the spot. It is almost the same as living in an active crater. Round about there is a continual seething and hissing and roaring and boiling, and the whole ground is warm. In the first night, the ground upon which I was lying, grew gradually so warm from below, despite the thick underlayer of ferns and despite the woollen blankets, that composed my bed, that I started from my couch unable to bear it any longer. To examine the temperature, I formed with my stick a hole into the soft clay soil, and placed the thermometer into the aperture. It rose at once to boiling-point; on taking it out again, hot steam came hissing out, so that I hastened to stop the hole up again. In reality the island Puai is nothing but a torn and fissured rock, which, boiled entirely soft in the warm lake, threatens every moment to fall to pieces. Hot water bubbles up all around partly above partly below the surface of the lake. On the South-side of the island is a boiling mud-pool; blocks of silicious deposit, scattered about, point to large hot springs existing in former times, and even now-a-days hot steam escapes from numerous fissures. No fire is required here for cooking; wherever we dug but a little into the ground, or cleared the existing crevices of the crusts formed on them, there we could cook our potatoes and meat by steam. In some places the crevices are covered with sulphurous crusts and a strong smell of sulphurous acid was observed; in other places I found under cakes of silicious deposit films of fibrous alum. East of Puai and separated from it by a channel only 40 feet wide, is a second island Pukura (red lump). It is of the same description as Puai, smaller in circumference, but higher by several feet, and has likewise several huts, which some of my Maoris chose for their dwelling place. On these islands we had our headquarters during two days, and from them we undertook our excursions round the lake.

I will give a brief account of my observations, describing the principal springs more in detail with a view of furnishing a guide to the numerous visitors of the lake.

The Rotomahana is one of the smallest lakes of the lake district, not even quite a mile long from South to North, and only a quarter of a mile wide. According to my measurement it is 1088 feet above the level of the sea. Its form is very irregular, on the Southside, where the shore is formed by swamps, three small creeks are meandering and discharging themselves into the lake, the Haumi from Southwest the Hangapoua from Southeast, the middle creek without a name. In many places of those swamps warm water streams forth; hot mud-pools are also visible here and there; and from the projecting points muddy shallows covered with swamp-grass extend almost as far as the middle of the lake. At its Northend the lake grows narrow, and where the Kaiwaka creek flows out, there are again on both sides nothing but grass-swamps and shallows. Only in the middle the water is deeper; and the shores East and West are high and rocky. It justly bears the name of "warm lake." The quantity of boiling water issuing from the ground both on the shores and at the bottom of the lake, is truly astonishing. Of course, the whole lake is heated by it. But on making attempts to ascertain the temperature of the water, it is soon found to be very different in various places. Where the rising of gas-bubbles indicates a hot spring at the bottom of the lake, the thermometer will be often seen to rise to 90 or 100° F. Near the mouth of the cold creeks, the water of which showed a temperature of 50 to 52° F., the temperature is found to be only 60 to 70° F.; but in the middle of the lake and near its outlet 80° F may be considered as the mean temperature of the lake. In bathing and swimming through the lake, the change of temperature is very easily felt; but care must be taken not to come too close to any of the hot springs. The water

[1] A detailed map of the lake, upon which the various springs are marked, I have published in Dr. A. Petermann's Geographishe Mittheilungen, 1862, as well as in the Topographical and Geological Atlas of New Zealand, 1863, and in the Geology of New Zealand, 1864.

is muddy-turbid, and of a smutty-green color; neither fish, nor mussel-shells live in it. On the other hand the lake is a favourite haunt of countless water- and swamp-fowls. Various kinds of ducks, water-hens, the magnificent Pukeko *(Porphyrio melanotus)*, and the graceful oyster-catcher Torea *(Hæmatopus picatus)* enliven the surface of the water. These birds have their brooding-places on the warm shores, while they have to seek their food in the neighbouring cold lakes. In certain seasons of the year the natives institute regular hunts; at other times, however, they refuse every body, even Europeans, the pleasure of shooting, declaring the birds tapu. In former years, natives are said to have constantly dwelled about the lake; but of late, they seem to shun more and more this dismal laboratory of subterraneous forces, where rocks are dissolved in water, and rocks again are solidified from the water; and consequently the shores of the lake are usually uninhabited. Numerous observations lead to the conclusion, that constant changes are going on at the Rotomahana, that some springs go dry; others rise; and especially the earthquakes, which are felt here from time to time, seem to exercise such a changing influence. The main interest is attached to the Eastshore. There are the principal springs, to which the lake owes its fame, und I will now describe the principal springs in their succession from North to South on the East-shore of the lake, such as they are found by the side of the road generally taken during a short visit to the lake.

First of all is Te Tarata[1] at the N. E. end of the lake with its terraced marble steps projecting into the lake, the most marvellous of the Rotomahana marvels. About 80 feet above the lake on the fern-clad slope of a hill, from which in various places hot vapors are escaping, there lies the immense boiling cauldron in a crater-like excavation with steep, reddish sides 30 to 40 feet high, and open only on the lake-side towards West. The basin of the spring is about 80 feet long and 60 wide, and filled to the brim with

---

[1] Te tarata is said to signify "the tattooed rock;" it therefore seems to have its name from the peculiar forms and figures formed by the silicious deposits of the terraces. But Tarata is also the name of a tree, *Pittosporum crassifolium.*

Hamel photogr

Te Tarata on the Rotomahana Lake

Boiling spring with terraces of siliceous deposits

A. Hermann sc.

perfectly clear, transparent water, which in the snow-white incrustated basin appears of a beautiful blue, like the blue turquoise. At the margin of the basin I found a temperature of 183° F., but in the middle, where the water is in a constant state of ebullition to the height of several feet, it probably reaches the boiling-point. Immense clouds of steam, reflecting the beautiful blue of the basin, curl up, generally obstructing the view of the whole surface of water; but the noise of boiling and seething is always distinctly audible. Akutina (Augustus), the native who served me as guide, asserted that sometimes the whole mass of water is suddenly thrown out with an immense force, and that then the empty basin is open to the view to a depth of 30 feet, but that it fills again very quickly. Such eruptions are said to occur only during violent easterly gales. The confirmation of this statement would be of great interest. If it be true, then the Tetarata spring is a geyser playing at long intervals, the eruptions of which equal perhaps in grandour the famous eruptions of the great Geyser upon Iceland. The Tetarata-basin is larger than the Geyser-basin,[1] the mass of water thrown out, therefore, must be immense. The reaction of the water is neutral; it has a slight salty, but by no means unpleasant taste, and possesses in a high degree petrifying, or rather incrustating qualities. The deposit of the water is like that of the Iceland springs, silicious, not calcareous, and the silicious deposits and incrustations of the constantly overflowing water have formed on the slope of the hill a system of terraces, which, as white as if cut from marble, present an aspect, which no description or illustration is able to represent. It has the appearance of a cataract plunging over natural shelves, which as it falls is suddenly turned into stone. I had these terraces, which are truly unparalleled in their kind, photographed and drawn from various points of view, and the annexed

---

[1] The basin of the great Geyser is 58 feet in diameter, and has a depth of 6 to 7 feet. From the middle of the bottom a cylindrical, shaft-like tube, 12 feet wide at the top, but growing narrower as it descends, runs down to a depth of $75\frac{1}{2}$ feet. In the eruptions, which generally take place every 24 hours, the water-spout is thrown out sometimes to a height of 100 feet and more.

colour-print is a copy of those views, as true to nature as possible. In order to receive the full impression of that marvellous freak of nature, it is necessary to have climbed those steps, and to have studied the details of its structure.

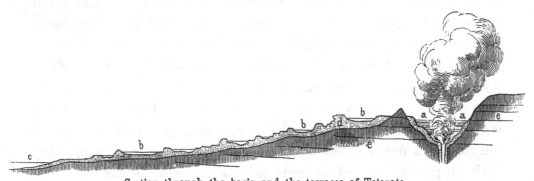

Section through the basin and the terraces of Tetarata.
a. Main-basin. *b.* Basins upons the terraces. *c.* Level of the Rotomahana. *d.* Silicious deposit. *e.* Ground consisting of decomposed rhyolite.

The silicious deposits cover an area of about three acres of land. For the formation of those terraces, such as we see them to-day, doubtless thousands of years were required. Forbes, judging by the thickness of the silicious deposits on the great Geyser of Iceland, which he estimates at 762 inches, and by the observation, that an object exposed to the discharge of the geyser-water for the space of 24 hours, is covered with a sheet of paper-thickness, has calculated the approximate age of the great Geyser at 1036 years. Similar calculations might be made also with regard to the Tetarata fountain, by examining the thickness of the silicious incrustations.

The flat-spreading foot of the terraces extends far into the lake. There the terraces commence with low shelves containing shallow water-basins. The farther up, the higher grow the terraces; two, three, some also four and six feet high. They are formed by a number of semicircular stages, of which, however, not two are of the same height. Each of these stages has a small raised margin, from which slender stalactites are hanging down upon the lower stage; and encircles on its platform one or more basins, resplendent with the most beautiful blue water. These small water-basins represent as many natural bathing-basins, which the most refined luxury could not have prepared in a more splendid and commodious style. The basins can be chosen shallow or deep, large or small, and of every variety of temperature, as the basins upon the higher stages, nearer to the main-basin, contain warmer water than those upon the lower ones. Some of the basins are so large and so deep, that one can easily swim about in them. In ascending the steps, it is, of course, necessary to wade in the tepid water, which spreads beside the lower basins upon the platform of the stages, but rarely reaching above the ankle. During violent water-eruptions from the main-basin, steaming cascades may occur; at ordinary times, but very little water ripples over the terraces; and only the principal discharge on the Southside forms a hot steaming fall. After reaching the highest terrace, there is a extensive platform with a number of basins, 5 to 6 feet deep, their water showing a temperature of 90 to 110⁰ F. In the middle of this platform, there arises, close to the brink of the main-basin, a kind of rock-island about 12 feet high, decked with Manuka, mosses, lycopodium and fern. It may be visited without danger, and from it the curious traveller has a fair and full view into the blue, boiling and steaming cauldron. Such is the famous Tetarata. The pure white of the silicious deposit in contrast with the blue of the water, with the green of the surrounding vegetation, and with the intensive red of the bare earth-walls of the water-crater, the whirling clouds of steam, — all together presents a scene unequalled in its kind. The scientific collector, on the other hand, has ample opportunity of filling whole baskets with the most beautiful specimens of the tenderest stalactites, of incrustated branches, leaves etc.; for whatever lies upon the terraces, becomes incrustated in a very short time.

From the foot of Tetarata fountain a path leads on the slope of the hill to the great Ngahapu (alias Ohapu). This fountain, shut up in by a thicket of shrubs, lies close by the margin of the shore, about 10 feet above the level of the lake. The colossal column of steam continually ascending from it betrays its locality from afar. The basin is oval, 40 feet long, 30 feet wide; the water in it clear and transparent, but nearly always in a terrific state of ebullition; it is but for some short moments, that the basin relapses into a calm repose; then it boils up madly anew, now more on this side, now on that, raising its foaming crest; the water is thrown up to a height of 8 to 10 feet and a terrific surf of boiling hot waves lashes the walls of the basin with wild up-roar, so that the observer stricken with awe shrinks back from the scene. But the elevated margin of silicious deposit prevents the free overflow of the water. The natives constructed an artificial outlet, conducting the water to several bathing-basins. The thermo-meter rose in the hot spring to 210⁰ F.; the water having a slight reddening effect on blue litmus-paper. The silicious deposits of the fountain have a smutty brown colour, and in the rear on the hill-side, where steam is rising from crevices and fissures, there also sulphur-crusts are found. The little Ngahapu, a basin in which turbid, muddy water is bubbling, lies farther up on the hill; but it is difficult of access.

Nearest to the great Ngahapu, more to the South, and close to the shore is Te Takapo, an incrustated basin, 10 feet long and 8 feet wide, with clear, gently boiling water of 206⁰ F. Some-times this fountain is said to rise in a jet 30 to 40 feet high. The numerous smaller springs, bubbling mud-pools, and tabular, lightly incrustated holes along the shore between Tetarata and Tetakapo bear no special names. Near Tetakapo there are some deserted huts, and a few yards farther there is a ravine, called Waikana-panapa (the coruscating water), extending in a N. E. direc-tion for a short quarter of a mile, in the background of which lies the Rotopunamu (green lake). The entrance to the ravine is overgrown with thicket, and rather difficult; it also requires con-siderable caution, because very suspicious places have to be passed, where the traveller is in danger of being swallowed up in heated mud. The inside of the ravine has the appearance of a volcanic crater. The bare walls, utterly destitute of vegetation, are terribly fissured and torn; odd-looking, rocky serratures threatening every moment to break loose, loom up, like dismal spectres, from red, white and blue fumarole-clay, — evidently the last remains of the decomposed rocks. The bottom of the ravine is formed by fine mud, and thick, burst and broken plates of silicious deposit lie scattered about, like cakes of floating ice after a thaw. Here there a big cauldron of mud is simmering, there a deep basin full of water is boiling, next to this lies a terrible hole emitting hissing jets of steam, and farther-on small mud-cones are seen, from two to five feet high, vomiting forth, volcano-like, from their craters hot mud with a deadened rumbling, and imitating on a small scale the play of large fire-volcanoes. Quite in the background, perhaps, 100 feet above the level of the Rotomahana, lies the green lake. It was a dirtyish-green water-basin, 40 feet in diameter; the reaction of the water was acid; its temperature 63° F. The basin

The mud-cones of the Rotomahana.

was surrounded by a flat, and partly broken shell of silicious deposit, and seemed to me to belong to an extinct fountain.

Towards South, at the mouth of the ravine, there lies quite picturesque between rocks and bushes, about 40 feet above the level of the Rotomahana, the fountain Ruakiwi (Kiwi-hole), a basin 16 feet long and 12 feet wide, with clear water of 210° F, which is in a constant state of gentle ebullition. The stony deposit, — lacking, however, the beauty of the Tetarata terraces, — extends as far down as the lake, at the shore of which there is another smaller fountain, Te Kapiti.

From the Waikanapanapa valley, opposite to the two islands Pukura and Puai, the shores of the lake become steep and rocky; hot springs gush out from below, under the surface of the lake, while above on the side of the hill the huts of the deserted settlement Ngawhana (or Ohana) lie scattered by the side of the spring bearing the same name. Here the natives, probably for bathing-purposes, have constructed square-basins of sinter plates and connected them by gutters with the springs on the hill-side. The flat stones, which are laid across hot places, are said to have served for the drying and roasting of Tawa-berries (of *Laurus tawa*). To the Ngawhana group belong several hot springs, which have all a share, more or less, in the silicious deposits covering the slope of the hill. Ngawhana itself is a quiet, hot water-basin without any special pecularities. Farther up, about 100 feet above the lake, is the Koingo (the sighing),[1] an intermittent fountain, from which discharges of water take place three or four times a day, alternating, as the natives say, with those of the neighbouring Whatapoho. The basin of that fountain, 9 feet long and 5 feet wide, is encompassed by a crust of silicious deposit. When I stepped up to the margin of the basin, the water was quiet and but slightly steaming. It was so low, that no discharge took place through the gutter laid by the natives. Suddenly, however, it commenced to stir again; the water rose, soon the whole basin was filled to the brim, and finally boiled over in a surf bubbling up to a height of 3 or 4 feet. This lasted about ten minutes; then the fountain began to subside again; a low rumbling was heard, as if the water were receding through a small tube, and the water in the basin stood again low and quiet as before. In this state it had a temperature of 202° F.

A few yards apart there is a fountain-basin about 16 feet deep, contracting like a funnel as it descends; this, years ago, is said to have been an intermittent geyser, but at the time of my visit it was entirely empty. Not far from it is the Whatapoho, one of the most remarkable points on the lake, part fountain, part solfatara, and part fumarole, or rather all three in one. From a deep, shaft-like aperture between brittle, ash-coloured rocks there streams, as from a steam-boiler, hot steam and sulphurous gas with a dismal, moaning sound. It being too dangerous to approach it quite closely, there is no chance of looking down into the depths of the shaft; but it may be easily seen, how the steam-jet is occasionally also throwing out spouts of water. The vegetation is withered

---

[1] So called from the sighing sounds heard on the water relapsing into the basin.

away all around, sulphur-crusts and silicious deposits cover the ground and the crevices of the rocks. Sometimes the Watapoho is said to throw out a boiling water-column to a considerable height.

The springs hitherto mentioned are only the principal springs on the Eastside of the Rotomahana; they all lie on the slope of a hill — rising about 200 feet about the level of the lake, overgrown with fern and Manuka — which steams on more than a hundred other places, and the rhyolithic rocks of which are totally decomposed by the hot vapours into a more or less ferruginous mass of clay, from which all traces of the original structure of the rock have been effaced. South of the steaming hill the shores are flat. Here, on the Southeast side of the lake lies the Wakaehu (water in motion), and a number of boiling springs bubbling forth from the sand and mud of the shore, partly with clear and partly with muddy water. There are also three lagoons in those flats: Rangipakaru, Te Ruahoata[1] and Wairake; and in the rear, there rises an isolated hill, Te Rangipakaru (broken sky), on the Westside of which, from a crater-shaped excavation, a powerful solfatara steams forth, depositing great quantities of sulphur. The Southside of the lake does not present a single spring of any noteworthy size.

On the West-shore, the great terrace-fountain Otukapuarangi[2] is a side-piece to Tetarata. The terraces extend as far as the lake, and the ascent is made, as it were, by artificial marble-steps, decked on both sides with Manuka, Manuwai, and Tumingi bushes. The terraces, although not so grand as at the Tetarata, are, on the other hand, neater and finer as to their structure. Besides, a light pink hue imparts an intrinsic charm to the marvellous structure. The platform is at a height of about 60 feet above the lake; it is 100 yards long, and equally broad. It contains neat basins three or four feet deep, full of transparent, sky-blue water of 90 to 110° F. In

[1] The hole of Hoata; Hoata is the name of one of the Taniwhas, who carried the sacred fire to the Tongariro.

[2] Otukapuarangi means cloudy atmosphere, from the continually ascending steam-clouds. Taylor writes Tutupuarangi.

the background, shut in by half naked walls tinged with various colours, red, white, and yellow, lies the large spring-basin, a great cauldron 40 to 50 feet in diameter, and probably very deep. The water is usually quiet, shining blue, merely steaming, but not boiling, and has a temperature of 179° F.; the ascending vapours smell of sulphurous acid. Round about the basin a light tincture of sulphur is also noticed, and here and there on the side-walls of the crater thick crusts of sulphur have been deposited. At the northern foot of the terraces is the great solfatara Te Whaka-taratara,[1] a crater-shaped hole, open towards the lake and full of hot, yellowish-white, muddy water, of a strong acid reaction, — a real sulphur-lake, from which a hot, muddy stream flows into the Rotomahana. In the crevices of the walls enclosing the sulphur-lake very fine sulphur-crystals are to be found.

A little beyond the lake, in a small side-valley, lies the Ate-tuhi; and in the marshy flats, at the N. W. end of the lake, the Te-Waiti fountain. Also at the outlet of the Rotomahana towards lake Tarawera, on both sides of the Kaiwaka river, there are still numerous Ngawhas observed, which have special appellations such as Te aka manuka, Te mamaku, Te poroporo, Tamariwi, Ma-krowa, Te karaka etc., which, however, I had no opportunity of examining more closely.

Altogether about 25 large Ngawhas may be counted on the Rotomahana; the number of smaller springs, coming to light at in-numerable places upon an area, occupying about two square-miles, I do not even dare to estimate. As those hot springs according to the experience of the natives have proved very effective in the curing of chronic cutaneous diseases and rheumatic pains, I have no doubt, that, at no very distant period, this remarkable lake will become the centre of attraction not only for tourists of all nations, but also a place of resort for invalids from all parts of the world. Moreover the Rotomahana has been repeatedly used as a climatic place of cure. Mrs. Spencer told me, that some years ago she took fifteen Maori children, affected with the whooping-cough, to

[1] The name refers to the broken, rugged appearance of the cliffs.

the Island Puai, and that in its uniformly warm temperature they all recovered in a very short time. I regret to say that my stay upon the island was too short, and I was too much engaged in other observations, to find time for making comparative thermo-metrical observations. [1]

Before leaving lake Rotomahana, the warm lake, however, I must also make mention of the Rotomakariri, or the cold lake, situated East of the Rotomahana at the foot the Tarawera moun-

View of the Rotomakariri (cold lake) with the Tarawera mountain.

tain. It is smaller than the Rotomahana and shows very remark-able circular coves reminding me of the circular tuff-crater basins in the vicinity of Auckland. However, having seen the lake only from the heights above the Rotomahana, I am not able to say, how that phenomenon is to be explained. East of the Rotomakariri, there is another little water-basin, surrounded by swamps, and the joint outlet of both is the Awaporohoe, which unites with the Kai-waka river between the Rotomahana and Tarawera. A third, much larger lake, about three miles long and one mile wide, by the natives called Rere-whaka-Aitu, is said to lie in the Kaingaroa plain about five miles East of Rotomahana; a lake, of which as

[1] The only observations contained in my journal concerning the atmospheric temperature upon Puai, measured in the shade in front of my hut, are as follows:

April 28. 5½ p. m. 54⁰ F.	*April 29.* 8 a. m. 50⁰ F.
6 p. m. 53⁰ F.	5½ p. m. 52⁰ F.
8 p. m. 53⁰ F.	8 p. m. 51⁰ F.
9 p. m. 53⁰ F.	*April 30.* 7½ a. m. 50⁰ F.

On the 29. and 30. April a heavy, rough Southwest wind was blowing.

yet no mention has been made in any account of travels, or on any map.

Our return-trip to Temu, on the 30. April, we made by water. The Kaiwaka,[1] the outlet of the Rotomahana, is the water-route into lake Tarawera. It empties into the South-East cove of the lake, called Te Ariki, which is very picturesquely surrounded by steep, rocky shores. Near the Pah Maura on a peninsula stretching far into the lake, the river turns into the great lake, and after a passage of one and a half hours we reached the landing at Temu, at the mouth of the Wairoa river. On our return we had the pleasure of becoming acquainted with Mr. Spencer, an American by birth; and the constant heavy rain preventing our departure, we spent two more days with that amiable, hospitable family. This delay was quite welcome to me, because by it I found time to finish the topographical sketches, deriving great benefits for my purpose from the excellent local knowledge of Mr. Spencer. My botanical collections were considerably increased by the kind contributions of the amiable daughter of the house.

On the 3. May we took leave, and set out to lake Rotorua, the second in size in the lake-district. Our intention to shorten the distance by passing across the Rotokakahi to the West-cove of that lake, we were obliged to renounce on account of the boisterous West-wind which churned the little lake into madness; we took the road by lake Tikitapu, a small, deep-blue lake of triangular shape, and about one mile long, picturesquely situated between steep, partly wooded heights. It is separated from the Rotokakahi only by a narrow range of hills, probably keeping up the connection with the same by means of a subterraneous water-course. On the North-side of the lake we came into bush for a short distance, then over fern-hills and through small, grass-covered valleys running in various directions, into the Waipa plain; thence we had to ascend a low, woodless range, and the picturesque Rotorua with the hot springs

---

[1] The word signifies canoe-eater, or canoe-destroyer, probably from the rapids which are to be passed in the river, as canoes there strike the ground and are easily damaged.

View of the lake Rotorua, Prov. Auckland.

of Whakarewarewa in front, lay spread before our eyes. Here I made halt for the purpose of making observations for the map, and committing the beautiful scenery to paper. The height bears the name of Te whaka-he kinga-whaka, and for the benefit of the lovers of interesting specimens of volcanic rocks I add, that the rocks along the road consist of obsidian, containing numerous, lavender-blue spherolites. This kind of rock is found over a large portion of the lake-district (the Ngongotaha mountain near the Rotorua also consists of it); but at the place above mentioned it is easy to knock off good specimens.

Rotorua means hole-lake or a lake lying in a circular excavation. With the exception of the southern bight, called Te arikiroa, it has an almost circular form with a diameter of about six miles and a circumference of twenty miles. Almost in the precise centre of the lake the island Mokoia is situated, formed by a conical hill rising about 400 feet above the level of the lake and with a Pah on its top. The circular form of the lake, the island in the middle, the white steam-clouds ascending along the shores, all this might easily induce the observer to take the Rotorua to have formerly been a volcanic crater, while in reality this lake, like all the other lakes of the lake-district, was produced by the sinking of parts of the ground upon the volcanic table-land. The depth of the lake is comparatively but small, perhaps at no place more than 5 fathoms; it has numerous shallow sand-banks, and the shores also, with the exception of the North-side, are sandy and flat. It is 1043 feet above the level of the sea, consequently of the same height with the Tarawera lake. On the Southwest-side, the wood-clad Ngongo-taha mountain towers up to a height of 2282 feet. This is the highest point of the range of hills encircling the lake. Among the numerous tributaries, the Puarenga river emptying into the lake on the Southeast-side near Whakarewarewa, is probably the most con-siderable. On the North-side the Ohau Creek forms the outlet of the lake to the Rotoiti, thus connecting the two lakes separated only by a low and narrow isthmus.

The principal native settlement is Ohinemutu, situated at the northwestern extremity of the lake; it is a famous old Maori Pah; famous for its inhabitants, and famous for its warm baths. At a distance of three miles from the village, my Maoris had kindled a big fire upon the height, the smoke of which, according to Maori custom, announced our coming to the inhabitants of the Pah, the chief of which had previously invited me by letter to pay him a visit. We hastened past the hot springs of Wakarewarewa, and made straight-way for the Pah. Pini te korekore, the gallant chief of Ohinemutu, accompanied by his special friends, came forth to meet me with due solemnity. He was dressed in a fine European suit, had a cloak wrapped about him, a straw-hat upon his head, and carried a white flag in his hand with the inscription in blue letters: *"Sancta Maria, ora pro nobis."* After we had approached each other within about 20 yards, he stopped. Bowing very low, and taking off his hat, he shouted to me with a loud voice the usual salutation: "haeremai" (come). I went, shook hands with the chief, exchanging some complimentary words, and then was conducted into the Pah. I found the tent pitched, the meal ready; for such is Maori custom. When they see from afar friends or strangers coming, the women of the village commence at once peeling potatoes, and by the time the ceremony of salutation, accompanied by the usual Tangi (cries of lamentation), is over, the dinner is ready.

Pini te korekore upon better acquaintance became extremely talkative. He had heard of the solemn reception, which the Roman Catholic Bishop Pompallier — his master — had been honoured with on the occasion of his visit on board the Novara in Auckland, and was very anxious to have me relate to my friends that he — a pupil of the Roman-catholic mission school — had come to meet me with the catholic banner. He was about thirty years of age, had only the lower part of his face tattooed, and had, both in his exterior appearance and manners, adopted a great many of the peculiarities of his French masters. Till late in the evening he continued telling me of the country and its inhabitants, and of the murderous wars and bloody fights, that were fought here.

May 4. — At breakfast-time, Pini te korekore again made his appearance in my tent, requesting with a solemn air my presence at a most peculiar ceremony. Round about in a circle sat the men of the village. In the middle was a fat pig tied to a post; by the side of it stood six baskets full of potatoes and kumaras, and a heavy sack of flour lay across them. The chief stepped forth into the circle, addressed some kind words to the assembled audience, and by his touching with a rod the pig, the potatoes and the sack of flour, at the same time pronouncing my name "Te Rata Hokiteta," they were dedicated to me and my companions, according to Maori custom, as a present offered in his own and his people's name. Captain Hay in the Maori language returned in my name our sincerest thanks for the hospitable gift.

After breakfast I ascended the Ngongotaha mountain and on arriving at the top I was greatly rejoiced to have an extensive view in all directions reaching to the shores of the Bay of Plenty, and as far as the volcanic island Whakari (White Island), which was seen emitting immense clouds of white steam. The afternoon was devoted to the inspection of the Pah and the examination of the hot springs.

Ohinemutu still bears to some degree the features of an old Maori Pah. The dwellings of the chiefs are surrounded with enclosures of pole-fences; and the Whares and Wharepunis, some of them exhibiting very fine specimens of the Maori order of architecture, are ornamented with grotesque wood-carvings. The annexed wood-cut is intended as an illustration of some of them. The gable-figure with the lizard having six feet and two heads, is very remarkable; at no other place in New Zealand have I seen a similar representation. The human figures are no idols, but are intended to represent departed sires of the present generation. The huts of the village are scattered over a considerable area on both sides of the Ruapeka Bay and on the slope of the hill Puke Roa arising to about 150 feet above the lake. Ohinemutu has a protestant and a catholic church. The place is noted far and near for its hot springs and for its excellent warm baths (Waiariki).

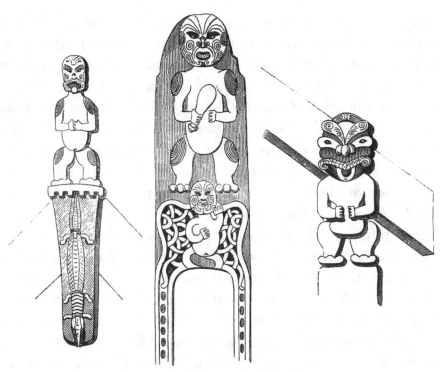

Carvings at Ohinemutu on lake Rotorua.

Ruapeka Bay forms the centre of the hot springs. There it seethes and bubbles and steams from hundred places. The principal spring is the Great Waikite at the South-side of the bay. The basin of the fountain communicates with the lake, and it is to the immense quantities of hot water issued forth here, that the whole bay becomes warm and forms an excellent bathing-place. By approaching the fountain more or less, any degree of temperature may be chosen. The water of the fountain is perfectly clear. For some short moments all is quiet in the large basin, only white steam-clouds ascending from it; then a powerful ebullition succeeds in raising the water to a height of four to six feet, sometimes even to ten and twelve feet. Little Waikite, a few yards above, forms a basin four to five feet wide, in which the water rises about every five minutes several feet high, sinking down again during the intervals to a depth of six to seven feet. The temperature I found to be 201° F. In going about between the countless pools of boiling sputtering mud and water, the greatest care has to be taken.

Whoever has once involuntarily bathed his feet in steaming water or boiling mud, will certainly remember it all his life. That even more serious accidents are of no rare occurrence, is proved by several monuments in the shape of figures carved of wood, which are posted in those places where persons have met with an untimely death.

From the Ruapeka Bay the hot springs continue in a south-western direction on the foot of the Pukeroa, along the Utuhina Creek as far as the small settlement Tarewa. In this direction there are moreover two small warm ponds, Kuirau and Timara, fed by hot springs, both favorite bathing-places of the natives. Also on the South- and East-sides of the Pukeroa steam is seen to ascend

The Pah Ohinemutu on the Rotorua.

from various places. Tabular blocks of silicious deposit, two to three feet thick, of a mass resembling milk-opal, lie scattered about over the slope and the base of the hill, indicating, that the activity of the springs in former periods especially on the East-side of the hill was still far more extensive than now, or that the springs change their place from time to time. The natives have special springs for bathing, for cooking and also for washing. On places, where only hot vapour escapes from the ground, they have established vapour-baths, and upon heated ground they have warm houses for the winter season, of which it is said, that no vermin of any kind is able to exist in them. The whole atmosphere in and about Ohinemutu is so constantly impregnated with watery vapours and sulphurous gases as to make them plainly perceptible to the sense of smell. This, however, seems only to improve the physical condition of the inhabitants; for they are known to be an extraordinarily robust set of Maoris. Haupapa, a renowned chief of Ohinemutu, who was absent at the time of my visit, but whom I happened to meet at Maketu, is a giant in height and a Hercules in strength.

May 5. With three hearty cheers we took leave of the kind and hospitable people of Ohinemutu. The gallant Te Pini, in a large war-canoe, took one half of my company to the Island Mokoia[1] and thence to Te Ngae, a former Mission station on the Northeast-side of the lake; myself, with the other half on foot went the circuit of the lake in order to have an opportunity for farther observations.

Two and a half miles distant from Ohinemutu in a south-easterly direction is the native settlement Whakarewarewa. Lying at some distance off the direct road from Ohinemutu to Lake Tara-wera, it is generally skipped over by tourists, but the springs here exceed those of Ohinemutu in variety and extent. The principal springs are on the right bank of the Puarenga Creek. Seven or eight of them are periodical geysers, having, however, their own, as yet unexplored caprices, as they are not always obliging enough to satisfy the curiosity of visiting travellers. It is said to happen

---

[1] The Island Mokoia was inhabited in 1859 by 40 Maoris; there are numerous hot springs on the island, on the S. W. side Waikimihia; on the Eastside Kaiweka and Kapoao; on the N. W. side Paipairau. On the N. E. side of the island a low isthmus is extending far into the lake. Mokoia, in the Maori wars, had always been a safe place of refuge for the dwellers on Lake Rotorua; but Hongi, who brought his own canoes with him from the Eastcoast overland to the lake, knew how to pursue his enemy even to that spot. The Ngawha of Hinemoa is pointed out by the natives as the spot, in which Hinemoa, the traditional ancestress of the Ngatiwhakaue tribe, warmed herself after swimming from the mainland to her lover Tutanekai, and on that account is held tapu to this day. A large stone — on which Hinemoa sat listening to Tutanekai playing his flute previously to her swimming to him on Mokoia — is pointed out on a headland on the eastern shore near the South-east corner of the lake.

now and then, that they play all together; the natives assert, that such is generally the case during heavy easterly gales. I was not fortunate enough myself, to witness such a grand spectacle, but had to content myself with observing a small eruption of one of them, the Waikite. It issues from the top of a flat silicious cone,

Waikite, intermittent fountain at Whakarewarewa on the Rotorua.

measuring 100 feet in diameter and 15 feet high, which rising between green manuka and fern-bushes, presents an extremely picturesque sight. The cone consists of white silicious deposit; it has numerous fissures and crevices, which are all incrusted with neat sulphur-crystals; the hot vapours, however, issuing from those fissures, smell neither of sulphurous acid, nor of sulphuretted hydrogen, but merely of sublimated sulphur. At intervals of about eight minutes the Waikite throw out a column of water two or three feet thick to a height of six to eight feet. But in January and February, Mr. Spencer told me, it shows itself in its full glory, spouting to a height of 30 to 35 feet. A little S. E. of the Waikite is the Pohutu. [1] Its basin is 12 feet wide; the masses of sili-

---

[1] Comp. the view of the Pohutu in chap. II. p. 44.

cious deposit surrounding it are very extensive, and piled up to a height of more than 20 feet, fissured and broken by numerous cracks; the sulphur-deposits are here still more distinct than on the Waikite. Parikohuru and Paratiatia are the names of the springs, supplying the large bathing-basins of 50 and more feet diameter, in which the natives, men and women promiscuously, bathe for hours, all cosily smoking their pipes and chatting together. The range of hot springs extends from Whakarewarewa along the course of the Puarenga river, a distance of one and a-half miles, to Tearikiroa Bay on lake Rotorua. The number of smaller springs, of boiling mud-basins, of mud-cones and solfataras, which are scattered over this extensive area, must be counted by hundreds and I will make special mention of only two points more. Near the peninsula Motutara on the West-side of Arikiroa Bay, there is a basin, 16 feet long and 6 feet wide, by the natives called Oruawhata, full of hot water with a temperature of 185° F. and a neutral reaction; close by, a cold waterbasin of 55° F., 80 feet long and 14 feet wide, contains yellowish-white water, acidulated with sulphurous acid, and of a strong acid reaction. The Arikiroa Bay also has yellowish-white water of an acid reaction; numerous sulphur-crusts, the yellow hue of which upon the white sand-beach of the shore is visible at a great distance, and a strong smell of sulphuretted hydrogen indicate from afar the solfataras on the shores and at the bottom of said bay. All these phenomena, however, cease after passing the mouth of the Puarenga Creek, and along the flat East-shore of the lake as far as Te Ngae there are no more hot springs to be met with. Two small lakes are situated on this side, the Rotokawa, a muddy basin in the immediate vicinity of lake Rotorna; and a little farther off between the wooded hills on the Northeast-side the Rotokawau of the size of lake Tikitapu.

Te Ngae, formerly the residence of the Rev. Mr. Chapman, who has of late settled at Maketu, we found deserted. The grounds around the old Mission house are planted with a great variety of fruit trees, and next to it there is a small settlement Waiohewa with a mill, turned by the Te Ngae Creek. The shores of the

lake near Ngae are formed by vertical sandstone and pumice-
stone crags, 20 or 30 feet high, which are more and more under-
washed by the heavy surf raised by West- and South-West-
winds. I came just in time, to see Te Pini's canoe, with my
travelling-companions on board, bravely struggling through the
waves of the lake lashed by the howling gale, and landing at
length in safety.

After one hour's repose and after a repeated solemn farewell
to Te Pini we set out from Waiohewa, taking a Northeast course
along a marshy valley, for the purpose of visiting the great Ngawhas
or solfataras situated upon the pumicestone-plateau between the
lakes Rotorua and Rotoiti. They are a peculiar group for themselves. Comparable to hideous carbuncles on the surface of a body,
those solfataras — holes, rotten more or less deeply into the ground,
surrounded with yellowish-white crusts, and diffusing an offensive
odour — lie bedded among the green fern-lands. Their list opens
with Tikitere, not merely a single pool of sulphur, but a whole
valley of solfataras, bubbling mud-pools and hot springs. In the
middle is a water-basin, 50 to 60 feet in diameter, called Huritini;
it seethes and boils and bubbles in all corners, the turbid and muddy
water sometimes rising to a height of 12 to 15 feet. Pumicestone-
sand cemented with silicious deposits, sulphur-crusts and black mud
form a very suspicious soil around, which can only be stepped upon
with the greatest caution. The atmosphere is impregnated with
sulphuretted hydrogen and sulphurous acid; dense clouds of steam
whirling up from the dismal haunt. North of Tikitere are the solfataras Karapo, Te Korokoro, Te Waikari and Te Terata; next,
Harakeke-ngunguru, Tihipapa and Papakiore; and finally Ruahine.

Ruahine (from rua, hole, and hine, wife) has the appearance
of an active crater. The crater-shaped basin lies on a declivity
sloping towards lake Rotoiti; at its bottom black mud may be seen
boiling, which by the rising and bursting steam-bubbles is sputtered
up into the air to a height of several feet. The column of steam
ascending here is designated by the natives Te Wata-Kai-a-Puna-
kirangi, meaning the place where the meal is hung up for Puna-

kirangi. Yellow masses of sulphur are sticking to the many-coloured clay-soil. Black, muddy water flows out of the mud-basin; the valley in front of the basin is covered with sulphur and sinter-crusts, emitting steam from more than hundred perforations. Here also great care is to be taken by the traveller, lest he breaks through into boiling mud. The apertures emitting pure steam the natives use for cooking.

The solfatara Ruahine on lake Rotoiti.

Towards evening we reached the southern shore of lake Ro-toiti, and pitched our tents near the Maori village Pukeko at the peninsula Te mihinga-a-terangi-tapu.

May 6. — Rotoiti or the "little lake," is of a very irregular shape, from West to East about six or seven miles long, and only from one to two miles wide. As regards the character of its scenery it is decidedly one of the most beautiful lakes. Picturesque pro-montories and peninsulas jutting far into the lake separate the various branches and inlets from each other. On the West-side it is separa-ted from the Rotorua only by a narrow isthmus scarcely half a mile broad, upon which the Pah Morea stands. The Ohua Creek flowing from the Rotorua into the Rotoiti, connects the two lakes. From the northwestern end of the Rotoiti issues the Okere river, which flows into the sea near Maketu on the East-coast. As towards

West from the Rotorua, so the Rotoiti is towards East separated by a low narrow isthmus from the Rotoehu or Roto-ihu (nose-lake). The Rotoehu is said to be three miles long and one and a half miles wide. Its outlet, the Waihi Creek, does not come to light until after a subterraneous course at some distance from the lake. North-East from Rotoehu there is moreover the small lake Rotoma (white lake), so called from its white sandy beach; it is about one mile long, and the last of the lakes in the lake-district, which I have to enumerate. The last hot springs, exclusive of the island Motu Hora and the island-volcanoe Whakari[1] in the Bay of Plenty, are said to be found on the shores of the Rotoehu.

The Fauna of all those lakes is identical with that of lake Taupo. They all are likewise characterized by their total lack of eels.

Having thus given a description of the principal ones of the Thousands of Puias and Ngawhas on the North Island, I will now say few words about general features and about the origin of the springs. We can distinguish three parallel lines of springs, striking in the direction of N. 36° E. One line connects the two volcanoes Tongariro and Whakari. On this line are situated the hot springs of Lake Taupo, the fumaroles of the Kakaramea mountain and the hot springs round Rotomahana. The second line is the line of the Puias of Orakeikorako and of the Pairoa-range; and to the third line belong the hot springs of the Rotorua and the solfataras of the Rotoiti. The chemical and mechanical processes, as

[1] Upon Motu Hora (Whale Island), 4 sea-miles from the coast, 467 feet high, and formerly inhabited, there are said to exist numerous hot springs. Whakari (White Island), 28 sea-miles from the coast and 820 feet high, is the noted point mentioned next to Tongariro among the active volcanoes of New Zealand. The crater, the bottom of which is level with the sea, has a circumference of 1½ miles. In its middle there is a hot lake, an immense solfatara, 300 feet in circumference, puffing in calm weather dense, white clouds of vapour into the air to a height of full 2000 feet. Round about there are numerous fumaroles, from which hot steam is escaping with an immense force and a hissing uproar; and the whole bottom of the crater is covered with sulphur-crusts. Polack (N. Z. Vol. I. p. 329) mentions, that in 1837 he saw also black smoke ascending from the Whakari, and that at night he had observed the glare of fire. There are, however, no accounts extant of lava eruptions.

exhibited by the New Zealand hot springs are throughout similar to those of the hot springs upon Iceland. The interesting results, to which the investigations of Krug von Nidda, Sartorius von Waltershausen, von Bunsen and others have led upon Iceland, may therefore be mostly applied also to New Zealand. As to the mechanical processes, the intermittent springs, or those which at certain periods display an increased state of ebullition, — raised sometimes to regular, geyser-like water eruptions, — may be distinguished from the permanent springs, the surface of which is in a state of constant repose, or in a uniform state of ebullition. The former may, as I have explained in the beginning of this chapter, be designated as Puias, the latter as Ngawhas. With this distinction corresponds moreover the chemical difference of alkaline and acid springs.

Both kinds of springs owe their origin to the water, permeating the surface and sinking through fissures into the bowels of the earth, where it becomes heated by the still existing volcanic fires. High-pressure steam is thus generated, which accompanied by volcanic gases, such as muriatic acid, sulphurous acid, sulphuretted hydrogen and carbonic acid, rises again towards the colder surface and is there condensed into hot water. The overheated steam however, and the gases decompose the rock beneath, dissolve certain ingredients, and deposit them again on the surface. According to Bunsen's ingenious observations, a chronological succession takes place in the cooperation of the gases. The sulphurous acid acts first. It must be generated there, where rising sulphur-vapour comes into contact with glowing masses of rock. Wherever vapours of sulphurous acid are constantly formed, there acid springs or solfataras arise. Incrustations of alum are very common in such places, arising from the action of sulphuric acid on the alumina and alkali of the lavas; another produce of the decomposition of the lavas is gypsum (or sulphate of lime), the residuum being a more or less ferruginous fumarole-clay, the material of the mudpools. To the sulphurous acid comes sulphuretted hydrogen produced by the action of steam upon sulphides, and by the mutual decomposition of the sulphuretted hydrogen and the sulphurous acid sulphur is formed, which in all solfataras forms the characteristic precipitate, while the deposition of silicious incrustations is either entirely wanting or quite inconsiderable, and a smell of sulphuretted hydrogen is but rarely noticed. These acid springs have no periodical outburst of water.

In course of time, however, the source of sulphurous acid becomes exhausted, and sulphuretted hydrogen alone remains active. The acid reaction of the soil disappears, yielding to an alkaline reaction by the formation of sulphides. At the same time the action of carbonic acid begins upon the rocks, and the alkaline bicarbonates thus produced dissolve the silica, which, on the evaporation of the water, deposits in the form of opal or quartz or silicious earth, and thus the shell of the springs is formed, upon the structure of which the periodicity of the outbursts depends. Prof. Bunsen — rejecting the antiquated theory of Mackenzie, based upon the existence of subterraneous chambers, from which the water from time to time is pressed up through the vapours accumulating on its surface, according to the principle of the Heron-fountain — has proved, in the case of the great Geyser, that the periodical eruptions or explosions essentially depend upon the existence of a frame of silicious deposits with a deep, flue-shaped tube, and upon the sudden development of larger masses of steam from the overheated water in the lower portions of the tube. The deposition of silica in quantities sufficient for the formation of this spring-apparatus in the course of years, takes place only in the alkaline springs. Their water is either entirely neutral or has a slightly alkaline reaction. Silica, chloride of sodium, carbonates and sulphates are the chief-ingredients dissolved in it. In the place of sulphurous acid, the odour of sulphuretted hydrogen is sometimes observed in those springs.

The rocks, from which the silicious hot springs of New Zealand derive their silica, are rhyolites and rhyolithic tuffs containing seventy and more percent of silica; while we know that in Iceland palagonite and palagonitic tuffs with fifty percent of silica are considered as the material acted upon and lixiviated by the hot water. By the gradual cooling of the volcanic rocks under the surface of the earth in the course of centuries the hot springs also will gradually disappear. For they too are but a transient phenomenon in the eternal change of everything created.

Appendix.

Chemical composition of the water and the silicious deposits of some of the hot springs; the analyses executed in the laboratory of Professor Dr. v. Fehling at Stuttgart.

1. Water.

No. 1. Te Puia-nui, near Tokanu on lake Taupo; reaction alkaline; analyzed by Dr. Kielmaier.

No. 2. Tetarata on the Rotomahana; reaction neutral; analyzed by Mr. Melchior.

No. 3. Ruakiwi on the Rotomahana; reaction neutral; analyzed by Mr. Melchior.

No. 4. Rotopunamu on the Rotomahana; reaction neutral; analyzed by Dr. Kielmaier.

In 1000 parts of the waters there were contained, of:

	in No. 1.	2.	3.	4.
Silica	0.210	0.164	0.168	0.231
Chloride of Sodium	4.263	2.504	1.992	1.192
Total residuum	4.826	2.732	2.462	1.726

Owing to the small quantities of water for analysis (one bottle of each), only silica, chlorine — computed as chloride of sodium — and the total amount of non-volatile ingredients could be quantitatively ascertained. Qualitatively, however, the presence of magnesia, lime, sulphuric acid and traces of organic substances has also been proved.

2. Silicious deposits of hot springs on the shores of the Rotomahana, analysed by Mr. Mayer.

No. 1. Tetarata, two samples, *a*) an earthy powdery mass; *b*) solidified incrustation.

No. 2. Great Ngahapu.

No. 3. Whatapoho.

No. 4. Otukapuarangi.

	1.		2.	3.	4.
	a.	b.			
Silica	86.03	84.78	79.34	88.02	86.80
Water and organ. subst. .	11.52	12.86	14.50	7.99	11.61
Sesquioxide of Iron . . . }	1.21	1.27	1.34 }	2.99	} slight indi-
Alumina . . . . }			3.87 }		cation.
Lime	0.45 }		0.27 }		} slight indi-
Magnesia	0.40 }	1.09	0.26 }	0.64	cation.
Alkalis	0.38 }		0.42	0.40 }	

I. Pattison (Philos. Magazine 1844. p. 495) and

II. Mallet (Philos. Magazine 1853. V. p. 285) give the following analyses of the silicious deposits on the hot springs of Lake Táupo, without, however, specifying the localities.

	I.	II.
Silica	77.35	94.20
Alumina	9.70	1.58
Sesquioxide of Iron .	3.72	0.17
Lime	1.54	Indic.
Chloride of Sodium .	—	0.85
Water	7.66	3.06
	99.97	99.86
specif. grav.	1.968	2.031

CHAPTER XIX.

The East Coast from Maketu to Tauranga, and Return to Auckland.

From the lake district to the East Coast. — Sepulchral monuments. — Rev. Mr. Chapman at Maketu. — The giant Haupapa. — The venomous spider Katipo. — Rev. Mr. Voelkner at Tauranga. — War and negotiations of peace on Tauranga-Harbour. — Trip into the interior. — Great waterfall Wairere. — The Waiho, Piako and Waikato. — The Waikato-bridge near Aniwhaniwha. — Peculiarities of the river-bed and of the river-banks. — Maunga-tautari. — The Maori-City Rangiawhia. — Rev. Mr. Morgan in Otawhao. — Visit to King Potatau at Ngaruawahia. — Return to Auckland.

Appendix. Table of altitudes in the southern part of the province of Auckland.

On the 6. May we left the lake district and turned towards Maketu, a noted Pah and a Mission station on the Bay of Plenty, East Coast. From the lakes Rotoiti and Rotorua, which lie 1040 feet above the level of the sea, the country gently slopes towards the coast, consisting of trachytic tuff and pumicestone. To the North-west the Otanewainuku, an old trachytic cone, is visible, commanding the table-land of the Patetere forest, towards Southeast the Putauaki (Mt. Edgumbe) rising from a low pumicestone plain, which forms the continuation of the Kaingaroa plain. The higher portions of the country are intersected by numerous water-courses and deep valleys, thus presenting a rolling, hilly surface, while towards the coast marshy flats extend. The woods have mostly disappeared. The road going up-hill and down-hill, lies chiefly across open fern-land, and in rainy weather is very slippery. The distance from the Pah Morea, situated upon the isthmus between the Rotorua and Rotoiti, to Maketu I estimate at 25 miles. As the whole distance can be travelled on horseback, this is the easiest and shortest route for visiting the lake district from Auckland. From Auckland

to the harbour of Tauranga on the East Coast, we have one or
two days sailing; thence along the sea-beach to Maketu one day,
and another day to lake Rotoiti. The only difficulty or inconvenience upon this route is a deep and broad swamp, which has to
be passed close by Maketu. A second road leads from Tauranga
across the Otawa wood-ridge direct to lake Rotorua, likewise two
days' journeys; but the road is much more difficult; swamps and
several creeks (Te Papapa, Mangakopekopeko, Terarenga, Mangarewa and Mangapore) are to be passed, and the traveller has to
plod for a whole day in the depth of the bush. By this route he
misses the picturesque lake Rotoiti. I, therefore, recommend to
future travellers the Maketu-route, although there is but little to be
seen along the road.

A Maori monument.

The carved Maori-figures, which are met with on the road,
are the memorials of chiefs, who, while journeying to the restorative

baths of Rotorua, succumbed to their ills on the road. Some of the figures are decked out with pieces of clothing or kerchiefs; and the most remarkable feature in them is the close imitation of the tattooing of the deceased, by which the Maoris are able to recognise for whom the monument has been erected. Certain lines are peculiar to the tribe, others to the family, and again others to the individual. A close imitation of the tattooing of the face, therefore, is to the Maori the same as to us a photographic likeness; it does not require any inscription of name. Not having set out from Rotoiti before noon, we were obliged to camp on the road. The only camping-ground, supplied with water and fuel, is just half-way, and is called Te rewarewa. Here we found a large party of Maoris encamped with their wives and children, who were likewise on a journey to Maketu.

The 7. May was an unpleasant, rainy day, and it was only with difficulty that we advanced upon the slippery, muddy road. A short distance from Maketu we were moreover obliged to wade knee-deep — in some places even up to the waist — through the Kawa-swamp, nearly half a mile wide. The plight, in which we arrived, may, therefore, be easily imagined. A messenger on horseback came to meet us, tendering us in the Rev. Mr. Chapman's name a cordial invitation to the Mission station. I remonstrated that it was impossible for us to enter a European house in the plight, in which we had come out of the mire and swamp; but the messenger brought word back, that we should come just as we were, and I consequently proceeded with my whole party numbering twenty-three persons. Rev. Mr. Chapman received me at his garden-gate with a most hearty welcome; the natives shouted their friendly "haeremai," and ere long we were all under comfortable shelter beneath the missionary's hospitable roof.

The 8. May, a Sunday, was to all of us a welcome day of rest. The name of the Mission station is Whare-Kahu, house of the falcon, or as we should call it, "the eagle's eyrie." Mr. Chapman, is one of the oldest and most deserving missionaries of the English Episcopal church, who has passed through the times of cannibalism and the reign of terror of the "New Zealand Napoleon" Hongi; and Maketu is the very place famous for native wars of the most cruel and barbarous kind. There is a stone in the yard of the Roman Catholic church, on which is inscribed the date at which hostilities ceased, viz. 16th September 1845. With what soothing contemplations such men can look back to their noble and eventful life, on comparing the Past and the Present!

The Pah of Maketu is situated upon an eminence on the South-east-side of the Kaituna river, close by its mouth. The dwelling houses and granaries (Pataka) are carved in the most elaborate and grotesque manner. The estuary of the river extending to a great distance inside of the bar is accessible only for the smallest coasting vessels. A stone-block on the beach is designated by the natives as the anchor of the canoe Arawa, in which, as the story goes, their ancestors had immigrated. Kumaras, maize, potatoes, taro and tobacco thrive splendidly in the fertile alluvial soil of the country and peaches are very plentiful. In the Pah I had the pleasure of meeting Hori te Haupapa, the celebrated chief from Rotorua. He told me, that in Auckland he had been on board the Novara, and that there the colossal dimensions of his body had been measured. He was of opinion, that, if the people in Europe were to judge by that measure, they would surely take all the Maoris to be giants. The headland of Maketu consists of alternating layers of trachytic tuff, sandstone and pumicestone. From the heights near the Pah towards N. E., the island Whakari (White Island), a distance of about 35 seamiles, can be distinctly seen, with its crater continually puffing powerful clouds of white steam. The island rises 863 feet above the sea. Towards North, at a distance of 6 miles from the coast, is the island Motiti, a low island, — its highest point being only 190 feet high, — which, however, is said to be very fertile and therefore inhabited.

On the 9. May we set out towards Tauranga. After several stormy and rainy days this was again the first pleasant day. At sunrise the thermometer stood at the freezing-point, and the puddles of rainwater in front of the house were frozen; but the sun soon shone warm, and we had one of the most delightful autumnal days I ever witnessed in New Zealand. Haupapa himself ferried me in his canoe across the Kaituna river, and finally carried me upon his gigantic shoulders over the muddy banks to dry ground. We parted with the kindest assurances of friendship and esteem.

The distance from Maketu to Tauranga is sixteen miles, which at a good pace may be travelled in six hours; to us, however, it appeared twice the distance, the road leading in uninterrupted monotony along the sea-beach; to the right the blue sea, and the booming surf, which wets the foot of the traveller; and to the left sand hills and swamps. The isolated cone at the entrance of Tauranga Harbour, the Maunganui (860 feet high), is continually in view, and on nearer approaching, seems more and more to recede. To judge from the character of the shores between Maketu and Tauranga one would never have thought of calling the extensive bay between the East Cape and the Mercury Islands "Bay of Plenty;" yet the shores Southeast of Maketu and North of Tauranga seem to correspond better with that name given by Captain Cook. As we were about to camp for dinner, we were cautioned by the natives against a small, black spider with a red stripe on its back, which they call Katipo (or Katepo). The spider is said to exist only here and about Otaki on Cook's Strait on the grass growing upon the sand-hills, and its bite to be so poisonous, that with sickly persons it has even caused speedy death. Farther in the interior, they say, it is nowhere met with. This is the only poisonous vermin in New-Zealand. But in spite of a long and careful search we could not find even a single specimen. Ralph[1] describes it as a real spider, of a very different appearance at different periods of its age; when full-grown it is black, with an orange-red stripe on its back. Ralph mentions also, that he had put the spider together with a mouse, and that the latter died after 18 hours in consequence of the spider-bite. We were therefore careful to move our camp farther inland, where a swamp furnished us the necessary water for making tea. It was not until evening that we reached Tauranga Harbour. From

[1] Ralph: On the Katepo, Journal Proc. Lin. Soc. Vol. I. Zool. 1856 p. 1—2.

the Mission station Te Papa we were still separated by a narrow branch of the sea We had first to signal our arrival by means of a kindled fire, calling a boat over from the other side to take us across. Our signal was responded to; the boat came, and on the opposite shore our worthy countryman, the Rev. Mr. Voelkner came to meet us with open arms.[1]

May 10. — Mr. Voelkner and his amiable wife had been expecting us for some days, and pressed by our kind host and hostess, we remained two days at Tauranga, during which we experienced the greatest kindness and hospitality. Te Papa (i. e. the plain), the residence of the Archdeacon Brown, who during our visit was absent, is a Mission station of the Anglican High-Church, situated upon an extremely fertile peninsula on the southeastern branch of the Tauranga Harbour; the buildings are concealed between beautiful fruit-trees and locust-trees. Mr. Voelkner conducted a school for boys and girls, which he had founded five years ago, raising it in very short time to a most flourishing condition. The school-children had collected for me, at his request, a great number of pretty sea-shells. Among them *Argonauta tuberculosa*, *Solemya australis*, *Bulla Zelandica* and *Trochus tigris*. The schooner Maiperi bound for Auckland, being then at anchor in the harbour, and happening to be just ready to set sail, I had an excellent opportunity to forward my collections at once.

The Tauranga Harbour is the only harbour on the East-coast between Mercury Bay and Port Nicholson, which is accessible to larger vessels, and offers shelter against every wind. The entrance is a narrow, serpentine channel, but perfectly safe, the isolated, truncated cone of the Maunganui on the East-side, which rises to a height of 860 feet above the level of the sea, serving as an excellent land-mark. The form of the mountain resembles very much a volcano, but a closer examination shows, that the hill is a remnant part of the volcanic table-land, consisting of horizontal

[1] Mr. Voelkner is the unfortunate German missionary, who, during the native wars in 1865, became in a most horrible way a victim to the brutality of a bloodthirsty horde of fanatical Maoris.

beds of lava and tuff. Farther out in the sea, East of the Maunganui, there is a small rocky island, Motu-Otau, upon which the
singular lizard, *Hatteria punctata* Gray, Ruatara of the natives,
18 inches in lenght, is said to exist, a specimen of which Dieffenbach has brought to London. The Tuhua island (Mayor island),
1100 feet high, 20 seamiles North of the entrance to the Tauranga
Harbour, is an extinct volcano with a large crater open towards
South-East. On the West-side huge blocks of the beautiful greenish-
black obsidian (Tuhua of the natives) are found, to which the island
owes its name, and of which a captain is said to have shipped a
whole load to Auckland, mistaking it for coal.

Formerly Tauranga was one of the most densely populated
parts on the East-coast. Dieffenbach in 1841 estimated the number
of the inhabitants still at 3000; now it will be scarcely found to
exceed 800 or 1000. Serious hostilities had broken out of late
between the two tribes dwelling along the harbour, the Ngatihokos
and the Ngaiterangis, and at the time of my visit they were at
open war with each other. At the Mission station we could distinctly hear the fire of musketry from the neighbouring scene of war.
The cause of the war were the claims of both tribes to the possession of a piece of land Southeast of Te Papa. Both parties endea-
vored to settle upon the contested ground, leaving their former
villages and plantings. Fortified camps (Pah Wawai) were built;
and in spite of the efforts made by the missionaries, the native as-
sessor, and even the Maori King Potatau, to settle the difficulty by
an equitable partition of the land in question, they took up arms.
After a few of them had been killed on both sides, the parties grew
more and more irreconciliable. Thus the conflict had lasted already
for three whole years. The opportunity to witness the Maori war
was too enticing; so we set out on the 11. May in company of
Mr. Voelkner to visit the seat of war. A short passage across
the Southeast branch of the harbour brought us to the spot. Red
flags with the English inscription "war" were waving from the
forts, as we arrived, and here and there shots were fired. I had
my white pocket-handkerchief tied to a long pole and this im-

promptu flag was borne in front by a Maori. We were soon observed. The Pah nearest to us hoisted instead of the red war-flag
the white flag of truce with the inscription "Rongopai" (glad tidings)
and the other Pahs soon followed its example. The firing ceased.
An old chief, named Hou, came forth to meet us, and conducted
us into the main-camp of the Ngaiterangis, to the Pah Tumatanui.
Here I was presented to the commanding general of the fort, Rawiri, a man with a defiant, frowning air. He invited me to step
into a very spacious hut, situated in the centre of the Pah, and
surrounded with earthworks; and ordered his adjutant to call his
men together. About forty warriors assembled gradually together,
all robust young men in the best years; more than a hundred double-
barrelled guns were leaning against the walls; and I soon found
myself through my friend Voelkner placed in the wholly unexpected
and to me rather strange situation of having to act the part of an
apostle of peace before the assembled combatants. The causes of
the conflict having been fully explained to me, I was to state my
opinion on the subject. I declared, that, whereas both parties
were already abundantly blessed with the possession of the most
fertile lands of the country, which they were not even able to
cultivate to their full extent, a war about a scanty patch of land,
for which formerly for years nobody had cared anything, was in
my opinion utterly unworthy of the wonted bravery and magna-
nimity of the Maoris, and that I should be greatly rejoiced to be
able to convey to the Governor at Auckland, whose guest I was,
the glad tidings that they had made peace. One of the warriors
then stepped forwards, and declared to me in a pathetic speech,
that the question with them was not the possession of the land,
but of their right, and that they would gladly make me a present
of the whole contested land, if I could prevail upon their enemy
to give their consent to it; that they would be proud to have me
settled among them, or if I should not be able to remain myself,
to send my friends there to settle. Thus I had actually a chance
of becoming a chief in New Zealand. I had some trouble in preserving countenance at this offer, but promised to use every means

to move also the opposite party in favour of peace. The negotiations, which had lasted for two hours, ended with a meal which
Rawiri had ordered for us.

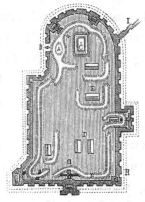

A. Terrace for the commandant.
B. Cannon.
C. Flagstaff.
D. and *E.* Casemates.
F. Ramparts.
G. Trenches.
H. Double raw of palisades.
I. Passage.

A fortified native camp
near Tauranga.

Thence we proceeded to the enemy's camp, to the Ngatihokos,
whom we found in their main-fort Tukiata. The annexed wood-
cut is intended to illustrate that Pah. Two brothers Manahira
and Tariha Kiharoa were the chief-officers in command. Theirs
was the weaker, but also the more obstinate party; they would not
listen to any proposols of peace, but gave vent to their bitter hatred
by the most violent abuse of their enemies. They told me, that
they were expecting the Ngatipukengas, the former owners of the
contested land, who were to decide the matter for them. My efforts
closed with the Ngaiterangis charging me with a memorial to the
Governor, in which they expressed their inclination to make peace,
and requested the mediation of the Governor to that effect. [1]

May 12. — Kind Mrs. Voelkner had baked some fresh bread
and prepared an excellent roast-pork for our benefit on the road;
and thus most liberally provided with food, we again parted from
a Mission house, the kind and hospitable inhabitants of which I
shall never forget. We crossed the Waikareao Creek in a boat, and
thence passed over a sandy plain through a deserted Pah in a north-

---

[1] Peace was really brought about through the mediation of a missionary; the
land was divided.

westerly direction along the shores of the harbour, which presents the character of a shallow estuary with many inlets and studded with numerous islands, separated from the sea only by a row of sand-hills, and extends, at an average breadth of two to three miles, in a N. W. direction, as far as the Katikati river, a distance of 15 miles. After a short hour's walk we reached the Otumoetai Pah (others spell it Otumoiti), one of the principal settlements on Tauranga Harbour, at the same time the seat of a Roman Catholic mission. The church of the place is of a very neat construction; there are also some Europeans settled here. On the beach lay a number of beautiful war-canoes; and next to them a schóoner, belonging to the Maoris, which they had bought for £500, leaving it afterwards to rot and decay. From Otumoetai we came over excellent, fertile alluvial plains to the Wairoa Bay. The Bay is very shallow, and the sandy rather than muddy ground so firm, that, although the tide had already half set in, we could without danger wade through to Peterehema (Bethlehem). From here we had to be ferried by means of canoes across a second, deeper branch, whereupon we found ourselves upon an abominable mud-bank, over which we were obliged to wade, the tide continually rising, for a whole mile sinking in more than knee-deep, until we came once more upon *terra firma* at a projecting sandy point. At 3½ p. m. we reached the village Potuterangi, from which we were hailed by shouts of "haeremai," so cordial and general that we halted here for dinner. Proceeding two miles farther along the strand we arrived at the Te Puna Bay and at eventide we crossed over to a small settlement Ongarahu, where we pitched our tents. We met only about twelve inhabitants, who exhibited a truly unfeigned joy at seeing Pakehas, and provided for our comfort to the best of their knowledge and ability. I for my part was most heartily glad, to have done with the mud-plains and creeks of the Tauranga Harbour, and to be once more upon dry ground. How very different everything here will look in after-years, when a European City shall rise on the Tauranga Harbour, and the beautiful country round about be dotted with flourishing farms.

May 13. — Rose early and pursued our route in a westerly direction, our destination being the Waikato Valley and the Middle Waikato Basin. We had to cross the Whanga range, the northern continuation of which is the Cape Colville peninsula on the East-side of the Hauraki-Gulf. From Tauranga, however, this range does not appear as a mountain-range, but the land rises gradually in a gently sloping, inclined plain to a wooded plateau, which to-wards the Waiho Valley breaks off abruptly; consequently it is only from the Waiho plains that the coast range appears as a steep and rugged mountain-range. It consists of trachytic rocks, and forms the northern continuation of the Patetere plateau. The road is a much frequented path for equestrians. For two hours we had to traverse woodless, fern-clad hills; then we came to a cool mountain-stream, the Waipapa, forming a few yards off the road a pretty cascade, called Taiharu, and about 30 feet high. Here the bush commences, at first interrupted by some intervening lawns, from which a charming view is had over the Bay of Plenty, from the Tuhua Island to the Whakari volcanoe; thence it grows denser and denser, until the traveller at last finds himself once more buried in the obscurity of a primitive New Zealand forest. Magnificent cobalt-blue mushrooms 2½ inches high, and the extremely frequent occurrence of the remark-able "vegetating caterpillar" in those woods deserve special notice. The Parapara creek — on the banks of which there are very pretty ferns (among them *Lindsœa microphylla* Presl) in luxuriant growth, — is the last creek running towards the Tauranga side. The water-shed to be crossed is about 1500 feet above the level of the sea. The road obstructed by the gnarled roots of the trees, and by deep mud-holes, was extremely difficult and tiresome, and towards even-ing we were obliged to encamp in the middle of the bush on the banks of a small creek running towards the Waiho side. Our camp-ing-ground on the height of the Whanga-plateau was 1414 feet above the level of the sea. The bright glare of our blazing fire illumi-nated the gigantic trunks of majestic Rimu and Rata trees; the moonlight with magic splendour poured through their lofty crowns, and the blustering night wind moaned in the tree-tops. I do not recollect a more romantic scene of a night in the bush, during all my rambles in New Zealand.

May 14. — We took an early start, our road continuing level along the wooded plateau, then for a short distance up-hill. There the woods opened, and we stood at the margin of the plateau breaking off abruptly towards the plains of the Waiho and Piako; on our right the great fall of the Wairere river, which noisily tumbles over the precipice into the depths below. In fair weather, the prospect from this point must be a glorious one; but the tempest chasing heavy clouds of mist over our heads, deprived us of that pleasure. The upper part of the mountain-side being densely wooded, it is not easy to find a suitable place from which a full view may be had of the fall. I therefore, before ascending the last height, proceeded along the left bank of the Wairere creek through the bush, advancing to a place where the river first falls over a bank 10 feet high, and then suddenly plunges down into the depths. The heavy Southwest wind drove the spray back, and I stood there like in a heavy shower of rain. On ascending from here the height on the left, I had quite a fair view of the falls. The river precipitates in three steps over columnar trachytic rocks, and the total height of the falls I found to be 670 feet. The descent from the plateau is very steep. Having reached the valley, we took a southerly direction up the valley along the foot of the range, and rested for dinner near the small settlement Okauwia. Not far from here on the left bank of the Waiho is the old Pah Pupunui, and half a mile above this Pah on the banks of the Waiho there are three warm springs, on the left bank the Paruparu and Ramaroa, on the right bank the Okahukura. The latter, of a temperature of 117° F., bubbles from the sandy bed of a small creek near its junction with the Waiho; the natives have constructed a basin for bathing by damming it up. Also farther North at the foot of the Aroha mountain warm springs are said to break forth. After passing numerous swampy creeks, we reached towards evening the settlement Whatiwhati, about ten miles South of the Wairere.

On the 15. May an easterly gale blowing and the rain pouring down in torrents, we were obliged to keep within our tents all day. It was not until evening that the sky cleared off.

The 16. May was bright and warm as a spring-day. The preparations, however, which my Maoris were making for our to-day's journey, packing shoes and trowsers into their bundles, led me to suppose, that despite the delightful weather we should have enough to do with water. We advanced in a westerly direction towards the Maungatautari mountain, an old acquaintance from the Middle Waikato Basin. The beginning of the road was tolerably good. It lay partly over low fern-hills, partly over grassy plains, the pumicestone-soil and the grass-vegetation reminding us of the terraced country about Lake Taupo. Towards South a white cloud was seen, which could be no other than a steam-cloud ascending from the Tongariro, although the volcanic cone itself was not visible. Judging from the colossal dimensions of the cloud, the crater seemed to-day to be in an extraordinary state of activity. We crossed the tributaries of the Waiho coming from the Patetere plateau, first the Rapurapu river, then the Waiomau, and six miles from Whatiwhati we struck the Waiho itself.[1] The river, which sixty miles farther North empties into the Hauraki-Gulf, bearing the proud name of the New Zealand Thames, which already Captain Cook gave the stream, is here, near its source in the Patetere plateau, only a small creek six to eight feet wide, which has cut its bed thirty to forty feet deep into the soft pumicestone-banks. A suspension-bridge most curiously constructed of branches leads across. The Maori geographers, however, seem to have taken an unusual liberty in this case; for scarcely two miles from the Waiho the traveller comes to a large rapid stream, running in a broad river-bed, which empties into the Waiho and bears the name of Oraka; this is evidently the main-river. From the Waiho to the Mangawhero, the last tributary of the Waiho — a distance of three miles — we were obliged to wade almost without interruption in stagnating water,

[1] According to the pronunciation of the natives the name ought to be written Waihou.

and it was not until reaching the Tirua range which separates the Waiho valley from that of the Piako, that we stepped once more upon dry ground, where we could rest for dinner. The Tirua range (the name Horokatoa has also been mentioned to me as belonging to the same range) is a low, woodless range of hills forming a spur of the volcanic table-land in the South and consisting of yellowish pumicestone-tuff. Beyond the Tirua range we had to cross first the Mangaokahu creek, and then the Waitoa, the principal tributary of the Piako. The Piako itself rises more to the Northwest in the Maungakawa range, which separates the Piako plains from those of the Waikato; and here we meet again the singular case that the Waitoa, the main-river with a longer course and a greater supply of water, must surrender its name to the Piako at their junction. The crossing of the Waitoa, a stream not very broad, but very deep and rapid, was difficult and dangerous. One of the Maoris lost his balance in the swift current and was in great danger of being carried down a fall 20 feet high, a few yards further; luckily, however, he succeeded in holding himself by some bushes on the bank, until we rescued him from his critical situation. At our place of crossing the valley was flat and swampy, but below the fall it changes to a deep gorge with vertical sides of pumicestone-tuff.

One and a half miles from the Waitoa runs the Waikato, separated from the former only by low hills. After passing through a marshy flat between the Hinuwera and the Tekopua ridges, we reached the topmost terrace of the Waikato valley, about 200 feet above the river, and thence descended upon the second terrace, covered with Manuka bushes. A mile farther up we came to the remarkable place, where the mighty stream, narrowing to within 30 feet, plunges roaring and foaming through a deep rocky gully; the only place narrow enough to admit of its being bridged over by the natives. The name of the place is Aniwhaniwha. It was already evening when we crossed the bridge, and pitched our tents on the left bank of the river upon the lowest terrace amongst a luxuriant growth of Manuka-bushes. Along a distance of about 18 miles we had on that day passed not less than eight rivers, without, however, having seen a single settlement. This I can only account for by the fact, that the whole country there is treeless, and that the natives in order to have firewood at hand prefer to locate their villages at the edges of the bush along the foot of the ranges bordering the plains and river-valleys.

May 17. — Aniwhaniwha is situated at the northeastern foot of the Maungatautari. The Waikato flows in a deep erosion-valley, on the West-side of which I counted seven terraces rising one above the other with the utmost regularity; near Mangatautari the river gradually works its way out of the volcanic table-land, which lies between the Taupo-district and the middle Waikato-Basin, and a

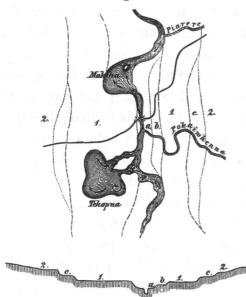

Sketch of the Waikato near Aniwhaniwha.
Terraces of the valley.

few miles farther-on it enters the plains. The formation of the river-bed near Aniwhaniwha is a very singular one. Above the gorge, across which the bridge is made, the river makes a futile attempt to escape to one side. One branch of it turns off, and has excavated the deep basin Tekopua; but the waters can find no outlet and rush with mad uproar, tumbling over large blocks of rock, back to the mainstream, in order to force in junction with the latter the passage through the rocks. Upon the rock-island girt by the two branches of the river, a Maori Pah is said to have stood in former times; but one may justly ask, how did those people make their way across the torrent to that island? The chasm or gorge, which the waters have to pass, is 400 feet long, 30 to 40 feet wide and, no doubt, very deep. The waters seethe and foam in the deep and narrow channel and dash from it into an extensive basin called Makiha, from which they continue their onward course smoothly and placidly. The rocks of the river-bed consist of a yellowish volcanic tuff, intermixed with pumicestone, and it is strange enough, that the brittle rock should withstand such a powerful swell of the water. The bridge consists of a Totara log hewn out in the shape of a flat canoe; it is safe enough even for horses. A few yards above the bridge, on the left bank of the river, and on a tabular rock, which in time of high-water is flooded by the rapid stream, a very striking phenomenon presents itself. In the bed of the rock there are numerous round holes, one to three and even four feet wide, and equally deep, appearing as if wrought by art; in each of those holes there is a stone, sometimes two or three, all round like cannon-balls, consisting of hard trachyt or dolerit. I stood for a moment wondering what it meant; but the interpretation of it is very simple. These pot-holes are evidently caused by a process of erosion, during the rapid flow of the river over the bed of the rock. Any obstacle causes the waters to move in a whirl and carry around pebbles or stones, and by this process the stones gradually grind their way into the rock. The hole being once started, it is bored deeper and deeper, the eddying waters keeping the pebbles and stones in a constant rotatory motion. Thus the stones grind a circular hole into the rock, and are themselves ground off into balls. The phenomenon is interesting enough to induce the traveller to walk a few yards up the river; to step upon the rocky bank, which is quite dry at an ordinary stage of water; and there to look at the pot holes. Just opposite the Pokaiwhenua (i. e. boring its way into the earth), empties through a narrow gorge into the Waikato. The height of the Waikato above the level of the sea near Aniwhaniwha I found to be 166 feet; consequently from lake Taupo to this place, a distance of about 70 miles, the river falls 1080 feet, forming rapids upon rapids; and only a few miles farther down it commences to be navigable for canoes.

Having scaled the pumicestone-terraces on the left bank of the river to a height of about 250 feet, we stood at the foot of the Maungatautari. Numerous, small tributaries of the Waikato have their source in the forests and ravines of the mountain. At a northern spur stands the Pah Whareturere, where we were kindly received

by the natives, and lodged in a house built in European style.
The Chief Tioriori being not at home, his wife, a portly Maori-
woman, acted the kind and obliging hostess, and ere long a pig
was killed and the tea ready.

May 18. — Maungatautari is a trachytic mountain like the
Pirongia on the Waipa, its highest peaks reaching a height of about
2600 feet above the level of the sea. Along its eastern foot runs
the Waikato. On its Northwest-side rises the Mangapiko, a considerable tributary of the Waipa, and towards West and North it
extends its branches far off into the plains between the Waikato
and Waipa.

From Tiorirori's Pah on the northern slope of the mountain
I dispatched one half of my party down the Waikato, which commences to be navigable in the vicinity of Horotiu below the Hau-
tupu and Paratoketoke rapids; myself and my friend Dr. Haast
taking a westerly course, pursuant to a friendly invitation on the
part of the Rev. Mr. Morgan to visit Rangiawhia and Otawhao.
In Kirikiriroa on the Waikato we were to meet again. It was a
bright and pleasant day. Upon the height of the Pukekura I met
a Maori messenger with two saddle-horses, sent for us by the Rev.
Mr. Morgan. We had to cross the Mangapiko creek, and were on
the other side most kindly received by Mr. Ireland, the teacher
of Otawhao, who had come to meet us with no less than nine
saddle-horses, which the natives of Rangiawhia had placed at our
service. It was a pleasing moment indeed, as we met there together,
mounted the horses and gallopped gaily along. The direct road
to Rangiawhia across the large swamps of the Mangapiko is impassable for horses. Mr. Morgan, therefore, had for years past
made a special path for horseback-travelling, which leads over the
gap running West from the Maungatautari and across the Whanake
mountain. It continues quite steep up-hill and down-hill; but on
arriving at the top a magnificent prospect over the Waikato country
from the Pirongia in the West to the Aroha in the East affords
ample indemnification for the fatiguing ascent; and even the Taupiri in the North we had the pleasure of hailing again.

At 5 p. m. we reached Rangiawhia, situated in the fertile plain between the Waikato and Waipa. Extensive wheat, maize, and potatoe-plantings surround the place, broad carriage-roads run in different directions; numerous horses and herds of well-fed cattle bear testimony to the wealthy condition of the natives; and the huts scattered over a large area are entirely concealed among fruit-trees. A separate race-course is laid out; here is a court-house, there a store; farther-on a mill on a mill-pond, and high above the luxuriant fruit-trees rise the tapering spires of the catholic and protestant churches. I was surprised on entering the latter sanctuary, at beholding a beautifully painted glass-window reflecting its mellow tints into my wondering eyes. Such is Rangiawhia — the only Maori-settlement, among those I have seen, which might be called a town —[1] a place, which by its central position in the most fertile district of the North Island, and as the central point of the corn-trade, bids fair to rise ere long to the rank and size of a flourishing staple-town. A large number of the inhabitants had assembled at our arrival, among them the pretty young wife of the chief Wiremu Toetoe, who had been one of the party on board the Novara on the return voyage to Europe. The magnificent sorrel, on which I was mounted, had been sent by Toetoe's spouse for my special accommodation; it was Toetoe's own favorite charger. I had to tell them of the Novara and her route to Europe; and was subsequently not only charged with letters and greetings, but the affectionate Mrs. Toetoe even sent a photograph by me to her distant husband. On a good road we had soon traversed the short distance between Rangiawhia and Otawhao, where we were received with a most hearty welcome by the Rev. Mr. Morgan and his family.

May 19. — Otawhao is one of the principal Mission stations of the Anglican High-Church. A church, a school, gardens, meadows and smiling fields have taken the place of an old dilapidated Maori Pah; and we could well appreciate the proud joy swelling the heart of our noble minded host, on seeing the lovely

---

[1] Otaki in the South of the North Island is said to have a still more city-like appearance.

group of blooming daughters, who are the ornament of the Otawhao
Mission station. The fleeting hours of the pleasant day I spent at
Otawhao, slipped away amid interesting conversations about the
country and the people, and promenades through the environs of
the place. In the old Pah I found among the rubbish of the huts

Missionary school at Otawhao.

one of those grotesque figures carved in wood, which in olden times
had graced the dwellings of the Maoris. It was 5 feet high and
still in a tolerable state of preservation. I did not hesitate to appropriate it and to take it home with me to Europe as a specimen
of Maori sculpture. Notwithstanding the greatest possible caution
on my part, however, it was rumoured among the natives that I
had packed up one of their ancestors; the figure was demanded
back from me on the plea that it was intended to grace the residence of the Maori king Potatau. Nevertheless it now stands safe
and sound in the Novara-Museum at Vienna; whence Potatau II is
quite welcome to reclaim it.

May 20. — Mr. Morgan very kindly lent us horses and ordered
one of his natives along to guide us a far as Kirikiriroa. The road,
a convenient horse-path, after passing the Mangahoe and Mangapiko, lies across an undulating hilly country, through a charming
pond-landscape with picturesque groves of Kahikatea pines. We

passed eight small lakes or ponds and several deserted Pahs. After a ride of four hours we arrived at Kirikiriroa, a settlement on the left bank of the Waikato, where we met the other half of our party, who had arrived an hour before. The Waikato here has dug its bed with two terraces about 50 feet deep into the pumicestone and sand-alluvium. On the upper river terrace are the plantations of the natives.

On the 21st May we proceeded down the Waikato. Captain Hay had hired a large canoe for us, hewn out of a rimu-log, 64 feet long and 4 feet wide, in which our whole party consisting of twenty-five persons was easily accommodated together with all our luggage. We had twenty paddles; the morning was bright and pleasant. At 9 a. m. we pushed off from the bank and at 11 a. m. we landed at Ngaruawahia in front of the residence of the Maori King. Favored by the current of the river, we had paddled the distance of about fifteen miles in two hours.

Ngaruawahia, residence of the Maori King Potatau in 1859.

Our object was to pay a visit to the King Potatau. In the first hut we came to, we met the king's private secretary Te Wetini Te Tekerahi, a tall stout man with his face beautifully tattooed, his whole carriage and expression indicating pride and determination. He received us very kindly, ordered forthwith dinner for the Maoris of my party, and went to acquaint the king with our arrival. We were soon informed by Te Wetini, that the king had granted the interview and was awaiting us. I therefore proceeded to the Royal residence. The king's palace — a spacious hut, built in the best style of Maori architecture with a flag-staff next to it — was situated upon the heighest point of the neck of land between the Waikato and Waipa, with a view of both rivers. Some smaller huts standing aside among ferns and shrubs were the first signs of the future capital of New Zealand designed by the king's party. The flag on the flagstaff had a centrepiece of white surrounded by a red border; on the white there was a blue cross, besides that the inscription "Nuitireni" and three diamond-shaped figures. In front of the royal residence stood a

The flag of the Maori King.

sentinel in blue uniform with red facings and shining brass-buttons. This was the guard of the palace. We entered through the door into a large room; in which there were about twenty persons assembled. On the right side, in a dark corner, sat upon a mat of straw and wrapped in a dark blue blanket a blind old man, his head bowed low with age, it was Potatau Te Whero-whero, King of the Maoris. The face of the old man tattooed all over, displayed handsome regular features; the deep scar upon his fore-head betokened the veteran warrior, who had fought through many a bloody cannibal-fight. Potatau returned our salutations on being presented to him only with slightly raising of his head. Captain Hay told him of our travels, but the old man spoke not a word. Two young men, however, setting near the king were very curious to receive from us information as to European history and geography, and kept up the conversation in his stead. It was evident, that the decrepit old man, who was on the very brink of the grave, merely lent his celebrated. name to the support of the royal title, and that other, younger men were at the head of the national movement, the so-called "king movement". A tall

---

[1] Potatau Te Wherowhero died in 1860. His son Matutaera was created King under the name of Potatau II.

young man of sullen mien was presented to us as the king's son; the daughter of the king was just busy washing. We were invited to dinner, and were regaled with dried shark as a special delicacy. I admired the appetite of the Maoris, without however being able to prevail upon myself to join in; I was truly glad, when after dinner the ceremony ended and I had once more an opportunity of inhaling the fresh air in front of the royal hut.

After a stay of two hours we paddled rapidly down the Waikato and enjoyed once more the fine scenery of the Taupiri range. After a short stop at the Taupiri Mission station, to greet the Rev. Mr. Ashwell, we proceeded down the river and took up our quarters for the night at Rangiriri.

On the 22ᵈ May a very heavy gale was blowing, which stirred up such waves upon the broad river, that we could not think of venturing to proceed in our heavily loaded canoe. The gale ended in the night with a violent tempest and heavy showers of rain. In the morning of the 23ʳᵈ the weather had cleared off so that we could continue our journey, and at 1 p. m. we landed at the Teia Creek, a little below the mouth of the Mangatawhiri. Thence we had to climb up and down a slippery hill, until we reached the house of an European settler at Mangatawhiri. After resting a short time, we thence set out to Drury. The Waipapa Creek near Mangatawhiri having risen to such a size, that we were obliged to build a temporary bridge over it, in order to cross, we were detained so long, that it was already quite dark when we reached the Great South Road. Hence forth, we thought we should have easy travelling; but how sadly were we disappointed! Never have I seen anything, that is called a road, in such a condition. The Great South Road had more the appearance of a river or a marsh, where we sank in knee-deep, than of a road; moreover a heavy tempest broke loose, the rain poured down in torrents, and only the frequent flashes of lightning served to illume the darkness of the night. Yet notwithstanding all this we were determined to reach the Drury Hotel that same evening. The plight we finally arrived there at 9 p. m., I need not describe.

However, "all's well that ends well." We found in the excellent hotel everything necessary to make us feel comfortable once more, and on the 24ᵗʰ we were all safely back in Auckland, and highly gratified with our three-months journey.

Appendix.

Table of altitudes in the southern part of the Province of Auckland.

† Altitudes taken from the New Zealand Pilot, and the Admiralty maps. — *b* Barometrical measurements by Dr. F. Hochstetter. — * Estimated.

	English feet.		English feet.
Auckland — Meteorological Observatory of the Royal Engineer Department .	140 †	Pilot Station	300 *
		Pukehuhu	690 †
		Omanawanui Peak . . .	1100 *
Claremont House, upper end of Princess-street .	130 *b*	Te Kaamoki, or Te Komoki Peak, near the Huia .	480 †
Kaipara Harbour, West Coast:		The Huia Peak	1280 †
Te Karanga hill, on the Otamotea river . . .	1440 †	Puponga, highest point .	390 †
Wakakuranga hill on the Oruawharu river . . .	476 †	Hill on the left border of the big Muddy Creek .	600 †
Opara hill	378 †	Hill near Whau Creek .	800 †
Auckland Peak, near Otau Creek.	1023 †	South Head, Mahanihani .	580 †
Koharanga, on the Kaipara river	326 †	*East Coast, from the Bay of Islands to the Waitemata Harbour (Harbour of Auckland):*	
Titirangi chain, between the Waitakeri and the Manukau Harbour:		Cape Tewara or Bream Head, Wangarei Harbour	1502 †
Mount Tea Wekatuku . .	1430 †	Summit between Bream Head and Home Point, Wangarei Harbour . .	1340 †
Pukematiku, Henderson's Bush	1300 *	Moto Tiri Island) Hen and (	725 †
Maungatoetoe, near Delworth's Farm	1200 *	Taranga Island) Chickens (	1353 †
Parera cliffs, West Coast .	700 *	Mount Hamilton, near Rodney's Point	1050 †
Manukau Harbour, West Coast:		Kawau Island, Mt. Taylor	510 †
Paratutai Island, Signal Station	350 †	Little Barrier Island (Houturu), Mt. Many Peaks	2383 †

English feet.

Great Barrier Island (Aotea), Mount Hobson . . 2330 †

The extinct Volkanoes near Auckland:

Rangitoto . . . . 920 †

North Head, Takapuna . 216 †

Mount Victoria, Takarunga 280 †

Heaphy Hill . . . . . 100 *

Mount Eden . . . . . 642 †

   „ Hobson . . . . 430 †

   „ St. John . . . . 400 †

   „ Albert . . . . . 400 †

   „ Kennedy . . . . 310 †

   „ Three Kings . . . 390 †

One Tree Hill . . . . 580 †

Mount Smart . . . . . 300 *

   „ Wellington . . . 350 *

Pigeon Hill . . . . . 110 †

Otara Hill . . . . . . 150 †

Mangere Hill . . . . . 333 †

Waitomokia . . . . . 120 *

Puketutu . . . . 263 †

Otuataua . . . . . . 300 †

Maunga lakelake . . . 300 †

Manurewa . . . . . . 300.†

Matakarua . . . . . . 300 †

Drury — Young's Inn (first storey) . . . . . 75 *b*

Brown coal shaft on Farmer's land . . . . . 356 *b*

Great South Road, between Drury and Mangatawhiri:

1. First hill at the entrance of the bush . . . 491 *b*

2. Highest point of the road 811 *b*

3. Waikohowheke, house on the road . . . . 598 *b*

4. Second height of the road on the point where the view of the Waikato opens 770 *b*

English feet.

Mangatawhiri, Maori settlement . . . . 77 *b*

Papahorahora, near Kupakupa, on the left bank of the Waikato, the brown coal seam . . . . 250 *

Taupiri, hill on the right border of the Waikato, opposite the Mission Station 983 *b*

Kakepuku, isolated mountain not far from the Mission Station on the Waipa 1531 *b*

Points between the Waipa River and the West Coast:

Toketoke creek, on the way from Whatawhata to Whaingaroa . . . . . 249 *b*

Highest point of the road from Whatawhata to Whaingaroa . . . . . 853 *b*

Whaingaroa Harbour, Captain Johnston's house . 93 *b*

Mill on the Oparau River, Kawhia Harbour . . . 97 *b*

Pirongia, highest point of the road from the Oparau River to the Waipa . . 1585 *b*

Pirongia, highest point of the mountain . . . . 2830 †

Waikato River:

1. Near Mangatawhiri . . 35 *

2. Near Rangiriri, pah on the right bank . . . 51 *b*

3. Near Taipouri, island in the river with a Maori village . . . . . 63 *b*

4. Near Tukopoto, Taupiri Mission Station . . . 75 *b*

5. Ngaruawahia, residence of the Maori King . . 85 *

English feet.

6. Kirikiriroa . . . . . 97 *b*
7. Aniwhaniwha, Waikato
   Bridge . . . . . . 166 *b*
8. Near Orakei Korako . 970 *b*
9. Outlet from Taupo lake 1250 *b*

Between the Waikato and Waipa:

Maungatautari, Maori pah 621 *b*
Otawhao, Mission Station 211 *b*

Waipa River and District:

Ngaruawahia . . . . . 85 *
Whatawhata, 20 feet above
   the river . . . . . 109 *b*
School-house at Whatawhata 112 *b*
Kaipiha, Mr. Turner's house 167 *b*
Waipa, at the junction of
   the Mangaweka . . . 143 *b*
Mission Station of the Rev.
   A. Reid, 25 feet above
   the bed of the river . . 173 *b*
Awatoitoi, Maori settlement
   on the right bank of the
   Waipa, 25 feet above the
   bed of the river . . . 185 *
Orahiri, on the left bank
   of the Waipa . . . . 186 *
Hangatiki, Maori settlement 195 *b*
Te Ana Uriuri, cave . . 204 *b*
Tauahuhu, Maori settle-
   ment on the left bank of
   the Wangapu . . . . 196 *b*
Mangawhitikau, Maori vil-
   lage . . . . 237 *b*
Puke Aruhe, hill . . . 877 *b*

Upper Mokau district:

Takapau, Maori settlement 823 *b*
Piopio, Maori settlement on
   the upper Mokau River 469 *b*
Mokau River, above the
   Wairere falls . . . . 420 *b*

English feet.

Pukewhau, Maori pah on
   the left bank of the Mokau River . . . . . 683 *b*
Mokauiti, between the Ma-
   ori settlement Huritu and
   Puhanga . . . . . . 473 *b*
Puhanga, Maori settlement
   on the Tuparue hill . . 937 *b*
Morotawha, place of en-
   campment on the 7th and
   8th of April, 1859. . . 570 *b*
Tarewatu mountain ridge,
   height of pass from the
   Mokau to the Wanganui
   district . . . . . . 1581 *b*
Tarewatu, highest point . 1790 *b*
Tapuiwahine, highest point
   on the way from Mokau
   to Wanganui . . . . 1933 *b*

*Upper Wanganui, Tuhua Di-
strict:*

Ohura, Maori village . . 917 *b*
Katiaho, Maori settle-
   ment, on the Ongaruhe
   river . . . . 650 *b*
Ngariha hill, on the On-
   garuhe river near Ka-
   tiaho . . . . 1551 *b*
Pokomotu plateau, highest
   point on the way from
   Katiaho to Petania . . 1386 *b*
Petania, Maori village on
   the Taringamotu river . 754 *b*
Takaputiraha chain, pas-
   sage from Petania to
   Taupo . . . . 1534 *b*
Pungapunga creek, on the
   road to Taupo . . . . 897 *b*
Puketapu, hill on the road
   to Taupo . . . . . . 2073 *b*

English feet.

Taupo District:

Moerangi, pumice - stone plateau on the west and south-west of Lake Taupo . . . . . . 2188 *b*

Whakaironui . . . . . 2175 *b*

Kuratao river on the road to Pukawa . . . . . 1719 *b*

Poaru, Maori settlement . 2289 *b*

Pukawa, pah on the southern bank of Lake Taupo 1399 *b*

Mission Station of the Rev. Mr. Grace, at Lake Taupo 1473 *b*

Koroiti plateau, on the south bank of Lake Taupo 1768 *b*

Taupo Lake (by Dieffenbach, 1337 feet) . . . 1250 *b*

Roto Aira, by Dieffenbach 1709

Roto Punamu, by Dieffenbach 2147

Tongariro and Ruapahu:

Tongariro, Ngauruhoe (by Dieffenbach 6200) . . 6500 *

Ruapahu, on Taylor's map 10,236 *

   „   Arrowsmith's map 9000 *

   „   Admiralty map . 9195 *

Pihanga . . . . 3500 *

Between Taupo Lake and the East Coast:

Oruanui, Maori settlement 1672 *b*

Plateau above Orakei Korako 2200 *

Orakei Korako, pah on the left bank of the Waikato river . . . . 1169 *b*

Boiling mud springs at the foot of Paeroa . . . . 1409 *b*

Waikite, hot springs at the Paeroa range . . . . 1241 *b*

Pakaraka, above Roto Kakahi . . . . 1801 *b*

English feet.

Roto Kakahi, Lake . . . 1378 *b*

Roto Mahana, Lake . . 1088 *b*

Tarawera, Lake . . . . 1075 *b*

Papawera plateau, between Roto Mahana and Tarawera . . . . 1867 *b*

Mission Station on Tarawera Lake, Rev. Mr. Spencer . . . . . . 1502 *b*

Rotorua, Lake . . . . 1043 *b*

Ngongotaha, mount on the southern bank of Rotorua . . . . . 2282 *b*

Rotokawa, small lake on the southern bank of Rotorua . . . . . 1098 *b*

Waiohewa, or Ngae, settlement on the north-eastern bank of Rotorua . . . 1103 *b*

Pukeko, on the Rotoiti . 1063 *b*

Omatuku, near Maketu . 1388 *b*

East Coast:

Major island (Tuhua), highest point . . . . . . 410 †

Monganui hill, at the entrance of the Tauranga harbour . . . . . . 860 †

Plate Island (Motunau), centrum . . . . . . 166 †

Whale Island, or Motu Hora, highest point . . 1167 †

White Island, or Whakari 863 †

Mount Edgcumbe, eastern summit . . . . . . 2575 †

East Cape (East Cape Islet) 420 †

Between the East Coast and the Waiho River:

Waipapa creek, on the road from Tauranga to the Waiho . . . . 803 *b*

	English feet.
Wairere river, immediately above the falls . .	1442 _b_
Height of the pass over the Whanga range, near the Wairere falls	1481 _b_
The height of the Wairere Falls	670 _b_

	English feet.
Waiho Flats, near Wairere Falls	573 _b_
Whatiwhati, settlement at the foot of the Patetere plateau	537 _b_
Castle Hill (Cape Colville range) near Coromandel Harbour	1610 †

CHAPTER XX.

Nelson.

Character of the surface of the province. — The Western mountain ranges. — The Eastern ranges. — The hills on Blind Bay. — Excellent climate on the shores of Blind Bay. — The town of Nelson, situation, foundation and development. — The Harbour. — The Boulder Bank. — The agricultural districts. — Ranzau and Sarau, German settlements. — Wood-cutters and shepherds the farthest out-posts of civilization. — The Mineral wealth of Nelson. — The copper and chrome-ore of the Wooded Peak and Dun Mountain.

———

The character of the surface is always more or less indicative of the geological structure of a country. Even to those who have not deeply studied the science the different forms, which mountain-ranges show, will indicate the difference of their geological structure. This difference in the external appearance of the country is most striking and surprising to the traveller on coming from the Province of Auckland on the North Island to the Province of Nelson on the South Island. In contrast with the comparatively low ranges of hills and table-lands, extending over the greater part of the northern Island, and broken only by high volcanic peaks, we find in Nelson high and steep mountain ranges with serrated peaks, striking in long parallel chains, separated by deep, longitudinal valleys, and broken at right angles by rocky gorges. The geological field presented here is consequently an entirely new one in comparison with that of the North Island.

From a central point forming the water-shed between the East and West coasts, and containing the sources of the boundary-rivers of the two provinces, Nelson and Canterbury, — the Hurunui running Eastward, and the Teramakau running Westward, — the Southern Alps send forth towards North two branches through the province of Nelson, the extremities of which are washed by the waters of Cook's Strait.

These branches present very different geological features. The Western Ranges, terminating in Separation-Point and near Cape Farewell, have an almost northerly strike. They consist of crystalline rocks and metamorphic schists, principally granite, gneiss, mica-schist and hornblende-schist, quartzite and clay-slate. To the auriferous character of those rocks, Nelson is indebted for its goldfields.[1] The peaks of these ranges ascending to a height of 5000 to 6000 feet above the level of the sea, and in winter-time covered with snow to a great extent, such as the picturesque Mt. Arthur, Mt. Owen and others, greet the traveller from afar on his arrival in Blind-Bay, and impart to the landscape about Nelson one of its most peculiar charms. The plains watered by considerable rivers, which interrupt the ranges, offer to the colonist extensive areas for agriculture, and to sheep-farmers very fine natural pasture grounds. Mr. T. Brunner was the first to traverse these regions under unspeakable difficulties; J. Haast, in his able report on the Western district of the province Nelson[2] was the first to publish a detailed account of them, giving names to ranges, mountains, lakes, and rivers not named before. My own name I had also the pleasure of finding in the report and on the accompanying topographical map attached to a mountain, a river and a lake near the sources of Grey river. For this token of friendly remembrance I am the more indebted to my worthy friend, as I have been honoured with a place in the most select and respected society, by the side of the mountains Werner, Herschel, Hooker, Albert and Victoria. A view of the mountain-range bordering the Grey-plains in the East, I have presented in the following chapter. The principal groups belonging to the Western Ranges, are to the North of the Buller river the Lyell and Marino

---

[1] See Chapter V. p. 99.
[2] Report of a topographical and geological Exploration of the Western Districts of the Nelson-Province; Nelson, 1861.

ranges and Mt. Owen, then the Tasman mountains and the Mt. Arthur ranges; finally in the North, bordering on Golden Bay, the Whakamarama, Haupiri and Anatoki ranges. South of the Buller is situated the isolated mountain-stock of the Paparoha range with the Buckland mountains, and the Victoria, Brunner and Mantell ranges. In the Victoria range the mountains attain a height of 7500 feet above the level of the sea. The most important plains and valleys are the Grey and Mawhera plains, the Buller or Kawatiri plains with their southern branches, the Matakitaki, Maruia and Inangahua plains; farthermore the Mokinui and Karamea (or Mackay) plains, the Wakapuia valley; finally on Golden Bay the broad valleys of the Aorere and Takaka rivers. The area of the Grey plains alone is estimated at about 250,000 acres; and we may take it for granted, that the Province of Nelson possesses all together in these Western mountain-ranges about half a million acres of available land.

The Eastern Ranges, stretching from Southwest to Northeast, consist of stratified sedimentary rocks of sandstones, red, green and gray clay-slates, with few limestone banks intervenig. The strata are highly inclined, all more or less vertical, and the parallelism of their strike from Northeast to Southwest continues with remarkable regularity. They are accompanied by an immense dyke of intrusive rocks, striking in the same direction from the northern extremity of d'Urville's Island across the French Pass, through the Croixelles by the Dun Mountain, Upper Wairoa and traceable as far as the Cannibals Gorge in the South of the province, a distance of 150 miles; thus constituing one of the most prominent geological features of the country. The nature of the rocks is very varying in the longitudinal extent of the dyke, the intrusive masses presenting themselves now as serpentine and olivin-rock (Dunit), now as syenite or diabase; at other points as diallage-rock and pyroxene-porphyry. To the serpentine and olivin range in the South of the city of Nelson belongs the famous Dun Mountain, the copper-ore and chromate of iron of which has for several years past given rise to mining enterprises.

On Cook's Strait those ranges terminate in numerous islands and peninsulas enclosing those fiord-like inlets and sounds (Pelorus Sound; Queen Charlotte Sound etc.), which already in Cook's time were noted as most excellent harbours. Towards South the mountains grow higher and higher. Ben Nevis and Gordon's-knob, visible from the heights about Nelson, rise already to a height of 4000 feet above the level of the sea; but then the mountain-range is interrupted by a pass leading from the Motueka valley along the Big Bush Road to the Wairau valley. On the southern shores of Lake Rotoiti the mountains rise again forming Mounts Travers and Mackey, and further in a southwesterly direction ascending in the Spencer mountains (Mt. Franklin and Mt. Humboldt) to a height of 10,000 feet, far beyond the limits of perpetual snow. These ranges are covered, to an altitude of 4000 to 4500 feet with dense forest, above which an Alpine vegetation begins; the summits form meadows of short smooth snow-grass, the whole reminding one strongly of the Alpine scenery of Switzerland. The group of the Spencer mountains forms the central knot, from which nearly all the principal rivers of the province of Nelson rise, the Wairau, Waiautoa (Clarence) and the Waiauua (Dillon) running towards the East coast; and the tributaries of the Kawatiri (Buller river) and Mawhera (Grey river), which empty into the sea on the West coast. It is very remarkable, that the most elevated heights of the province consist of sedimentary rocks, the strata of which have been upturned nearly vertically. The sandstones and slates, however, being destitute of fossils, it has as yet been impossible to ascertain their exact geological age. From other reasons we may infer that in those mountains we have to do with strata corresponding partly with Silurian and Devonian, partly with Triassic formations in Europe.[1] The eastern-most portions of this mountain-system, beginning at Pelorus-Sound — including the Wairau plains and the broad longitudinal valleys of the Wairau, Awatere and Waiautoa (or Clarence) river, and likewise comprising the seaward and landward Kaikoras, with the towering peaks bearing the names of Scandinavian deities, such as

[1] See Chapter III. p. 56—57.

Odin, 9700 feet high (the Maori name is Tapuenuka), Thor (8700 feet) and Freya (8500 feet) — have been separated from the province of Nelson since 1859 as the new province of Marlborough. [1]

Between the Eastern and Western Ranges is the deep indentation of the coast which forms Blind Bay, and from the southern extremity of which the land rises gradually towards South to a height of 2000 feet above the level of the sea, forming an undulating hilly country, which extends along the lakes Rotoiti and Rotorua, past Mt. Murchison to the point, were Southeast of the Spencer

View of Lake Rotoiti in the Province of Nelson.

mountains the Eastern and Western Ranges, converging in the line of their strike, meet together. The hills near Nelson bear the name of the Moutere hills. They are intersected by numerous rivers running in deeply channelled, terraced valleys. The Motueka and Waimea rivers empty into Blind Bay, while the Buller river takes a transverse direction from the Eastern to the Western Ranges

[1] The Capital of this new province is Picton on the Waitohi Bay, one of the inmost recesses of Queen Charlotte Sound, with about 800 inhabitants. The province contains good, arable soil and pasture-grounds (about 200,000 acres), exporting already from the ports Underwood and Queen Charlotte Sound considerable quantities of wool. The landward Kaikoras, according to the statements of some colonists in the vicinity, are said to be of volcanic origin. Perhaps those colossal mountain-cones consist of porphyry, or, if bearing a more recent date, of trachyte and andesite.

and towards the West coast. The hills are composed of irregular and imperfectly stratified beds of shingle, gravel, sand and clay, resting upon tertiary strata. These beds are of quaternary age, and being part of the generally diffused drift formation, which fills up all the principal valleys and covers all the flats amongst the mountains, afford evidence of the great action of water and ice upon the surface within a comparatively very recent period. The large dam composed of colossal rugged blocks of rock, which borders the lower end of lake Rotoiti, may be looked upon as the moraine of a former glacier.

It is, no doubt, in consequence of the peculiar configuration of the country just described that the shores of Blind Bay are favored with the excellent climate, for which they are famous. However heavy the gale in Cook's straits may be, Blind Bay is always calm. From the swell of the sea the Bay is protected by the land projecting to a great distance near Separation Point and d'Urville's Island; while the mountain-ranges converging towards the South form a regular wedge warding off the violent atmospheric currents from the South. Ships, therefore, find allways shelter in Blind Bay from the dreaded gales that rage in Cook's Straits; and the town of Nelson, situated on the Southeast shore of the Bay at the immediate foot of the Eastern ranges, unlike other cities on the coast of New Zealand, which are rather too much subject to wind, enjoys a soothing calm, which combined with a clear and rarely clouded sky renders its climate the pleasantest in New Zealand. Nelson, therefore, has been justly styled "the garden of New-Zealand."

The town of Nelson was founded but few years after Wellington, and was the second settlement formed by the New Zealand Company in Cook's Straits. In February 1842, the first vessel arrived with a number of immigrants, and the 25th May of the same year is marked in the annals of the city as the ever memorable day, when the first plough penetrated the virgin soil of the new colony. Notwithstanding sore and heavy trials, which the infant-colony had to undergo — already in 1843, in a fatal encounter with the natives, who under the command of Rauparaha and Rangihaiata opposed the colonization in the Wairau, they lost a number of their best men — it has steadily gained ground; and when upon a closer exploration of the country, coal, copper-ore, chrome-ore, plumbago and gold were discovered, the fame of Nelson was at once established as being the principal mineral-country of New Zealand. In 1862 the province numbered about 10,000 inhabitants, 5000 of whom reside in the town and its vicinity. The town lies close by the foot of the mountains, being built upon a kind of alluvial delta, which is formed by the confluence of two small streams, named the Maitai and Brookstreet creek, extending also up their valleys, and along the ranges of hills lining the harbour. On account of its beautiful site and its delightful climate, Nelson is justly considered one of the most pleasant places of sojourn in New Zealand. The impression made by the snug little cottages, surrounded by beautiful gardens, is an extremely cheerful one. As the rows of houses in the principal streets are already closing up more and more, and larger buildings are growing up, the place gradually improves in city-like appearance. On the 26th August 1859 the laying of the corner-stones to new Government buildings took place with all due solemnity, and on that occasion I was honored by the inhabitants with the office of laying the corner-stone to an edifice designed for the noble purposes of art and science, to the Nelson Institute. Certainly a most cheering and memorable epoch in the history of the development of the young colony, when the enterprising pioneers, — after the toils and labours of their first settling down had succeeded, after their houses had been roofed over, and fields and meadows put in due order, — now direct their attention also to the nobler purposes of life, to the nursing of the blossoms and fruits of our civilization, of art and science! The Protestant Church is situated upon a commanding elevation in the centre of the town. There are two bridges across the Matai river, a suspension-bridge and a wooden-bridge; and Nelson can even boast of a rail-road, the first constructed upon New Zealand soil. It is the work of the Dun Mountain company, constructed for the purpose of developing the mines of chrome-ore on the Dun Mountain, and leads from the harbour through the town along the Brookstreet valley.

The harbour of Nelson is safe, but small and difficult of access to larger sailing-vessels. It owes its formation to a most singular "boulder-bank" which extends eights miles along the coast, forming a natural dam, behind which there is a narrow and shallow arm of the sea, which grows deeper at its southern extremity, where it communicates with Blind Bay, and here forms the harbour. The

Entrance to the Harbour of Nelson,
with part of the Boulderbank.

entrance to the harbour is between the southern extremity of the boulder-bank and the mainland, but is narrowed so much by the Arrow rock,[1] a rock rising in the middle of it, that the navigable channel is only 50 yards wide. Owing to the extremely swift current of the tide in this narrow channel and its shallowness, larger vessel can pass in and out only in time of high-water, and are moreover obliged to improve the tide in coming or going. These unfavorable circumstances would greatly disparage navigation, but for the excellent anchoring places outside the harbour, which by

[1] The Arrow rock consists of altered schists streaked with veins of quartz. It is covered all over with *Mytilus* and *Balanus* as far as highwater-mark.

the sheltered situation of Blind Bay are safe in almost any kind of weather. During northwesterly gales the neighbouring Croixelles Harbour offers a perfectly safe place of refuge. The boulder-bank is one of the natural curiosities of Nelson. It consists of rounded pebbles on boulders. In time of high-water a large portion of it is under water; at low-water it is dry throughout its whole length. The largest and heaviest boulders are towards the sea side; on the harbour side the boulders grow smaller; and at a point close by the entrance to the harbour they are so small, that vessels there can drive on the strand without any damage, thus using the place as a natural dry-dock in consequence of the great difference of the water level between ebb and flow.[1] The boulders consist nearly all of one and the same kind of a syenite, containing blackish-green hornblende, flesh-coloured feldspar and a small quantity of iron-pyrites. On following the narrow bank from South to North, it is easily observed, that the boulders towards North grow larger and more angular, and originate from a precipitous bluff of syenite, called "Mackay's knob" which abuts upon the sea a little beyond Drumduan, the residence of Mr. Mackay. The fragments constantly falling from the cliffs are gradually rolled towards South by the heavy northerly swell combined with a strong current of the sea, passing in time of springtide with considerable velocity along the coast.[2] The reason of their

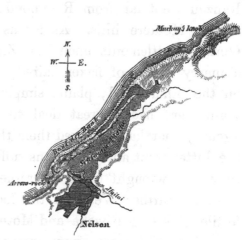

The boulder-bank, Nelson Harbour.

[1] The springtide in Nelson Harbour rises 14 feet.

[2] In a similar manner the remarkable sand-bar, Cape Farewell Spit, extending from Cape Farewell 22 miles into the sea, owes its origin to the currents of the sea running on the one side along the West coast, on the other along the shores of Golden Bay. The direction of the sand-bar is the resultant of the direction of those currents and of the moving force of the westerly and north-westerly winds prevailing on the West coast.

being deposited on the existing line is, that in all probability a submarine reef underlies them, of which the Arrow Rock, in the entrance of the harbour may be regarded as the southern termination. The boulder-bank in front of Nelson is a rich field for the zoologist. At neap-tide the sea-side teems with all sorts of fishes, sea-pads, sea-urchins, muscles and snails, while upon the bank itself beautiful spongias are found, which are cast out by the surf.

An excellent road leads from Nelson in a S. E. direction, through the agricultural districts of the Waimea and Waiiti plains covered with the most luxuriant meadows and fields. Upon the most fertile alluvial soil there is here farm joining farm, and smaller and larger boroughs are springing up. There is a Richmond with a "Star and Garter Hotel," the proprietor of which is striving to establish the well-earned reputation of that title, so renowned on the banks of the Thames, also among the Antipodes; furthermore Stoke, Hope, Spring-Grove, Wakefield, and what more names there are of villages and boroughs. We also find two German villages: Ranzau, not far from Richmond, and Sarau situated farther East on the Moutere hills. As far as I know, these two are the only German settlements upon New Zealand. At Sarau I was hailed by a merry crowd of flaxen-haired, blue-eyed children; the old folks on the other hand, plain, simple peasants from Mecklenbourg and Hannover, had a great deal to relate about the rascality of the agents, that had decoyed them thither; of the sad disappointments, the bitter want and privations suffered in the first years, until later they had wrought out a tolerable existence by the sweat of their brows. Farther West, at the foot of the Western ranges are the fertile plains of Riwaka and Motueka, which, but fifteen years ago a perfect wilderness, now present the most charming sight: luxuriant meadows with magnificent cattle grazing upon them; thriving fields and orchards, interspersed with the dwellings of the settlers. The white glistening snow-peaks in the back-ground remind us of the most charming valleys of our Alps. In order to conceive a correct idea of the amount of labour required to transform those plains into smiling fields and meadows, let the traveller proceed up

New-Zealand, South Island.

A. Campbell del

M. Arthur-von post

A. Mermann. sc.

Motueka-Valley near Nelson.

the valleys. A single day's journey in that direction will suffice to advance from the cultivated parts on the shores of Blind Bay in a southerly direction into those districts, where wood-cutters and shepherds form the farthest outposts of civilization, and a wilderness commences, the virgin-soil of which man has scarcely ever set foot upon, — a wilderness of bush, swamp, and rock. Each wood-cutter's or shepherd's hut, however scanty, which offers a hospitable shelter to the way-worn traveller, is valued as highly at these borders of an uninhabited wilderness as the Oases in the desert or a solitary island in the ocean. With peculiar feelings one leaves the last inhabited hut, for the purpose of exploring unknown regions, thence taking a direction, where there is no path to lead one on; and whereever the eye roves into the distance over hill and dale, no vestige of human beings is seen. With difficulty one works a way through woods and thicket, and follows the river-banks across uniform grass-plains; with great trouble, and even with danger one crosses rapid mountain-streams, climbs over rocks and mountains, and in short, has all sorts of obstacles to encounter. No one can tell to where one will come, and with joyful surprise one views from open heights the novel landscape. Mountains, valleys and rivers are as yet without names; they are named according to the accidental notion or taste of the explorer, according to recollections from the dear old home, or after distant friends and acquaintances; he fancies himself living in future periods, when all those plains and valleys will be inhabited even to the remotest snow-mountain, the hoary peaks of which are looming up on the horizon; when commodious roads and lanes will enable the traveller to reach in one day the same point, to arrive at which weeks of toilsome and dangerous journey are required at present.

Having already given a detailed account of the coal-beds and gold-fields of the Province in former chapters, I shall here limit my remarks respecting the mineral treasures of the province to the copper and chrome-ore mines on the Dun Mountain, a few miles Southeast of the town. On approaching the harbour of Nelson from the high sea, a bare mountain ridge is seen rising to a height of about

4000 feet which owes its name "Dun Mountain" to the rusty-brown (dun) colour of its surface. It consists of a very peculiar kind of rock, of a yellowish-green colour when recently broken, but turning rusty-brown on the surface when decomposing. The mass of the rock is olivine, containing fine black grains of chromate of iron interspersed; it is distinguished from serpentine for which it was formerly taken, especially by its greater hardness, and its crystalline structure. I have called it Dunite.[1] The copper-mines, however, which are worked by an English Company (the Dun-Mountain Copper-Mining Company, residing in London), do not lie on

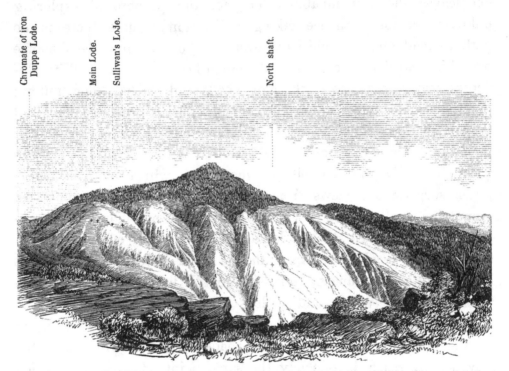

View of the Wooded Peak.

[1] Analysis of the Dunite, by R. Reuter (Labor. of the polyt. Institute of Vienna)

Silica	42.80	sp. Gr. 3.30
Magnesia . . .	47.38	
Protoxide of iron	9.40	
Water	0.57	
	100.15	

The Olivin-rocks are of rare occurrence. In Europe they are found principally at Lake Lherz in the Pyrenees, and therefore called Lherzolite.

the Dun Mountain proper, but on the opposite declivity of a mountain-ridge, the highest point of which is named Wooded Peak. In 1856, sixteen tons of excellent copper-ores, principally red oxide of copper and native copper were shipped to England, only as at kind of sample as it were, but the enterprises of the company established in consequence of it were not crowned with the eagerly expected success; and at the time of my stay at Nelson the Dun Mountain was the subject of a hot controversy, as the enthusiastic anticipations and promises of an old Cornish miner, whose vivid imagination beheld in every superficial trace of ore the richest lodes, were in direct contradiction with the unsatisfactory results, obtained by the technical leader of the enterprise. I regret much to say that the results of my examinations were not such as to corroborate sanguine hopes. The escarpments of Wooded Peak consist of serpentine traversed by dykes of diallage-rock. The mass of the larger dykes is coarse-grained, and very fine specimens of diallage may be knocked off. In this serpentine mountain, over a line of about two miles running from South almost due North, traces of copper-ores are here and there met with in the shape of green and blue silicates of copper (rarely malachite), forming thin crusts upon the crumbling serpentine. Nearly in every instance, where such indications appeared upon the surface, farther researches at a greater depth produced smaller and larger nests of red oxide of copper, and of native copper; sometimes also bunches of copper-pyrites, of purple-copper and of copper-galena, which, however, soon disappeared again. I could not convince myself of the existence of a number of parallel lodes, so as to justify the various names which have been given, and which appear to designate different lodes such as Sullivan's lode, Main lode, Windtrap-Gully lode, Duppa-lode etc. The copper ore does not occur in a regular lode; by which I mean a metalliferous dyke of different mineral composition from that of the rock of the mountain. As is usual in serpentine, the copper ore occurs only in nests and bunches. The richer deposits of ore form lenticular-shaped masses, which, when followed, may increase to a certain distance, but then disappear

Lenticular mass of ore (9 feet long, 2 feet broad), in the serpentine of the Wooded Peak near Nelson.

a. Iron-ochre and decomposed serpentine.
b. Nests of copper-ore.

again in a thin wedge. Where these nests are large and rich, one alone may sometimes make the fortune of a mine. The richest found on the Dun Mountain appears to have been that of the Windtrap Gully, from which pieces of native copper, some of them weighing as much as eight pounds, were extracted. The green and blue silicates of copper are surface minerals, which are only of value by showing the direction of the fissure in which the real ore may be looked for at a greater depth; at a certain distance below the surface they disappear entirely, and it is only by the broken and softened character of the serpentine that the miner is enabled to follow the fissure from one deposit of metal to the other. The occurrence of the best indications of copper ore on the surface over a continuous line of about two miles, affords good ground for supposing that considerable quantities of ore are contained in the mountain; but, on the other land, owing to the manner in which the ores occur in isolated bunches, mining operations in such a region are always attended by less certain profits than where the metal is deposited in a regular lode.

The Dun Mountain Company, therefore, has within the last years devoted all their attention and energy to the systematic and successful working of the chromate of iron which occurs in the same serpentine mountain in large quantity. Whole groups of rocks on the sides of the Wooded Peak consist of almost pure chrome-ore; and there is no doubt, that the lodes of this ore continue with more regularity than the copper-ores.[1] In order to facilitate the hitherto so very difficult and expensive transportation from the heights of the mountain to the harbour of Nelson, the company has constructed a railroad, which runs from the harbour through

---

[1] The chromate of iron of the Dun Mountain is but little inferior to the best ore from the mines in the vicinity of Baltimore in North America. The price of a ton of ore is in England about £10.

the Brookstreet valley and in numerous curvatures to the height of the mountain.

In the northern continuation of that extensive serpentine mass which is met with on the Dun Mountain, traces of copper-ores have also been found on the Croixelles Harbour, on Current Basin and upon d'Urville's Island. But on those points, also, the experiments made have as yet been without any striking success. Nelson, however, may well be satisfied. Its beautiful agricultural and pasture-land, its coal-mines, and its gold-fields, are rich sources of wealth and prosperity, and the discoveries of future years[1] may perhaps even realize in their full extent the past and present anticipations of great quantities of copper and other ores existing in that province.

---

[1] The Broth. Curtis in 1861 opened extensive beds of plumbago near Pakawau.

CHAPTER XXI.

The Southern Alps.

The Southern Alps proper. — Explorers. — Dr. Haast's merits. — Sacrificed lifes. — Summits of the central range: Kaimatau, Mt. Tyndall, Mt. Cook and Mt. Tasman. Mt. Aspiring. — Structure of the central chain. — Strike of beds. — Jointed structure of the rocks. — Two systems of valleys. — Passes. — Geological features. — Two sections through the South Island. — The glacier regions. — Rangitata valley. — Terraces. — Enormous mass of detritus. — The Forbes glacier. — Red snow. — The Clyde glacier. — The Ashburton glacier. — Lake Te Kapo. — The Godley glacier. — Lake Pukaki. — The glacial region near Mt. Cook. — The great Tasman glacier. — The Hourglass glacier. — Glaciers on the West Coast. — The Francis Joseph glacier. — Difference between the climate of the West Coast and East Coast. — Signs of enormous glaciers of the pleistocene period. — Causes of the pleistocene glaciation. — Extensive distribution of the drift deposits. — The Canterbury plains. — Their formation. — The rivers of the plains.

Appendix. The Otira road, the first road across the Southern Alps.

The Southern Alps proper commence South of the saddle between the Teramakau[1] and Hurunui rivers, on the boundary between the provinces of Nelson and Canterbury. Here, in the middle of the Southern Island, the mountains attain their greatest height; and as far as Haast's Pass on the boundary of the Province of Otago, leading from Lake Wanaka to the West Coast, — a distance of 200 miles — they form in the direction from N. E. to S. W. a chain of towering mountains, which as to the height of their summits,[2] and as to size and extent of their snow-fields and glaciers, rival with the Pennine and Rhaetian Alps. The first navigators on the coast of New Zealand looked already with wonder at those

[1] *Alias* Taramakau or Teremakau.
[2] Between 11,000 and 13,000 feet above the level of the sea.

magnificent Alpine heights clothed in perpetual snow, the giant summits of which now bear the names of Cook and Tasman. The wild forms of the huge rocky masses on the West Coast towering to the skies, and bidding defiance to the terrific breakers, were always an object of deep admiration to the sailors visiting these shores; but up to our times this mountain region remained a wilderness, untrodden by the foot of man. On the discovery of New Zealand it was uninhabited, — for the natives shunned this solitary mountain wilderness, — and it has remained uninhabited to this very day; incontestably one of the most remarkable and the grandest objects, which has been reserved for the physico-geographical and geological investigations of our time. It is only within the last ten years, — since European colonists from the rapidly rising and prosperous settlements of Lyttleton and Christchurch, situated in the neighbourhood of the excellent harbours of Banks Peninsula, have taken possession of the fertile plains at the eastern foot of the Alps, — that any attempt has been made to penetrate the unknown mountain districts. Some few squatters have advanced with their runs to the grassy valleys and downs at the foot of the Alps; and of late years intrepid men, inspired by a bold spirit of discovery, have penetrated to the icy glaciers of the highest mountain masses, while the discovery of the rich goldfields on the West Coast has accelerated the construction and opening of the first road across the mountains.[1] Thus we see now the chaos of this grand alpine system become, through geographical and geological explorations, more and more disentangled and its formation and structure brought to light.

Foremost among the alpine explorers stands the name of my energetic friend, Dr. J. Haast, who as Government geologist of the province of Canterbury penetrated, in 1861, to the sources of the Rangitata, in 1862 undertook the task of exploring the headwaters of the Waitaki in the neighbourhood of Mount Cook, and in the last years repeatedly has crossed the dividing range to the West Coast. His animated descriptions and interesting communi-

[1] The Otira Road See Appendix to Chapter XXI.

cations form the main part of this chapter, and we shall follow his footsteps into the very heart of the Alps. But in mentioning first of all the name of my friend Haast, I must not omit to mention also the names of the other explorers, who for the sake of science and knowledge were induced to undergo hunger, fatigue, cold and all the dangers connected with New Zealand alpine explorations, and some of whom by their bold spirit of discovery were even led to sacrifice their lifes. [1] To the geologist Dr. Hector, to the botanist Dr. Sinclair, to the surveyors and civil engineers Messrs. Brunner, J. T. Thomson, E. Dobson, G. Dobson, A. Dobson, W. T. Doyne, H: Whitcombe, Charlton Howitt, Rob. Park, Browning, J. Rochfort, James and Alexander Mackay, J. Burnett and many others, to all of these a tribute is due for their merit of contributing to the knowledge of the New Zealand Alps.

Three colossal peaks, towering up to a height of 11,000 to 13,000 feet above the level of the sea, are most prominent in the central range of the Southern Alps. In the North, the colossal snow-pyramid of the Kaimatau (lat. 42° 58'; long. 171° 35'), the ice-fields of which are feeding the sources of the Waimakariri; further South, Mount Tyndall (lat. 43° 20'; long. 170° 46) with its neighbours Mount Arrowsmith, Cloudy Peak and Mount Forbes 9000 to 10,000 feet high, the glaciers and snow-fields of which give rise to the Rangitata; finally Mount Cook (lat. 43° 36'; long. 170° 12') with its giant neighbour Mount Tasman [2] and the adjoining peaks of Mount Petermann, Mount Darwin, Mount Elie de Beaumont, Mount De la Beche and Mount Haidinger, from which the headwaters of the Waitaki rise. Although some of the last named peaks are almost equal in height to Mount Cook, yet the

---

[1] Dr. Sinclair found his death, while crossing the Clyde river in 1861; Mr. Whitcombe was drowned by the upsetting of his canoe on the Teramakau in 1863; Mr. Howitt was lost on lake Brunner in 1863; Mr. G. Dobson was murdered on the West coast in 1866.

[2] Mount Tasman has been named by Dr. J. Haast in 1865, its height estimated at 12.300 feet. On the Admiralty maps the northern peak of the Mount Cook range (probably Mount Tasman) is noted at 12.200 feet, the southern peak (Mount Cook proper) at 13.200, while Mr. Thomson, Chief Surveyor of the Otago Province in his reports to Mount Cook designates an altitude of only 12.460 feet.

Mt. Alexander. Mt Haast. Mt. Hooker.   Black Hill.   Grey River.   Mt. Hochstetter.   Mt. Herschel.

The Western Ranges of the Province of Nelson.

View from the junction of the Grey and Mawheraiti towards East.

Mt. Cook.

The Southern Alps.

View from the mouth of the River Grey, from sketches by Dr. J. Haast.

latter surpasses by far all the other peaks in grandeur. The gigantic snow and rock-pyramid of this mountain terminates in a curved sharp ridge, the northern point of which is about 600 feet higher than the southern, and the sides of which are so keen and steep, that an ascent of it seems impossible. The first attempt to ascent the Mount Cook Range was made by Dr. Haast in 1862 and is described by him, as follows. [1]

"The weather, which for several days had been very tempestuous, cleared up at last, and on the 12th of April at daylight, we started to ascend Mount Cook Range. It was a cold but sunny morning, and with great expectations we climbed through the fagus forest, which, for the first six or seven hundred feet, intermixed with sub-alpine shrubs covers the side of the range. After leaving the forest, we came to alpine vegetation, becoming still more characteristic about 1800 feet above the valley amongst the rocks, where we climbed along to the crest of the mountain leading towards Mount Cook proper. But, although the ridge, as seen from the valley, seemed quite smooth, it consisted of huge rocks, broken up into very sharp prismatic fragments lying loosely upon each other, often with deep precipices on both sides, where one false step would have cost life or limb. Soon patches of snow appeared which were remaining from the last storms, over which we worked our way higher and higher. The view became every moment grander, and, having reached an altitude of 6500 feet, I established my first station. Although the sun shone brilliantly from a cloudless sky, it was extremely chilly in the shade amongst the rocks, where we went to shelter ourselves from the icy blowing winds. The thermometer at 11 o'clock stood below freezing point. Again on our road, the rocks became more and more broken; hitherto they had consisted of dioritic sandstones, but now slates made their appearance, and about 7500 feet above the sea we came upon a precipice of about 10 feet wide and 30 feet deep, the vertical stratum of clayslates between two others of dioritic sandstone having

[1] Dr. J. Haast, Report on the headwaters of the River Waitaki, Christchurch 1865.

been here removed; and as it was impossible to round it, and we had no ladder with us to throw across, we were obliged to retreat. The view from this point was admirable in the extreme. The bold tent-like form of Mount Cook proper occupied the foreground, surrounded by many peaks of every conceivable shape. Deep below us the great Tasman glacier carried slowly but steadily its heavy detritus load down to its terminal face, whilst towards the South the large watershed of Lake Pukaki appeared on the horizon."

From the main sources of the river Rakaia to Mount Cook Range the Alps consist of one chain sending off divergent branches, which become gradually lower till they reach the Canterbury plains; but near Mount Holmes at the southern termination of the Mount Cook Range and its branch, called the Moorhouse range, an important change in their orographical structure occurs. The Southern Alps divide into two almost equal chains, of which the eastern one extends along the western bank of the river Hopkins to Mount Ward, whence it turns in a south-west by west direction towards Mount Brewster. The western chain commences also near Mount Holmes, losing near the sources of the Hopkins somewhat in altitude, but afterwards rises again to a great height, being formed by magnificent ranges, of which Mount Hooker and the Grey range are the most conspicuous. It runs in a south-west direction till it is broken through by the river Haast, after the junction of the Clarke, the broad valley of the latter occupying the space or basin between those two alpine chains. After this second break the Alps rise again, on the left side of that river, to a considerable altitude covered with vast fields of perpetual snow, both chains uniting at Mount Stuart, and running down in one longitudinal chain towards Mount Aspiring (lat. 44° 25′, long. 168° 49′) a mountain situated upon the frontier-line of the provinces Canterbury and Otago, and represented as a colossal cone of very regular shape, 9135 feet high.

It is a singular fact that the three principal peaks above named are all situated precisely in one and the same straight line, striking in the direction from E. 35° N. to W. 35° S. This line, passing towards Southwest over the Pembroke Peak (6710 feet) intersects the West Coast near Milford-Haven, at the point where the striking fiord-formation on the Southwest Coast of the Island commences. Towards Northeast it strikes the coast between Cape Campbell and Queen Charlotte-Sound on Cook's Strait, thus intersecting the South Island diagonally and forming the dividing line between the West and North Coast on one side and the East and South Coast on the other side.

The explorations of Dr. Haast at the head-waters of the Molyneux, the Waitaki, the Rangitata, and the Rakaia, and those of the Provincial Engineer Mr. E. Dobson in the upper valley of the Waimakariri, have fully established the fact that throughout the entire length of the Southern Alps in the province of Canterbury there are only three real passes, viz, the Hurunui Saddle, dividing the sources of the Hurunui and Teramakau; Haast's Pass at the head of Lake Wanaka, which leads over a very low saddle into the valley of the Haast river, which falls into the sea near Jackson's Bay; and Arthur's Pass, which is nothing more than a great fissure, running in a tolerably direct line from the valley of the Waimakariri to that of the Teramakau. The so-called North Rakaia Pass has no real claim to the title; its eastern face being simply a wall rising abruptly from the valley to a height of 1500 feet, and being at so great an elevation as to be buried deep in snow during eight months in the year.

This absence of passes throughout so great a distance can only be accounted for by the very peculiar structure of the central chain, as clearly explained by Mr. E. Dobson in his able report upon the Passes through the dividing range of the Canterbury Province. [1] The first point, says Mr. Dobson, to be noticed in regard to the central chain is, that it does not, as is popularly supposed, present an unbroken line of watershed, but rather a series of peaks and broken ridges separated from each other by deep ravines, and for the most part perfectly inaccessible. The clue to this

---

[1] E. Dobson, C. E., Report upon the Gorge of the Otira and upon the Character of the Passes through the dividing Range of the Canterbury Province. Christchurch 1865.

system of ravines and ridges is to be found in the fact that the palæozoic rocks forming the main range, have been at a very early period subjected to intensive pressure, the effect of which has been to crumple them up into huge folds, the upper portions of

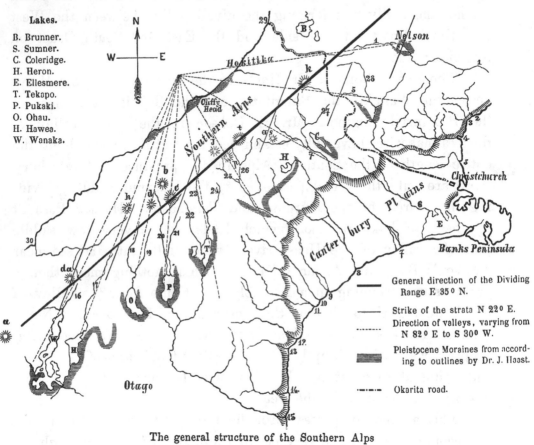

Lakes.

B. Brunner.
S. Sumner.
C. Coleridge.
H. Heron.
E. Ellesmere.
T. Tekapo.
P. Pukaki.
O. Ohau.
H. Hawea.
W. Wanaka.

General direction of the Dividing Range E 35° N.

Strike of the strata N 22° E.

Direction of valleys, varying from N 82° E to S 30° W.

Pleistocene Moraines from according to outlines by Dr. J. Haast.

Okarita road.

The general structure of the Southern Alps
reduced from a map by E. Dobson C. E

Mountains.		Rivers.	
k. Kaimatau.	1. Hurunui.	11. Opihi.	21. Jollie.
as. Arrosmith.	2. Waipara.	12. Pareora.	22. Cass.
t. Tyndall.	3. Kowai.	13. Otaio.	23. Godley.
j. Keith Johnstone.	4. Ashley.	14. Waihao.	24. Macaulay.
p. Petermann.	5. Waimakariri.	15. Waitaki.	25. Havelock.
c. Cook.	6. Selvyn.	16. Makarora.	26. Clyde.
b. Beaumont.	7. Rakaia.	17. Hunter.	27. Harper.
d. Delabeche.	8. Ashburton.	18. Hopkins.	28. Poulter.
h. Hooker.	9. Rangitata.	19. Dobson.	29. Teramakau.
da. Dana.	10. Orari.	20. Tasman.	30. Haast.
a. Aspiring.			

which have been removed leaving the remaining portions of the strata standing up on edge, either in a vertical position, or at very steep inclinations. The strike of the beds, corresponding with the direction of the axes of the foldings is tolerably regular, being generally about N. 22 deg. E. (true), thus differing from the general direction of the dividing range by 33 degrees. At the same time it is important to observe that the rule which has been found to prevail in other mountain chains of similar formation, appears to hold good also in the central chain, viz., that the greatest amount of denudation has taken place along the original ridges which are now occupied by valleys, whilst the existing peaks and ridges are on the sides of former depressions.

The next feature to be noticed is the jointed structure of the rocks. Although the joints cross each other in all directions, apparently without order, there are two prevailing systems of joints, which have an important influence on the configuration of the surface. These are: first, a system of vertical cross joints at right angles to the stratification and running in unbroken lines for great distances with such regularity that they might easily be mistaken for planes of stratification, were it not for the frequent occurence of beds of trap rock, the outcrop of which marks unmistakeably the true bedding; secondly, a system of joints more or less inclined to the horizon, not running in parallel planes but arranged in a series of curves radiating from a common centre.

The effect of this system of jointing combined with the strike of the beds or the direction of the axes of folding is to produce two distinct systems of valleys in the central chain, the direction of which i very remarkable. The one radiates from a common centre, situated about 50 miles North of Mount Darwin in the sea near Cliffy Head. This system includes all the principal valleys, from the Teramakau on the North to the Makarora on the South, their direction varying from N. 82° E. to S. 30° W., giving the idea that the country has been starred, just as a mirror is starred by a violent blow; or, as in rock blasting, a set of radiating fissures is sometimes produced by a single shot. To the other system belong the valleys of rivers and watercourses running either on the strike of the beds or in the direction of the cross joints, or in a compound zigzag course, following alternatly these two directions, like a line struck diagonally across a chess board, but following the sides of the squares, and giving to the cliffs which bound these valleys a peculiar rectangular appearance, resembling ruined masonery on a gigantic scale.

With the exception of the Hurunui valley none of the radiating valleys runs directly across the main chain, which at the heads of the Rakaia and Waimakariri, stands up like a wall barring all further progress. Haast's Pass, the lowest, and probably the easiest of all, does not extend across the northern branch of the chain, but leads to the coast by following the westerly course of the Haast river.[1] Arthur's Pass does not, as it were, cross the range in a direct line as does that by the Hurunui, but leads along it from one radiating valley to another; the Waimakariri and the Teramakau overlapping each other to the extent of about 20 miles. Thus it will be understood that the three passes occur under three distinct sets of conditions. Haast's Pass, at the head of the Wanaka Lake, is both in the line of one of the great radiating valleys and also in the direction of the axis of the great foldings of the strata, these two causes in combination having formed an unusually low gap in the mountains. The Hurunui Pass, on the other hand, is one of the fractures running directly across the range, whilst Arthur's Pass[2] is a fissure parellel to the planes of stratification, from which the rock already bruised and shattered, when the surface of the country was crushed up into the huge foldings before referred to, has been gradually removed by glacial action and by the weathering process constantly going on over the whole face of nature.

The general features of the Geology of the Southern Alps have been established by Dr. Haast's and Dr. Hector's researches in the provinces of Canterbury and Otago[3] and may be illustrated by the accompanying sections.

---

[1] Haast's Pass has not yet been opened, even for foot traffic; but it will probably one day become a very important line of communication between the West Coast of Canterbury and the northern part of the Otago Province.

[2] 1866 a bridle road was opened over Arthur's Pass, descending into the Teramakau by the gorge of the Otira. See Appendix.

[3] Dr. J. Haast, Report on the Geological Survey of the Pronvince of Canter-

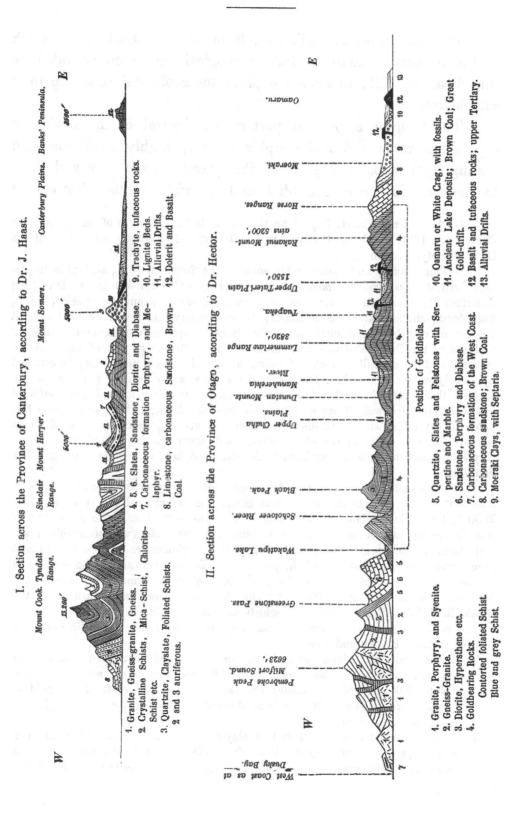

I. Section across the Province of Canterbury, according to Dr. J. Haast.

1. Granite, Gneiss-granite, Gneiss.
2. Crystalline Schists, Mica-Schist, Chlorite-Schist etc.
3. Quartzite, Clayslate, Foliated Schists. 2 and 3 auriferous.
4. 5. 6. Slates, Sandstone, Diorite and Diabase.
7. Carbonaceous formation Porphyry, and Melaphyr.
8. Limestone, carbonaceous Sandstone, Brown Coal.
9. Trachyte, tufaceous rocks.
10. Lignite Beds.
11. Alluvial Drifts.
12. Dolerit and Basalt.

II. Section across the Province of Otago, according to Dr. Hector.

Position of Goldfields.

1. Granite, Porphyry, and Syenite.
2. Gneiss-Granite.
3. Diorite, Hypersthene etc,
4. Goldbearing Rocks. Contorted foliated Schist. Blue and grey Schist.
5. Quartzite, Slates and Felstones with Serpentine and Marble.
6. Sandstone, Porphyry and Diabase.
7. Carbonaceous formation of the West Coast.
8. Carbonaceous sandstone; Brown Coal.
9. Moeraki Clays, with Septaria.
10. Oamaru or White Crag, with fossils.
11. Ancient Lake Deposits; Brown Coal; Great Gold-drift.
12. Basalt and tufaceous rocks; upper Tertiary.
13. Alluvial Drifts.

The formations are following from West to East pretty much in the natural sequence of their geological age; sections taken in this direction will, therefore, explain the geological structure most satisfactorily.

The western slope and part of the central chain consists of crystalline rocks and metamorphic schists, highly auriferous and resting upon a basis of granite, that presents itself here and there to the view in the rugged bluffs and declivities on the West coast.[1]

bury 1864. — Dr. J. Haast, Report on the Geological Exploration of the West Coast Canterbury 1865. — Dr. J. Hector, On the Geology of Otago (Quart. Journ. of the Geol. Society, London 1865.

[1] This granitic and metamorphic zone of the Southern Alps, so highly important from the auriferous character of the rocks (See Chapter V.) is, as Dr. Haast describes it, composed of varieties of granite, syenite, pegmatite, gneiss, mica-, hornblende-chloritic and talcose schists, dioritic porphyries, trap rocks and a great variety of clayslates and quartzites. Dr. Hector explains the general geological features of the Otago Province as follows: "The southwestern part of the Province is composed of crystalline rocks, forming lofty and rugged mountains, intersected by deeply cut valleys, which are occupied on the west by arms of sea, but on the east by the great lakes. The base-rock of this system is a foliated and contorted gneiss. Associated with it are granites, syenites and diorites. Wrapping round this batch of crystalline strata and sometimes resting at great altitudes (say 5000 feet) on its surface are a series of hornblende-slates, soft micaceous and amphibolic gneiss, clay-slates and quartzites, associated with felstone-dykes, serpentine, and granular limestone.

To the eastward of these two formations the country is traversed by a depression that widens towards the south, and enters the mountain-chain by a pass only 2000 feet in altitude above the sea; this is the Greenstone Pass. Along the line of this depression and resting on the last mentioned slates etc. (unconformably?) are well bedded sandstones, shales and porphyritic conglomerates, together with soft greenstone-slates and diabase-rock in patches. — To the eastward of the depression we have a great development of the auriferous schistose formations, shaped as a flattened boss, and rising to 4000 feet. The strike of the auriferous schists is on the whole N. N. W. I have divided the schists into three parts.

1. Upper. A grey arenaceous, almost slaty rock containing but little quartz in the form of veins and laminæ.
2. Middle. Soft blue slates, often highly micaceous and intersected with quartz-veins of small size. The thickness of this formation is not more than from 100 to 200 feet and is probably the same from which most of the gold in Western or Lake-goldfields has been derived by the direct erosion of glaciers and mountain torrents.
3. Lower. Contorted schist. This is a clay-schist, foliated, not with mica, nor with felspar, but with quartz. It is often chloritic and in the upper part the quartz is nearly wanting. Gold occurs segregated in the interspaces of this

To the eastward of the crystalline zone, stratified sedimentary rocks appear, such as slates, sandstones, conglomerates, indurated shales interstratified with trappean rocks of a dioritic or diabasic nature. These compose by far the greater part of the eastern side of the central chain, exhibiting everywhere there huge foldings. This extensive formation of sandstones and slates in some places is overlaid uncomformably by a carbonaceous system. The relation of these sedimentary formations one to another, and their exact geological age, as compared with the European series of formations, has at yet not been distinctly made out.[1] The extensive development of limestones, such as are peculiar to the European Alps, is totally lacking; and a simple glance at the above sections shows farthermore, that only the eastern half of a complete mountain-system has been preserved, while the western half is buried in the depth of the main. The eastern foot of the mountains is formed by tertiary and alluvial deposits broken through by volcanic rocks. The period of volcanic energy was one of upheaval; and since it closed we see no evidence of there having been any submergence of the island on the East side, whilst on the West coast, the evidence derived from the mountains rising directly from the sea and penetrated by the fiords, indicates rather a gradual submergence.

Another very remarkable feature of the magnificent snow clad range of the Southern Alps, is the enormous amount of glacier surface and glacier action, and moreover, the constant traces of those glaciers having at no remote period extended far below their

contorted schist, but is rarely found in situ. Quartz-reefs are confined to the upper schists.

External to all the above formations we have a series of Tertiary rocks; the lowest of these, however, may possibly be of the Upper Mesozoic age. The series consists of coarse conglomerates, sandstones and shales, containing estuarine shells, and associated with thick deposits of brown coal of excellent quality. The shales with the coal contain ferns and dicotyledonous leaves. This carbonaceous formation is generally tilted at considerable angles and is unconformably overlain by the newer Tertiary rocks. These consist of two series, the one freshwater, occupying basins or depressions in the schistose rocks of the interior, the other marine, confined to the coast-line and low altitudes."

[1] See Chapter III.

present limits. I shall here follow the footsteps of my friend Dr. Haast, who, in 1861, undertook the task of exploring the headwaters of the river Rangitata, and on that occasion first examined and described the beautiful glaciers of the New Zealand Alps.

In ascending the Canterbury plains along the course of the Rangitata we reach near the gorge of the Rangitata the place, where the river enters the mountains, or, properly speaking, issues from the ranges and enters the plains by a deep rocky gorge about 260 feet deep. On the other side of the gorge the river-bed widens into a basin-like valley, several miles broad and about twenty miles long. The sides of the valley rise with steep walls, whilst the bottom is filled to a depth of more than 1000 feet with immense masses of boulders and sand, into which the river has excavated its channel by terraces. The regularity and the great number of those terraces — in some places as much as twenty-eight have been counted — is truly astonishing; and in fact, no greater contrast can be conceived than the long horizontal lines of the terraces on the sides of the river-bed, their steps ascending the valleys with a gradient of 1 to 3 degress like broad, artificially laid out roads, and the broken outlines of the wild jagged rocky peaks above them. The river-bed itself has still, even here, the considerable breadth of one or two miles. The banks are covered partly with dense scrub, partly with Fagus forests, and partly with a growth of grass and with the "spear-grass" of the settlers (*Aciphylla grandis* Hook.). At the upper end of this deeply excavated valley, the terraces disappear gradually, and the river divides itself into two branches, the Havelock and the Clyde. At the junction of the Clyde with the Havelock, the bed of the river lies about 2200 feet above the level of the sea. The broad valley of the Rangitata is here divided into two narrow defiles running higher and higher to the flanks of mountains covered with perpetual snow and ice, and bear in

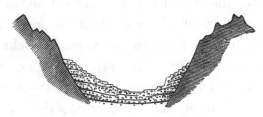

Terraces in the Rangitata valley.

their upper-parts the character of wild mountain gorges. The masses of debris filling these valleys take more and more the character of the ordinary mountain detritus, and are detached from the steep and precipitous mountain sides in the shape of sharp-edged fragments of rock, which are carried further along and rounded by the rushing waters of the river. The enormous mass of detritus in the mountains is truly astonishing. Haast says that mountain sides rising 5000 or 6000 feet above the valleys are often covered from their tops to the bottom with one unbroken talus of debris, so that not one projecting rock is to be seen. Avalanches, numerous watercourses, frequent rainfalls together with violent storms and frost, are, combined, the chief causes of this enormous denudation among the easily crumbling sandstones and slates. At altitudes above 3000 feet the temperature at night for at least six months in the year, is below the freezing point, while during the day it is so warm that generally a continued freezing and thawing takes place. The beds of the rivers have for a long course upwards a width of nearly a mile, but masses of detritus of every description, and gigantic blocks of rock (old moraines), which appear at times to shut up the whole valley, make it very difficult to proceed; even the vegetation assists in obstructing the way. He who has not seen the subalpine vegetation of New Zealand, writes Haast, can form no idea, how difficult it is to pass through it. The growth of scrub is in some places so dense, that it is necessary to walk literally on the top of it; the natural consequence being to break through, and then only releasing oneself with the greatest trouble and exertion. This occurs chiefly where the wind has bent the branches in one direction, giving them the appearance of clipped hedge rows. There is very indifferent feed for horses in these high alpine valleys; snowgrass and the leaves of the *Celmesia coriacea*, called "cotton plant" by the settlers, are often the only feed procurable.

Deep mountain ravines issue from the sides into the main valleys, and the higher the traveller ascends, the more grandly the great majestic character of this high mountain region becomes developed. From ten to fifteen miles from the already mentioned junction, both branches divide into numerous streams which spring from the glistening ice-portals of great valley glaciers (glaciers of the first order), plunging with deafening roar over a chaos of broken rocks, whilst high above, from the crest of the mountains, cascades fall down from the high-glaciers (glaciers of the second order), which hang like gigantic icicles from the vast snow fields; wild torrents are gushing, which like ribbons of silver, come down between the bar walls of rock, or spring from crag to crag in picturesque water-falls, dissolving into spray and falling into the valley in the form of gentle rain.

Dr. Haast writes as follows about his first glacier journeys in those distant mountain regions: — "On the 14 March (1861) I broke up camp in the Havelock valley and followed up the first stream coming in on the right, which I named after the disting-nished English naturalist Forbes. I had to cross numerous moun-tain streams, which issuing from high glaciers plunged down the steep declivities of the mountains with a noise of thunder, and after a laborious scramble of several hours over the fallen rubbish and enormous blocks of rocks, I came at last in sight of the first valley glacier, which I named the Forbes glacier. At its lower end the breadth was 600 feet, and the height I estimated at 100 feet, consisting of well stratified masses of ice, the layers of a thickness from three to five feet, concave and apparently adopt-ing the form of the valley. The ice itself was dirty, and the glacier completely covered with fragments of rocks, some of them of an astonishing size. From a glacial cave twenty feet high and equally wide, issued a discoloured glacial stream, which sought foaming and roaring for a course among the huge blocks of rocks, as they fell continually from the top of the terminal face of the glacier. I climbed in to the ice-grotto, where I found protection from the tumbling pieces of rocks and ice. A beautiful azure-blue twilight shone through the grotto; but the walls of ice were so loose that at a single blow from the hammer shattered large masses into a thousand pieces. But I was not allowed to stay there very long. Seeing that a part of the vault was giving way, I retreated, and being warned by the call of my companions, I had to stop behind a large rock, whilst an enormous fragment fell down into the river with a tremendous crash. The temperature of the water near its issue from the ice-portal was 32° Fahr.; three miles lower down from the glacier it was already 39° Fahr. By joint observations with my aneroid and the boiling water apparatus, I found the altitude of the glacier to be 3837 feet above the level of the sea. As far as I could see, the glacier was wholly covered with debris. The rocky cliffs on both sides of the glacier were about 500 feet high, almost vertical, and showed plainly streaked slips, a proof that the glacier must have been much higher in former times.

A few hundred yards below the origin of the Forbes a second glacier stream joins the former. There being no chance for me to cross the stream, I had to content myself with observing the second glacier, that gives rise to the latter stream, in the distance. It consisted of pure white ice; only a few blocks lay scattered over its surface. Extensive ice and snow-fields were looming up from the glacier to gigantic rock-pyramids. The uniform surface of the ice-field appeared at the starting-point of the glacier fissured, and a grand ice-fall was to be observed, on which the glacier-ice was not only broken into the most differently shaped crags, needles, towers and walls, but also glistened in all shades from the brightest blue to the deepest green, few of the icicles appearing dyed in a deep pink.[1] The terminal face of this glacier lay about 200 feet higher than that of the former, and the shorter descent seems to be chiefly caused by the circumstance, that this glacier was not protected from the rays of the sun by piles of stones and rubbish like the other.

---

[1] In the European Alps the so-called "red snow" is found only upon the snow-fields, but not upon real glacier-ice. It occurs generally some few lines below the surface of the snow-field, and consists chiefly of the breed of small infusoriæ of the genus *Discerœa*, especially the *Discerœa nivalis*, which, when grown, is not transparent and varying from brown-red to purple; likewise of the germs of *Protoccocus nivalis* and *Giges sanguineus*. Haast likewise mentions such red snow. He observed it upon the snow-fields of the M'Coy Creek valley. But as to the nature of the red glacier-ice, future explorations will be necessary.

It was with intense admiration that I gazed on the indescribably magnificent panorama before me. All around the patriarchs of the alps were enthroned, veiled in solemn, silver coverings of dazzling snow with venerable heads, whence flowed their long icy beards. Only the thunder of the avalanches, or the mournful cry of the large alpine parrot broke through the stillness of virgin nature. A deep feeling of veneration and holy overcame me, when I reflected that on the spot where I stood in the solitary wilderness, human foot had never pressed before."

The exploration of the Havelock valley was followed by an exploration of the second branch of the Rangitata, the Clyde. The first tributary from the left, or from North, was named Lawrence. It springs from Mt. Arrowsmith, a mountain-colossus about 10,000 feet high, from the slopes of which huge glaciers descend in all directions. Like the Lawrence so also the other tributaries of the Clyde, St. Clair-Creek, and M'Coy Creek, are glacier streams. Following the main valley to its upper end Dr. Haast discovered the principal glacier descending from Mt. Tyndall,[1] which he named the Great Clyde glacier. The end of this glacier is 3762 feet above the sea; it forms an ice-wall 1300 feet broad and 120 feet high. The river springs from a large glacial cave; and above the cave, about twenty feet below the surface of the glacier, a second stream issued forth from a round hole as from the gutter of a house. The glacier was entirely covered with stones, so that the ice was scarcely visible on the surface.

The third region of glaciers, which Dr. Haast visited, was that of the Ashburton river. The Ashburton glacier lies at the eastern foot of Mt. Arrowsmith. It descends to a height of 4823 feet, and at its terminal face is only 300 feet broad, and 30 to 40 feet high. It pushes a large moraine ahead of it, and appears very remarkable by its clean surface, which bears no stones whatever, as well as by the clear stratification of its ice. Its inclination near the terminal face amounts to 7 or 8 degrees; farther up the valley an ice-fall is to

[1] Named by Dr. Haast after our much esteemed friend Prof. J. Tyndall in London.

be seen, from which the ice, traversed by deep fissures and clefts, ascends in terraces up to the snow-fields. Great portions of this glacier having melted off on the sides of the valley, even its lower surface could be seen resting upon large rounded boulders as upon rollers, so that it was possible to creep between the larger boulders

The Ashburton glacier with Mt. Arrowsmith.

under the glacier. The commencement of winter put an end to Dr. Haast's investigations. Till end of May, my friend wrote to me, the weather was delightful, despite the very cold nights, in which the thermometer fell as low as 24° Fahr., but now winter set in with fogs, rain and snow; it was time to retire to the lower regions; but it was most reluctantly that I parted from that grand nature and from the majestic scenery of the Southern Alps. [1]

[1] The joyful satisfaction over these discoveries was sadly dimmed by the death of my lamented friend Dr. A. Sinclair, who found a premature death during this first New Zealand glacier expedition. It was but in April 1861 that I received a letter from him, dated Christ-Church, 18th February 1861, in which he communicated to me with youthful enthusiasm: "I now proceed with our common friend, jolly, joyous Haast, to the glaciers of the Southern-Alps." But already on the 26th March he found his death while crossing the Clyde river, in the swift floods of that torrent. In him science lost the man, who had been called to create a work

Still grander glacier-regions has Dr. Haast discovered in the summer of 1862 at the headwaters of the Waitaki river on Mount Cook and on the neighbouring gigantic heights of the Alps. The Waitaki (or Waitangi), one of the principal rivers of the East coast, is formed by three branches, springing from three mountain-lakes on the S. E. foot of Mount Cook, the Tekapo, Pukaki and Ohau. Lake Tekapo lies 2468 feet above the level of the sea; it measures 15 miles in length and about 3 miles in breadth. The water of the lake is turbid of a milky colour and it is only after months of cold and dry weather, that the water becomes somewhat clearer. The difference between high and low water amounts to 8 feet. The lake abounds in large eels, which, however, disappear in the month of April with the setting in of winter, and do not reappear until October. It is plainly visible that the lake is filling up rapidly, as the river deltas are extending far into the lake. The shore on tho lower end of the lake seems to have been formed by the colossal moraine of an old glacier, through which the outlet of the lake gradually forced its way. The principal tributary of the lake from the North was called Godley, the northeastern one the Cass.[1] Both spring from large glaciers descending from the mountains North of Mount Cook, the Godley from the great Godley glacier and the Classen glacier, the Cass from two glaciers situated about 20 miles from the lake on the slopes of Mount Darwin. The great Godley glacier — at its terminal face one and a quarter of a mile in breadth, but farther up, where it receives the Grey glacier about three miles broad — may be designated as the "Mer de Glace" of New Zealand. The view from its middle moraine upon the ice- and snow-fields[2] of Mount Tyndall, Mount Petermann, and the Keith-Johnston range Dr. Haast pictures as the grandest scenery he ever beheld in the Alps.

Lake Pukaki, measuring 10 miles in length and 4 miles in

on the geography of the plants of New Zealand, for which he had collected the materials through many years.

[1] To the memory of Mr. Godley the founder of Canterbury and Mr. Cass the Chief Surveyor of Canterbury.

[2] Red snow is said to be a very frequent occurrence upon those snow-fields.

breadth, has an elevation of 1717 feet above the sea,[1] its water likewise being milky-turbid. It is shut in by an old terminal moraine attaining a height of 186 feet above the lake. The view from the outlet of the lake towards the Southern Alps with Mount Cook in the centre and a wooded islet in the foreground is described as sublime in the extreme, and when we imagine villas and parks around its shores, the Lago di Como or Lago Maggiore would not bear comparison with it. The main-tributary from the North is the Tasman river, a sluggish stream, separated into numerous branches — Dr. Haast in one place counted 34 of them — and flowing through a broad valley, which is noted for swamps and dangerous tracts of quick-sand. This valley leads to one of the most extensive glacier-regions of the Alps.

Five large glaciers extend from the slopes of Mount Cook and of the adjoining peaks, — which Dr. Haast has designated by the names of Mount Haidinger, Mount de la Beche, Mount Elie de Beaumont and Mount Darwin, — in a southerly and southeasterly direction far down into the valleys. The Great Tasman glacier, the main source of the river Tasman, of a length of 18 miles, and a breadth of nearly two miles at its terminal face is the largest glacier hitherto observed in New Zealand.[2] For a distance of three miles upward this glacier is entirely covered with an enormous load of debris, so that the ice is only now and then visible in transverse and longitudinal crevasses and in large holes 100 to 150 feet deep. About nine miles up the valley it receives from a western side-valley a glacier one mile broad, which descends in two arms from the slopes of Mount Cook, Mount Tasman and Mount Haidinger, and which my friend Haast was pleased to designate after my name — Hochstetter glacier. The streams from the Tasman glacier flow on both sides of the glacier. The Murchison glacier,

---

[1] According to Mr. Thomson 1377 feet.

[2] The Tasman glacier was, till within the last years considered the largest glacier in any temperate region, surpassing in magnitude by far those of the Himalayas and European Alps. Since then Captain Godwin Austin, of the Indian Survey, has explored some glaciers in Thibet of a length of thirty-six miles, and conclusively shown that they extended formerly as far as one hundred miles down the valley.

East of the Tasman glacier, brings the ice masses down from Mount Darwin and Mount Malte Brun. The river streaming from it issues from a glacial-cave on the extreme East-side of the glacier. The Hooker glacier comes in two branches from the southern foot of Mount Cook; whilst the ice masses of the Mueller glacier descend from Mount Sefton and the Moorhouse range.[1] Near the head of this glacier is a pass over a wall, formed of nearly perpendicular rocks about 100 feet above the glacier, which would bring us upon the Selwyn glacier, forming one of the sources of the River Dobson, falling into Lake Ohau (or Ohou).[2] About a quarter of a mile down the Dobson river valley another glacier, of a very peculiar formation, comes down the principal peak of the Naumann range, which runs from the Moorhouse range between the two main branches of the Hopkins river to their junction. Owing to the great steepness of the mountain sides, a large portion of the ice is pushed before it can melt, at many spots over the perpendicular ledges, and falling down with a tremendous crash, is again cemented together to expand to a new glacier; at the same time a narrow channel between the upper and lower glaciers brings down the remains of the upper ice masses, which have escaped being thrown over. Dr. Haast named this glacier the Hourglass glacier, from its peculiar form. The elevation of its terminal face is 3816 feet.

What an impassable wall of ice and rock the Southern Alps proper present, is still more apparent at the West Coast, where also glaciers scarcely less inferior in size to their eastern neighbours descend towards the West Coast plains. It is there where the existence of a glacier is stated, the terminal face of which lies only 705 feet above the sea, with a rich and varied vegetation close to

[1] In honor of the Superintendent of the Province of Canterbury.

[2] According to Mr. Thomson lake Ohau has an elevation of 1498 feet, measuring 12 miles in length and 2½ miles in breadth. The other lakes on the eastern foot of the Southern Alps are Hawea, measuring 12 miles in length, and 2 in breadth, having an elevation of about 1000 feet above the sea; Wanaka, 1036 feet above the sea, 14 miles long, 4 miles broad, Taieri, having an elevation of 906 feet, and measuring 1½ miles in length and one mile in breadth; lake Wakatipu on the upper branches of the Clutha, finally North and South Anau at the sources of the Waiau river, which empties into Foveaux Strait.

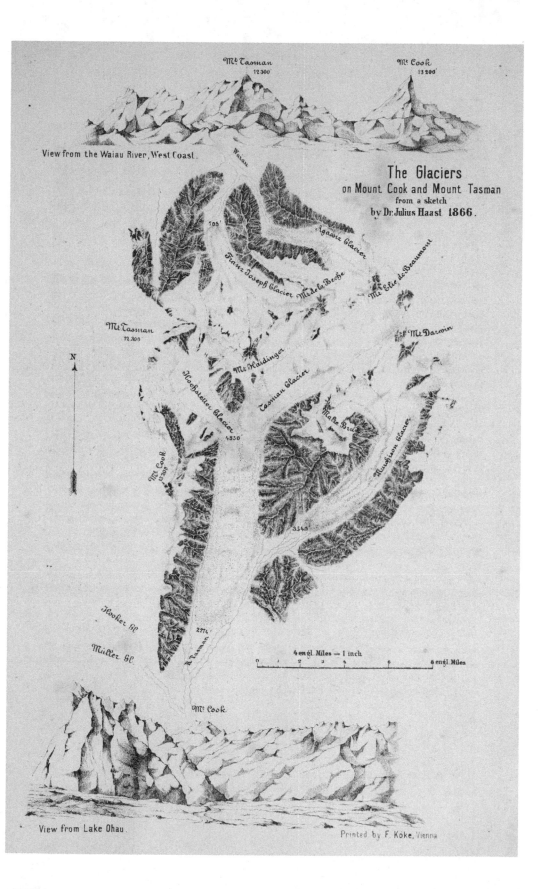

View from the Waiau River, West Coast.

The Glaciers
on Mount Cook and Mount Tasman
from a sketch
by Dr. Julius Haast 1866.

Waiau

705'

Agassiz Glacier

Franz Joseph Glacier

M.de la Beche

M.t Elie de Beaumont

M.t Tasman
12,300

M.t Darwin

M.t Haidinger

Hochstetter Glacier

4350'

Tasman Glacier

Malte Brun

Murchison Glacier

M.t Cook
13,200

3548

2774

Hooker Gl.

Müller Gl.

R. Tasman

4 engl. Miles = 1 inch

0   1   2   3   4   5   6   8 engl. Miles

M.t Cook

View from Lake Ohau.

Printed by F. Koke, Vienna.

it, and is was again Dr. Haast, who during a journey along the West Coast between Hokitika and the Waiau river, in 1865, has discovered that very remarkable glacier and named it Francis Joseph glacier in honor of His Majesty the Emperor of Austria and in remembrance to the Novara Expedition, which had been sent out by His Majesty.[1] The Francis Joseph glacier comes down from

---

[1] Dr. Haast (Lecture on the West Coast of Canterbury, Christchurch 1865) gives the following description of the glorious panorama from the lake Okarita on the West Coast, and of the discovery of the Francis Joseph glacier. "The contrast between the ever restless sea — the gigantic waves coming and going without intermission — and the quiet watershed of Lake Okarita, with its numerous islands, surrounded by luxuriant forest, was most striking. Above the forest plains rose low hillocks, also clothed with the same intensely green West Coast vegetation, over which the Southern Alps appeared a mass of snow, ice, rock, and forest. As far as the eye could reach, mountain appeared behind mountain, all clad in their white garments, with which they are covered during the whole year almost entirely, becoming apparently lower until they appeared only as small points over the sea horizon — half cloud, half ghost, as a modern philosopher has said so well. But what struck me more than anything was the low position reached by an enormous glacier, descending north of Mount Cook from the ranges, appearing between the wooded hillocks at the foot of the Alps. The sun being near his setting, every moment new changes were effected; the shades grew longer and darker, and whilst already the lower portion lay in a deep shade, the summits were still shining with an intense rosy hue. Turning towards the sea, the same contrast of colours was exhibited, the sea being deep blue, whilst the sky was of such a deep crimson and orange colour, that if we could see it faithfully rendered by an artist, we should consider it highly exaggerated. But the beauty of the magnificent scene did not fade away even after the large orb of the sun had disappeared, because, as the night advanced, the full moon threw her soft silver light over the whole picture, and lake and sea, forest and snowy giants still were visible, but assuming, apparently, other dimensions, shapes, and colours. It was late at night before I could leave this glorious view, and my heart swelled with such a pure delight as only the contemplation of nature can offer to her admirers." — The next morning, the 14th of June (1865) Dr. Haast continued his journey along the Coast to the Waiau river. — "Te view from the mouth of the Waiau river is most magnificent, as the valley, being straight and nearly two miles broad, allows us to gaze at the Southern Alps from foot to summit, having in the foreground the enormous ice masses of the Francis Joseph glacier appearing between a rich forest vegetation. The Waiau is a shingle river, flowing in several branches through its wide valley, the semi-opaque bluish colour of its waters at once revealing its glacier origin. Owing to the cold nights it was very low, so that we could easily cross, it being scarcely above our knees. We kept on the southern side of its bed, travelling partly in dry channels, over grass-flats, or sometimes through dense bush. This forest consisted either of pines intermingled with arborescent ferns, the whole interlaced by climbing plants, or — and what was still worst — of shrubs

the snow-fields of Mount Tasman, Mount Haidinger and Mount de la Beche and feeds the Waiau river. The position of the glacier

the branches of which were not only grown to dense masses and towards the ground, but were still more closely united by bushlawyers (*Rubus australis*) and supple-jacks (*Ripogonum parviflorum*). It was a herculean task to pass through bush of the last description, often only a few hundred yards long; and we generally did not reach the river-bed without having left part of our garments or skin in our battle with that pleasant West Coast vegetation.

After two miles the bed enlarged still more, the river flowing in two principal branches with a large wooded island in the centre. Towards evening we camped about seven miles from the coast, near a grove of pine trees and arborescent ferns. During our journey up the river we had occasion to observe what a great abundance of animal life is there existing; there were large numbers of woodhens, and my Maori companions soon made a sad havock amongst them. Next morning the same fine weather favoured us, and after four miles we arrived at the foot of the mountains. Here the main river turned towards south and an important branch joins it from the south-east, coming also from a large glacier, which I called after Professor Agassiz, the illustrious naturalist. The valley of the main river narrows here considerably, and rocky points are washed by its water on the right side, consisting of micaslate full of garnets; but even here, close to the glacier, the fall of the river is inconsiderable, so that it easily could be crossed on foot. The vegetation at the same time continued to have the same lowland character. Turning a rocky point, we had at once the white unsullied face of the ice before us, broken up in a thousand turrets, needles and other fantastic forms, the terminal face of the glacier being still hidden by a grove of pines, ratas, beeches and arborescent ferns in the foreground, which gave to the whole picture a still stranger appearance. About three-quarters of a mile from the glacier we camped, and, after a hasty meal, started for its examination. The same vegetation still continued, and it was in vain that I looked for any alpine, or even sub-alpine plants. From both sides numerous water-courses come down, mostly forming nice falls over large blocks of rocks. Before we reached the glacier, the valley expanded again, the left side having hitherto been formed by an ancient moraine, more than a hundred feet high, the river flowing in two channels, with a wooded island, from which huge blocks rose between; but, owing to the very low state of the river, the southern channel was nearly dry, and only received, on that side, the contents of numerous small water-falls from the outrunning spurs of the main chain. Before we reached the glacier itself we had to cross a moraine, mostly consisting of small detritus, denoting, by its mineralogical character, that it came from the very summit of the snowy giants before us. My whole party had never seen a glacier, and some of the Maoris had never seen ice; thus, the nearer we came, the greater was their curiosity, and whilst I stopped a few hundred yards from the terminal face to take some bearings, the whole range, owing to the clear sky, being well visible, they all ran on, and I saw them soon ascend the ice, which, with the exception of a few small pieces of *debris* in the centre, was perfectly spotless, and presented a most magnificent sight. Having finished my work, I followed them, and stood soon under the glacial cave at the southern extremity, forming an azure roof of indescribable beauty. On both sides of that glacier, for a is about lat. 43 deg. 35 min. S., long. 170° 12′ E. of Gr., corresponding in the northern hemisphere with that of Montpellier, Pau and Marseilles in France, and Leghorn in Italy, where already the orange and lemon tree, the vine and the fig tree, are covered with juicy fruits, and where palm trees raise their graceful crown into the balmy air. Even in the European Alps, which lie some degrees further north, the average altitude of the terminal face of the larger glaciers is about 4000 feet, whilst we have to go twenty degrees more to the north, till we find, in Norway, glaciers descend to the same low position, and to about 67° north, before the terminal face reaches the sea. Thus, we see, that the glaciers of New Zealand in proportion to the heights of the mountains are much larger, than the glaciers of the European Alps.

We may, therefore, justly ask for the reasons of the enormous size of the glaciers in the Southern Alps, considering the narrowness of the central chain, and for the causes to which we must attribute the remarkable occurrence of a glacier in such a low position, as the Francis Joseph glacier on the West Coast. We may best explain that by regarding the equal and humid oceanic climate of New Zealand, and the great difference of the climate on the two coasts. From the first explorer, who ever set his foot on the West Coast, to the discovery of the gold-fields, the difference of rainfall on the two coasts has always been a topic of great interest. While on the East-side the country was languishing for rain, on the other side the exploring parties suffered day after day from deluging rain and heavy mists. Fortunatly we have now some data to go upon to determine that difference more accuratly. According to the very valuable observations of Mr. J. Rochfort[1] during

---

good distance, the mountains are covered with a luxuriant vegetation, amongst which fern trees, rimu, totara, rata and fuchsia are most conspicuous.

[1] I Rochfort, Results of meteorological observations taken at Christchurch, Canterbury, for the year ending 31st December 1865, Christchurch. We extract the following table:

1865	at Hokitika	at Christchurch
May	17,128 inches	4,225 inches
June	14,530 ,,	2,674 ,,
July	4,380 ,,	3,417 ,,

the period from the month of May to end of December 1865 the rainfall at the West Coast was more than five times as great as at the East Coast, amounting at Hokitika to 96,136 inches, and at Christchurch only 17,395. In the southwestern part of the island Dr. Hector, in 1863, during a period of seven months from the first of June to end of December had observed 87 inches, whilst in Dunedin the rainfall was only 23¼ inches. Thus, it is obvious, as the perpetual snow-line, owing to the equable and humid climate, is at the West Coast very low, probably about 6500 feet near Mount Cook, and as the fall of snow and condensation of moisture must still be greater in those higher regions, where the equatorial currents come in contact with the cold surfaces of the Alps, that all necessary conditions exist not only for the formation of large glaciers, but also for their descent to much lower regions than at the East Coast, where the line of perpetual snow in a latitude from 43 to 44 degrees probably only reaches 7500 feet. The difference between the eastern and western side of the central chain is well exhibited by the great Tasman glacier, which, although of much larger dimensions than the Francis Joseph glacier, yet descends only to 2774 feet above the sea-level, whilst the latter reaches more than 2000 feet lower, namely to 705 feet above the sea. It is true that particular circumstances, as, for instance, a large cauldron-like basin, sheltered from the·sun's rays by Mount de la Beche and its out-running spurs, in which these enormous snow-masses can accumulate, are very favourable for allowing the glaciers to descend to such a low position above the sea level, where arborescent ferns, pines, and other low-land trees are growing. But if we compare its position with others in South America, whe shall find that, from ranges which are not so elevated as the Southern Alps, even in latitudes corresponding with the northern end of Stewart's Is-

1865	at Hokitika		at Christchurch	
August . . .	7,890	inches	0,833	inches
September . .	14,471	„	2,347	„
October . . .	11,172	„	2,201	„
November . .	12,426	„	1,024	„
December . .	14,085	„	0,674	„

land, enormous glaciers descend in latitude 46 degrees 50 minutes, according to Darwin, to the level of the sea, their terminal face being ultimately washed away and carried along as huge icebergs. Thus the conditions for the lowering of the snow-line and of the excess of moisture must still be greater in that part of America than in New Zealand, where the neighbourhood of Australia and Tasmania will certainly exercise some moderating influence, which in Terra del Fuego does not exist. From observations made in those and other regions, it is clear that the lowering of the snow-line does not depend on the mean temperature of the year, but on the low temperature of the summer.[1]

Although, as we have seen, the size of the principal glaciers in the Southern Alps is enormous, considering the height and narrowness of the central chain, the astonishment of the geological observer increases on meeting every where the signs of a still more considerable glaciation during former periods. There are perhaps, as Dr. Haast states, nowhere in the well-known regions of the earth so easy of access, and in such comparatively low positions above the level of the sea, such clear and fresh signs of the existence of enormous glaciers during the pleistocene epoch, than upon the Southern Island of New Zealand. Far below the moraines at the terminal face of the present glaciers old moraines are met with, striking across valleys several miles broad and overgrown with a dense subalpine vegetation, which is nearly impenetrable to horse and man. Combined with them the strait courses, the great depth and breadth of the valleys, which are in no proportion to the rivers now running through them, and the rounded outlines of the lower mountains *(roches moutonnées)* are unmistakable indications, that the

---

[1] The mean summer temperature of Christchurch, from observations made in 1864, is $61\frac{1}{4}°$, enjoying generally a clear and cloudless sky; but it is evident that at the West Coast it is much less, owing to the over-cast state of the atmosphere and the frequent rainfalls. This will account for such a lowering of the snow line on the western side of the Alps. But the mean temperature of the year will, nevertheless, not be lower than in Christchurch, where it was $53\frac{3}{4}°$ during 1864, with an average temperature of $61°$ for the summer, and $44\frac{1}{4}$ for the winter months; the warmer winter at the West Coast compensating, without doubt, for the lower summer temperature.

valleys have been excavated by glacier-action, and that glaciers of great dimensions had filled them and polished their sides.[1]   On the western coast the valleys have been scooped out to a depth which is now at least 1800 feet, in some cases, beneath the present sea-level,[2] while on the East-side, where the depression of the land has not been so great, these valleys are occupied by lakes, the surfaces of which have a mean altitude of 1000 to 2000 feet, but the bottoms of which are partly considerably below the sea-level.   The borders of those lakes are formed by the enormous lateral and frontal moraines of glaciers of earlier date, while the lateral moraines are now represented as fringing shelves along the sides of the valleys, just like the terraces skirting the valley of a river, which is changing its course from side to side of a gradually deepening channel.

In inquiring into the causes of the pleistocene glaciation of the

[1] Dr. Haast (Notes to a sketch-map of the Province of Canterbury, showing the Glaciation during the Pleistocene and Recent Periods as far as explored. Quart. Journ. Geol. Soc. 1864) mentions the following distances of the farthest terminal moraines of the Pleistocene glaciers on the Eastside from the points in the central chain from which the present glaciers take their rise: Wanaka glacier 48 miles, Hawea glacier 50 miles, Waitaki 78 miles, Rangitata glacier 40 miles, Rakaia 52 miles.   On the West Coast at the northern boundary of the province near lake Brunner the moraine accumulations are situated 15 miles from the sea; they approach it towards South, so that in the Arahura they are seven, in the Hokitika five miles, and in some localities even nearer to the coast.   More towards South, between the Mikonui and Waitaha rivers, they reach the sea level, forming bold headlands in the nearly vertical cliffs, of which magnificent sections are offered to the geologist for the study of these stupendous moraine accumulations.   From here these ancient moraines cover uniformly the whole country to the very base of the central chain, and in only few instances small hills consisting of granitic or metamorphic rocks strech their glacialized rounded summits above the ice-born beds.

[2] At the Milford Sound, according to Dr. Hector, the sea now occupies a chasm that was in past ages ploughed by an immense glacier.   The lateral valleys join the main one at various elevations, but are all sharply cut of by the precipitous wall of the Sound, the erosion of which was no doubt continued by a great central glacier long after the subordinate and tributary glaciers had ceased to exist. The precipices exhibit the marks of ice action with great distinctness and descend quite abruptly to a depth of 800 to 1200 feet below the water level.   Towards its head the Sound becomes more expanded and receives several large valleys, that preserve the same character.   A great ice lake must have existed in the upper and expanded portion of the Sound, from which the only outlet could be made through the chasms which form its lower part.

South Island we need not resort to the hypothesis of a general ice-period caused by cosmical influences, and which was supposed to have covered the surface of the globe from the poles to the torrid zone with snow and ice, but find the must reasonable explanation in physical causes now in existence, if we only suppose, that the Southern Alps during the pleistocene period differed somewhat from their present features, forming higher and more plateau-like ranges, while the climatical influences as well as the action of water and ice at that period were the same as at present. My views in this respect are perfectly in accordance with the views advocated by my friend Dr. Haast.[1]

From the extensive tertiary strata, covering large areas upon the South Island even to an elevation of more than 2000 feet we may infer, that during the tertiary period the South Island was to a great extent submerged below the level of the sea. The country emerging again the physical feature was a high, plateau-like mountain mass, but with depressions existing before the tertiary submergence — now partly obliterated. After the mountains had risen above the line of perpetual snow, the accumulation of snow and ice fields (*névés*) began. It would be however a mistake, to estimate the size of glaciers generated merely by the altitude of the mountain region, as it is truly the area which in the districts is elevated above the snow-line together with the climate, that determines their extent. The névés, considering the insular position and the oceanic climate of New Zealand — its principal backbone running from S. W. to N. E., thus lying at a right angle to the prevailing air currents, the equatorial north-west and the polar south-east, both bringing moisture with them — must soon have attained an enormous extent, even had the land not been raised, at it is probable, to a higher elevation than at present. The consequence was, that glaciers were formed descending down the natural out-

[1] Dr. Haast, Report on the formation of the Canterbury plains, Christchurch 1864. Note on the Climate of the Pleistocene epoch of N. Z. Notes on the causes which have led to the excavation of deep Lake-Basins in hard rocks in the Southern Alps. Quarterl. Journal Geol. Soc. London 1864.

lets and grinding down the rugosities of bottom and sides. Thus the scooping and ridge making action of ice and water began. If this be the case, the area of the snow-fields, as the mountains began to become more and more eaten into by the eroding action of the descending ice and the wearing away of the heights by atmospheric influences and sharp ridges and peaks were more and more formed, must have been rapidly diminishing; and in the same measure as the valleys were scooped out, the glaciers must have retreated from the coast towards the centre of the mountains and their extent diminished. Thus we have the most simple and natural explanation for the fact, that the glaciers of New Zealand since the pleistocene period have been more and more retreating into the deep alpine valleys to their present limits; and we need not suppose, that the pleistocene glaciers, although generally larger than the present ones, attained such an enormous length, as the distance of their terminal moraines seems to indicate. Judging, however, from the structure of the Sounds on the West-side of the mountains, and that of the Lake districts on the East-side, it seems probable that the opposite sides of the mountain range have besides undergone repeated and alternate oscillations to the extent of at least 1000 feet in either direction from a nominal point; and that the western district being at present near to the period of greatest depression the re-elevation of the land to the other extreme would be also sufficient to extend again the glaciers to farther limits.

From the present features of the New Zealand Alps we can easily imagine what enormous masses of rock must have been destroyed in course of time by the combined action of atmospheric influences, of ice and of water. We need only look upon the immense extent of drift, partly glacial, and partly alluvial throughout the Alps, and as far as the borders of the sea, to know what has become of them. The rocks desintegrated into boulders shingle sand and mud, and the colossal moraines of the glaciers have furnished the material for the drift-formation, which on the foot of the mountains forms extensive plains, filling the valleys up to a height of more than 1000 feet, and which extends from valley to valley across the mountain-passes, reaching sometimes the height of 5000 feet, thus enabling the traveller to proceed from range to range, from valley to valley, without once setting foot upon the native rock. The most prominent example of drift-formation, formed partly by the accumulations of the old pleistocene glaciers, partly by the deposits of the torrents falling from the ranges to the sea, are the Canterbury Plains, the formation of which has been so clearly shown by the combined labours and investigations of Messrs. Haast, E. Dobson, Beetham, Doyne, and several others.[1] In concluding the chapter of the Southern Alps, I may be allowed to dwell upon those plains, which, by their position, nature and general characteristics, form such a prominent feature upon the South Island.

The Canterbury Plains extend from Double Corner, South of the mouth of the Hurunui to the dolerite plateau of Timaru. Their length from N. E. to S. W. is 112 miles. Their breadth from a few miles at both extremities, North and South, augments as we advance towards their centre, having their greatest lateral extension near Banks' Peninsula, where in a direction from East to West they stretch a distance of more than fifty miles to the base of the mountains. A long uninterrupted shore-line, called the Ninety miles Beach, streches from Timaru towards the volcanic system of Banks' Peninsula which rises so conspicuously from that low shore, and to the existence of which a great portion of the loose strata composing the plains owe mainly their preservation from the destructive agencies of the waves and currents. The plains rise from the East Coast towards the mountains, reaching at the foot of the hills a height of 1500 feet. A general section, taken in a direct line from the sea to the hills may be described as a curved line, differing but little from dead level near the coast, but rising at a gradually increasing gradient until it reaches the hills, which, in most places, rise abruptly from the plains. The material as found upon the surface of the plains, and for a depth, as proved by wells sunk

---

[1] Dr. Haast, Report on the formation of the Canterbury Plains 1864. — W. T. Doyne, Report upon the Plains and Rivers of Canterbury 1864. — W. T. Doyne Second Report upon the River Maimakariri and Lower Plains of Canterbury 1865.

of nearly 200 feet below it, from a few miles within the coast line to the foot of the hills, is altogether composed of hard grey shingle and gravel filled up with variable proportions of sand and fine silicious silt. It is usually quite permeable to water; but nearer the sea, where the finer mud has been deposited in lagoons, deep beeds of impermeable soil are found. Nowhere over the whole extent of the plains has a trace of marine deposit been discovered.

The plains are intersected by numerous rivers, which the settlers divide into three classes: snow rivers, rain rivers and leakage rivers. The snow rivers, to which the Waimakariri, Rakaia, Ashburton and Rangitata belong, have their sources in the central chain. They are wild torrents issuing from the deep rock-bound gorges of the Alps, fed by the glaciers and melting snows, and are consequently subject to sudden and powerful freshes. They never dry during the autumn or summer months. The so-called rain rivers are those which have their origin in the outrunning spurs of the Alps, as, for example the Selwyn, Hinds, Ashley and others. One of their characteristic features is, that they generally dry up during the summer and autumn months, that to say, that they disappear in their course below the shingle, of which the plains are composed. The leakage rivers consist of those smaller water-courses which take their rise on the plains themselves, as for instance the Heathcote, Avon, Styx, Little Rakaia and many others. They are merely formed by the waters of the rivers of the former two classes, which, losing themselves among the shingle till they find impermeable beds to arrest their further sinking, and on which they flow downwards, form underground streams till forced to the surface by meeting a change of the strata. The water flowing from each of these outlets is remarkably uniform throughout the year, and is always beautifully clear.[1]

[1] Great bodies of water are also seen to leave the channels of the main rivers, which entirely diappear under the surface and have no outlet any where on the plains. There is no doubt, that for instance the Waimakariri has many undergrouud streams travelling for long distances at different depths. One of them feeds the artesian wells in and around Christchurch. The water in all these wells has been reached at about 67 feet below high-water mark and rises to about 26 feet above

The snow rivers are those, by the action of which the surface of the plains has derived its present form. In their passage through the mountains they carry with them the desintegrated materials they meet with. These are of every variety of size, from impalpable mud to large boulders; they are deposited along the beds of the streams, the heaviest particles highest up and the remainder over the whole length to the sea. Thus the beds of the streams become built up to, and above the level of the already illdefined banks; the waters leave the old courses and cut out new ones; these they rapidly widen till the water spreads and rambles over miles in width, and again deposits the materials brought down. Each stream thus roams over a large area, its various courses radiating from a centre at or near the gorge through which it debouches on the plains. It builds up, by layer on layer, that portion of the plain which it is destined to work over, and is ever thus travelling from north to south, and south to north, over a distance averaging about forty miles. It thus leaves within its field of operations a nearly level surface, slightly corrugated in lines radiating from a centre, and closely resembling a lady's fan, by which name it has been most appropriately designated in Dr. Haast's report. This operation has been in action since the pleistocene period, and no doubt, the present rivers repeat simply on a smaller scale the action of the large pleistocene torrents which formed the upper part of the plains. The present rivers have, in consequence, according to their present size and position of sources cut more or less deeply into the old fans at the foot of the mountains, whilst they raise their beds at a greater distance from the mountains and forme there the new fans near the sea coast. Thus, the river beds themselves rise from the sea to the mountain gorges at a tolerably regular slope of from 20 to 30 feet per mile; running from the gorges between terraces of great height, which gradually diminish until they die away altogether, leaving the rivers to run on the surface of the plains for a short space, after which they begin, instead

it. The depth of the bore-holes in the streets of Christchurch averages about 83 feet and the water rises to about 10 feet above the surface.

of excavating to fill up and raise their beds, and run to the sea upon elevated dams, similar to the Po and the Adige in Upper-Italy, and between high cliffs of shingle, whose height varies with that to which the edge of the plains rises above the sea beach. With few exceptions, therefore, every one of these rivers presents a point at which it may be crossed on the general level of the country, below which it is either inaccessible on account of the cliffs by which it is bounded, or difficult to cross on account of the number and depth of the channels into which it spreads on the surface of the plains, and above which it can only be approached by long sidling descents cut in the terraces.[1]

Such are the laws which guide these rivers, by the action of which extensive and fertile plains have been formed at the foot of an inhospitable mountain range. These plains, which will one day be the centre of a rich, industrious and large population, are now the home of an energetic and high minded class of settlers, in the name of which Mr. Dobson, the Vice-President of the Philosophical Institute of Canterbury at the annual meeting on November 5, 1866, could indeed proudly say: "With scanty means, and a comparatively small population, we have succeeded in introducing amongst us most of the great inventions of the civilised world. We have our telegraph through the country, and our submarine

---

[1] It may be interesting, Mr. Dobson says, to glance at the nature of the channels through which the great rivers find their way from the hills to the sea — as they all possess, to a greater or less extent, the same features — which governs the selection of points of crossing, and, as a consequence, the direction of the main lines of road running parallel to the eastern seaboard. Between Christchurch and the Waitaki, a distance of 143 miles, the position of the southern lines, both of road and railway, has been determined by considerations of this nature. The main route to the South forms a tolerably accurate line of division between the swampy and well watered belt of agricultural land on the sea-board and the dry shingle plains, which are only suitable for pasturage. In many places the agricultural land does not extend up to the road. Thus, whilst for a distance of twenty-five miles from Christchurch, along the Leeston road, the country is fenced in and mostly under cultivation, producing largely both grain, dairy produce and live stock, the Southern Railway, which is laid out so as to cross the Rakaia river as near to the sea as practicable, is yet two or three miles from the edge of this cultivated district, and runs for miles across a desolate looking plain — without water, trees, or human habitations.

cable connecting our capital with the seat of Government in the Northern Island. We have our great tunnel in construction, and our road across the New Zealand Alps. We have our goldfields, our coal mines, our founderies, our broad acres tilled with the steam plough, our clipper steamers, our mail coaches, and our locomotive railways, and we have all this in a country which, fifteen years ago, was an almost unknown land, but which is now, by God's blessing, the happy home of prosperous thousands of our fellow men."

Government Buildings at Christchurch, Province of Canterbury.

Appendix.

The Otira or Arthur's Pass road, the first road across the Southern Alps. [1]

Amongst the public works of the Province of Canterbury stands prominent the new road, just constructed, by the gorge of the Otira, across the New Zealand Alps, connecting the City of Christchurch with Hokitika and Greymouth, the ports of the western goldfields. The Otira road is a remarkable work, in every point of view; whether we consider the grandeur of the scenery through which it passes, the geological interest of the Alpine districts which it traverses, the engineering difficulties attendant on its con-

[1] The present state of applied science in the Canterbury Province (Adress, delivered at the annual meeting of the Philosophical Institute of Canterbury, on November 5, 1866 by Mr. E. Dobson, Prov. Engineer, and Vice-President of the Institute.

struction, or the hardships manfully endured by those engaged on the undertaking, it is in every way a work reflecting credit, not only on the Canterbury province, but on all New Zealand.

Up to the commencement of the year 1865, there was no road from the Canterbury plains to the West Coast, except a very rough and dangerous path, cut across the Hurunui saddle by Mr. Charlton Howitt, in 1862, by means of which, at considerable risk, horses could be taken as far as Lake Brunner. Mr. Howitt was engaged at the time of its death, in 1863, in cutting a track from Lake Brunner to the mouth of the Greenstone Creek, but it was not practicable for horses. Up to the date last mentioned, but little attention had been paid to the fact that, notwithstanding the inaccessible nature of the country, a very large number of diggers had found their way into Westland, and were pursuing their vocation with considerable success. About the beginning of 1865, however, the reports sent by the miners to their friends were of such a favourable character, that a violent rush set in from Eastland to the new El Dorado, and the attention of the Government was directed to the best method of opening up a communication with the goldfields. To this end Messrs. Edwin and Walter Blake were sent to improve Howitt's track by the Hurunui and Teramakau to Lake Brunner, and to explore for a line of road in continuation across the country between the Teramakau and Hokitika. At the same time, Mr. George Dobson and the Provincial Engineer were charged with the examination of the Waimakariri and its tributaries, to ascertain whether there were any passes which might afford greater facilities for constructing a road across the Alps than that by the Hurunui saddle. The results of these explorations were published in a report from the Provincial Engineer, dated May 15, 1865, amply illustrated with maps and sections.

In accordance with the recommendations of this report the Government at once took steeps for the construction of a bridle road over Arthur's Pass, descending into the Teramakau by the gorge of the Otira. It was at first intended to construct a mere bridle track, but the importance of the new goldfields developed itself so rapidly, that within a few weeks of the commencement of the works it was decided to construct a coach road throughout. From the date of this decision the works along the whole line of road from the plains to the sea beach were put in hand as rapidly as possible, and pushed forward with such energy, that by the 20th of March, 1866, the road was open for traffic from end to end, and has been regularly travelled ever since by four-horse coaches running twice weekly each way; the distance of 150 miles between Christchurch and Hokitika, being completed in 36 hours, including a night stoppage of 12 hours at the halfway station.

It is very difficult by a verbal description to give any idea of the obstacles that presented themselves to the construction of this road. Perhaps the greatest of all arose from the inaccessible character of the country; the only way of getting tools and stores to the central portion of the work being either by pulling canoes up the Teramakau from the beach, or by packhorses travelling over the Hurunui saddle, from the edge of the plains — a journey of seventy miles, and, moreover, this had to be done in a densely timbered country, in the depth of the winter.

No pen can describe the sufferings endured by both man and beast during that terrible winter, exposed to sleet and snow and bitter frost, hardly lodged and scantily fed, whilst the working parties were liable at any moment to be cut off from communication with each other by the rising of the rivers. By the end of July, however, a pack-horse track was opened through the Otira Gorge, which enabled supplies to be taken into the Teramakau valley with comparative ease, and the works in the latter valley were greatly facilitated by the use of drays, which were carried in pieces across Arthur's Pass, and put together in the Teramakau river bed which was used as a temporary road whilst the bush clearings were being made. As with the opening of the tracks greater facilities were given for the conveyance of stores to the works, the number of men employed was increased, until it amounted to upwards of a thousand.

Since the opening of the road the work has gone steadily on, and may now be said to be completed, although, from the nature of the country through which it passes, it will always require constant attention to keep it in repair, especially in the valley of the Teramakau, which is periodically visited by dangerous floods.

The total distance from Christchurch to Hokitika, by the Otira route, is 150 miles as above stated, of which about one hundred miles of road from the eastern foot of the hills to the sea beach at the mouth of the Arahura have been made and metalled between May 1st, 1865, and October 31st, 1866, at coast, in round numbers, of £145,000, or something under £1,500 per mile.

The engineering works upon this line are of a very varied nature. In some places the cliffs are scarfed out for a portion of the width of the road, the remainder being carried on timber brackets in the fashion of Trajan's celebrated road on the bank of the Danube; in others, the line is carried across ravines on embankments faced with walls made of timber cribbing, filled with blocks of stone.

The fords in the rivers have been protected by wing dams formed of large trees, backed with boulders; whilst in many places the mountain torrents have been made passable by building timber weirs across them, and filling up their beds to a uniform level with stones and gravel. Through the swampy forest the ground has been drained and fascined for many miles, whilst the whole length of the road has been thoroughly metalled. Amongst the bridges, that over the Taipo, 270 feet long, built upon pies with steel shoes, driven into a mass of granite boulders, deserves mention, as being a difficult work successfully executed, and which has, up to the present time, resisted the heaviest floods, although the stream has been at times blocked with drift timber from bank to bank.

Altitudes along the Arthur's Pass road.

Cook's Accommodation House	217
White's Accommodation House	623
Southern base of Little Race Course Hill	995
M'Cray's Accommodation House	1327
Riddle's Accommodation House, foot of Porter's Pass	2091
Summit of Porter's Pass	3234
Lake Lyndon	2814
Watershed between Lake Lyndon and River Porter	2852
Springs, sources of River Porter	2609
Riverbed of River Porter, crossing of road	2270
Summit of terrace on its western side	2473
Stables of Cobb and Co., near Mr. Enys's station	2548
Top of terrace of Broken river	2413
Bed of Broken river, where crossed by road	2168
Top of Cragieburn saddle	2843
Cragieburn Accommodation House	2221
Lake Pearson	2095
Roches moutounées, near River Cass and Lake Grasmere, eastern base	2157
Road crossing of River Cass	1879
Saddle between the Cass crossing and the River Waimakariri	1938
Bed of Waimakariri between the two cuttings	1863
Bealey Township, Police reserve	2155
Southern foot of Arthur's Pass (Mr. Smith's camp)	2534
Arthur's Pass, highest summit	3038
Southern foot of Moraine	2697
Summit of Moraine	2865
Mr. Wright's camp	2238
Riverbed of Otira, near first bridge	2026
Junction of the two branches of the Otira near the stockyard	1449

Bed of Teramakau river, near junction of the Otira where the road
joins the former 769
Western bank of River Taipo, near the junction of the Teramakau . 356
Waimea, where the road leaves the Teramakau 148
Foot of terrace between the Rivers Teramakau and Kawhaka . . . 121
First terrace 292
Summit of highest terrace · 647
M'Clintock's store, near Kawhaka Creek 410
Junction of Kawhaka with Arahura 145

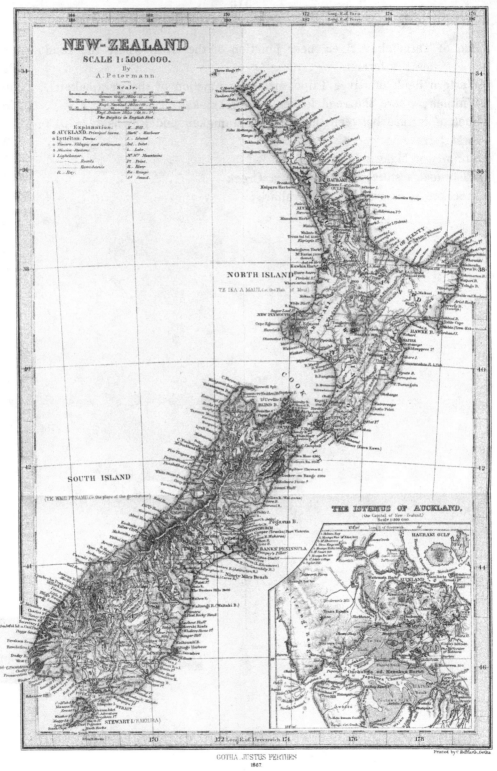

The material originally positioned here is too large for reproduction in this reissue. A PDF can be downloaded from the web address given on page iv of this book, by clicking on 'Resources Available'.

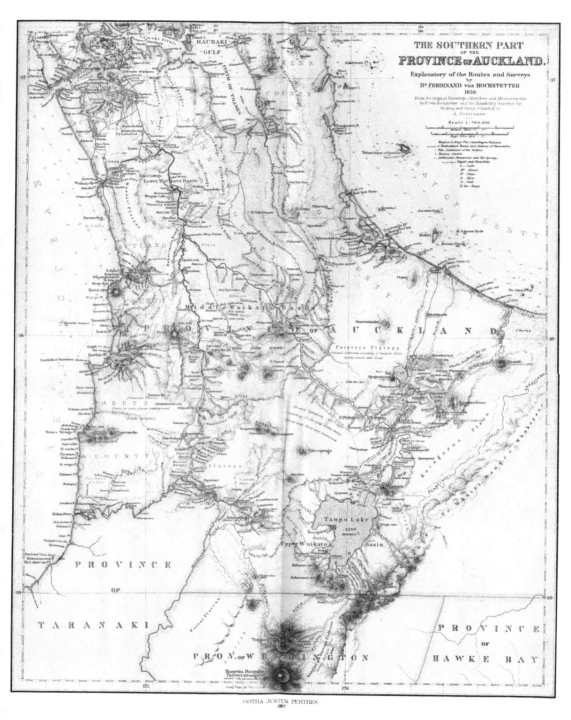

THE SOUTHERN PART
OF THE
PROVINCE OF AUCKLAND.
Explanatory of the Routes and Surveys
by
Dr FERDINAND VON HOCHSTETTER
1859.

GOTHA JUSTUS PERTHES

The material originally positioned here is too large for reproduction in this reissue. A PDF can be downloaded from the web address given on page iv of this book, by clicking on 'Resources Available'.

Printed in the United States
By Bookmasters